Android and IOS Mobile Forensics

Leveraging Blockchain, Machine Learning, and Deep Learning for Digital Investigations

Ravi Sheth
Keshav Kaushik
Chandresh Parekha
Narendrakumar Chayal

Apress®

Android and IOS Mobile Forensics: Leveraging Blockchain, Machine Learning, and Deep Learning for Digital Investigations

Ravi Sheth
Gandhinagar, Gujarat, India

Keshav Kaushik
Greater Noida, Uttar Pradesh, India

Chandresh Parekha
Gandhinagar, Gujarat, India

Narendrakumar Chayal
Ahmedabad, Gujarat, India

ISBN-13 (pbk): 979-8-8688-1747-2
https://doi.org/10.1007/979-8-8688-1748-9

ISBN-13 (electronic): 979-8-8688-1748-9

Copyright © 2025 by Ravi Sheth, Keshav Kaushik, Chandresh Parekha, Narendrakumar Chayal

This work is subject to copyright. All rights are reserved by the Publisher, whether the whole or part of the material is concerned, specifically the rights of translation, reprinting, reuse of illustrations, recitation, broadcasting, reproduction on microfilms or in any other physical way, and transmission or information storage and retrieval, electronic adaptation, computer software, or by similar or dissimilar methodology now known or hereafter developed.

Trademarked names, logos, and images may appear in this book. Rather than use a trademark symbol with every occurrence of a trademarked name, logo, or image we use the names, logos, and images only in an editorial fashion and to the benefit of the trademark owner, with no intention of infringement of the trademark.

The use in this publication of trade names, trademarks, service marks, and similar terms, even if they are not identified as such, is not to be taken as an expression of opinion as to whether or not they are subject to proprietary rights.

While the advice and information in this book are believed to be true and accurate at the date of publication, neither the authors nor the editors nor the publisher can accept any legal responsibility for any errors or omissions that may be made. The publisher makes no warranty, express or implied, with respect to the material contained herein.

> Managing Director, Apress Media LLC: Welmoed Spahr
> Acquisitions Editor: Miriam Haidara
> Desk Editor: James Markham
> Editorial Project Manager: Jessica Vakili

Distributed to the book trade worldwide by Springer Science+Business Media New York, 1 New York Plaza, New York, NY 10004. Phone 1-800-SPRINGER, fax (201) 348-4505, e-mail orders-ny@springer-sbm.com, or visit www.springeronline.com. Apress Media, LLC is a Delaware LLC and the sole member (owner) is Springer Science + Business Media Finance Inc (SSBM Finance Inc). SSBM Finance Inc is a **Delaware** corporation.

For information on translations, please e-mail booktranslations@springernature.com; for reprint, paperback, or audio rights, please e-mail bookpermissions@springernature.com.

Apress titles may be purchased in bulk for academic, corporate, or promotional use. eBook versions and licenses are also available for most titles. For more information, reference our Print and eBook Bulk Sales web page at http://www.apress.com/bulk-sales.

Any source code or other supplementary material referenced by the author in this book is available to readers on GitHub. For more detailed information, please visit https://www.apress.com/gp/services/source-code.

If disposing of this product, please recycle the paper

I dedicate this book to my beloved mother, Purnima Sheth, whose love and support have shaped my journey. To my wonderful wife, Kinjal Sheth, whose unwavering companionship has been my greatest strength. I am also blessed by my loving daughters, Kyara Sheth and Nyara Sheth, whose smiles and presence bring endless joy and inspiration to my life. Their collective love has been the driving force behind this work.

—Dr. Ravi Sheth

Dedicated to my beloved parents Shri Vijay Kaushik and Smt. Saroj Kaushik, my wife Priyanka, daughter Kashvi, and son Harshiv.

May god always bless us, Har Har Mahadev!!

—Keshav Kaushik

I dedicate this book with heartfelt gratitude to my elder brother, late Kirit Parekh, and to my loving family—my wife, Parul Parekh, and my two sons, Dr. Kathan Parekh and Rishi Parekh—whose constant love, support, and belief in me have made this journey possible. This book is as much yours as it is mine.

—Dr. Chandresh Parekha

This book is dedicated to my parents, my sister, my brother Vivek Chayal & sister-in-law Dr. Hetal Chayal, and family who gave me the motivation to write these chapters. It is also dedicated to my mentor and research guide, my faculties, teachers, colleagues, and my friends whose guidance, encouragement, and unwavering support have been a constant source of inspiration throughout this journey. Your belief in me made all the difference.

—Narendrakumar Chayal

Table of Contents

About the Authors ....**xvii**

About the Technical Reviewer ....**xxi**

Chapter 1: Introduction to Mobile Forensics 1
    Evolution of Mobile Forensics 1
        Introduction to Mobile Forensics 3
        Why Do We Need Mobile Forensics 6
        Mobile Forensics Approaches 8
        Mobile Phone Evidence Extraction and Digital Investigation Process 12
        Forensic Examination of Digital Evidence—A Guide for Law Enforcement 18
        Challenges of Mobile Forensics 22
    Summary 25
    References 26

Chapter 2: Android Forensics Fundamentals 27
    Android Device Architecture 28
        Overview of Android System Components 29
    Types of Data on Android Devices 42
        User Data 42
        System Data 44
        App Data 45
        Locations of Key Artifacts 46
    Forensic Acquisition Methods 47
        Physical Acquisition 47
        Logical Acquisition 48
        File System Acquisition 48
        Challenges in Acquisition 49

TABLE OF CONTENTS

Forensic Tools and Techniques 51
    Overview of Popular Forensic Tools 51
    Steps for Logical, Physical, and File System Acquisition for Android Mobile 52
    Data Recovery Techniques 54
    Handling Encrypted Data 55

Android File System and App Analysis 55
    File Systems Used in Android 56
    Analyzing Android Package Files (APKs) and App-Specific Data 56
    Key Tools for File and App Analysis 57

Summary 59

Reference 59

Chapter 3: iOS Forensics Fundamentals 61

iOS Device Architecture 62
    Introduction to iOS System Components 62
    Secure Enclave and Encryption 63
    Runtime on iOS and App Sandboxing 63
    System Security and Permissions 64

Types of Data on iOS Devices 64
    User Data 65
    Places of Key Artifacts in iOS 67
    Forensic Acquisition Methods 68
    Physical Acquisition Method 68
    Logical Acquisition 69
    File System Acquisition 70
    Cloud-Based Acquisition 71
    Acquisition Challenges in iOS Devices 72

Forensic Tools and Techniques 74
    Process to Be Followed for the Data Acquisition 75
    iCloud Data Acquisition 81
    Analyzing iCloud/iOS Backups 84
    Data Recovery and Decryption Techniques 85

iOS File System and App Analysis 88
    iOS File Systems: APFS and HFS+ 88
    Analyzing App Data and iOS Apps (IPA Files) 89
    Tools for iOS File and App Analysis 89
Summary 90

Chapter 4: Leveraging Blockchain for Mobile Forensics 91

Introduction to Blockchain 91
Blockchain Applications for Mobile Device 92
    Axie Infinity 92
    PancakeSwap 93
    Uniswap 93
    Compound 93
    OpenSea 93
Blockchain-Related Cybercrimes 94
    Case 1: The DPRK's DMM Bitcoin Exploit 94
    Case 2: Gainbitcoin Cryptocurrency Scam—Multilevel Marketing Scheme 94
Forensic Analysis of Blockchain Data on Mobile Devices 95
Key Concepts in Blockchain Forensic Analysis on Mobile Devices 95
Steps for Forensic Analysis of Blockchain Data on Mobile Devices 96
Tools for Blockchain Forensic Analysis on Mobile Devices 99
Key Blockchain-Related Artifacts on Mobile Devices 100
Challenges in Blockchain Forensic Analysis on Mobile Devices 101
Decentralized Application and Investigation 102
What Are Decentralized Applications (dApps)? 102
Investigating Crimes Involving dApps 103
Cryptocurrency and Wallet Investigations 107
    Understanding Cryptocurrency Wallets 107
Types of Wallets 107
    How to Identify the Wallet 108
Investigating Cryptocurrency Transactions 108

TABLE OF CONTENTS

Wallet Analysis 109
Techniques for Linking Wallets to Individuals 109
Challenges in Blockchain Mobile Forensics 110
    Pseudonymity and Anonymity 110
    Decentralized Nature of Blockchain 110
    Data Encryption and Privacy Features 110
    Volume and Complexity of Data 111
Opportunities in Blockchain Mobile Forensics 111
    Transparent and Immutable Data 111
    Integration with Mobile Forensic Tools 112
    Smart Contract and dApp Investigation 112
    Real-Time Monitoring of Blockchain Transactions 112
Summary 113

Chapter 5: Investigating Mobile Banking and Financial Applications 115

The Evolution of Mobile-Based Applications 116
Security Features of Banking Application 117
Cyber Crimes and Fraud Associated with Mobile Banking 126
    Malware Categories 127
Forensic Evidences and Acquisition of Mobile Devices 129
Methods of Mobile Forensics Acquisition 129
    Physical Acquisition 129
    Physical Acquisitions Process 129
    Examine the Collected Data 130
    Logical Acquisition 131
    Full File System Mobile Extraction 131
Timeline Analysis 131
    Common Data Sources for Analysis for Banking Application Investigation 132
Examine the Timeline for Trends and Insights 133
Examination of Mobile Banking Data or Malicious Applications Using a Case Study ... 134
Dynamic Analysis Process 140

Identifying Behavior of Application	141
IPhone Malware	145
Summary	147
Reference	148

Chapter 6: Examination of Social Media and Messaging Applications ... 149

Types of Data Generated by Social Media and Messaging Applications	150
Forensic Acquisition Methods for Social Media and Messaging Data from Desktop/Laptop or Mobile	154
Forensic Acquisition Procedure	155
Forensic Imaging/Cloning of a Drive	155
Forensic Artifacts or Evidence	157
Internet Browsing History	157
Prefetch Documents	158
Restoration of Erased Social Media Artifacts	159
Social Media Artifacts from Volatile Memory	160
Tools for Acquiring Memory Dumps	161
Why RAM Dumps Are Important for Social Media Forensics	161
Common Social Media Artifacts Identified in Memory Dumps	162
Standard Methods for Analyzing RAM Dumps	163
Tools for Analyzing RAM Dumps	164
Extraction of Social Media Artifacts from Mobile Devices	164
Social Media Artifacts from the Cloud	167
Cloud Storage, Synchronization, and Forensics	167
Common Characteristics	168
Challenges in Cloud Storage Investigation	170
Cloud Storage Forensics: Methods for Acquiring Digital Evidence	171
Forensic Tools for Data Retrieval and Examination	171
Forensics Investigation Challenges in Social Media and Instant Messaging Apps	172
Summary	175

TABLE OF CONTENTS

Chapter 7: Location-Based Data Analysis and Geolocation Artifacts 177

What Is Location-Based Data? 178

Geolocation Artifacts 178

Forensic Methods for Collecting and Analyzing Geolocation Data 180

Location Data Storing Path from Android and IOS Device 184

Location Data from Social Media Applications 187

IP Address Geolocation 187

Forensic Challenges in Analyzing Geolocation Data 187

Analysis Techniques for Geolocation Artifacts 189

Privacy and Security Implications 191

Data Security Risks 193

Ethical Issues in Geolocation Data Analysis 194

Summary 195

Chapter 8: Mobile Device Network and Forensics 197

Overview of Mobile Networks and the Importance of Forensic Analysis in Mobile Network Environments 197

    Introduction to Mobile Networks 197

    The Growing Complexity and Security of Mobile Networks 198

    Importance of Mobile Device Network and Forensics 199

    Challenges of Mobile Network Forensics 201

Mobile Network Architecture (3G/4G/Wi-Fi/Bluetooth) 202

    Mobile Networks Overview: Cellular Network Architecture 202

    3G and 4G Networks 203

    Mobile Wi-Fi Networks 206

Capturing Network Traffic from Mobile Devices and Infrastructure 213

    Packet Sniffer 214

    Network Tap Devices 216

    Tools and Software for Traffic Capture 217

    Capturing Infrastructure-Level Traffic 218

Types of Network Forensic Artifacts in Mobile Devices 219
    Log Files 219
    IP Address 220
    Subscriber Identity Module (SIM) and International Mobile Subscriber Identity (IMSI) Tracing 221
    Evidence Preservation 222

Mobile Network Attacks and Threats 223
    Common Mobile Network Threats 223
    Advanced Persistence Threats (APTs) 225
    Case Study: High-Profile Attacks 226

Incident Response and Investigation 228

Summary 231

Reference 231

Chapter 9: Mobile Browser Forensics 233

Introduction to Mobile Browser Forensics 233
    Overview of Mobile Browser Forensics and Digital Investigation Relevance 233
    Common Mobile Browsers and Platforms 234
    Importance in Criminal, Civil, and Corporate Investigations 235
    Forensic Acquisition Challenges: Sandboxing, Encryption, and Permissions 236

Analysis of Browser Artifacts 237
    Types of Browser Artifacts 238
    Artifact Storage Paths on Android and iOS 239
    Volatile vs. Persistent Artifacts 240
    Tools and Methods for Extraction 242

Recovery of Deleted Data 244
    Techniques for Browser Data Recovery 244
    Challenges in Recovering Incognito or Cleared History 245
    File System Behavior and Data Remanence 246

Identification of Web-Based Threats 250
    Mobile Browser Threat Vectors 251
    Artifact-Based Threat Detection 251

TABLE OF CONTENTS

    Indicators of Compromise (IoCs) 252

    Timeline Correlation of Malicious Behavior 252

Cross-Browser Forensic Analysis 254

    The Need for Multi-browser Correlation 254

    Structural and Encryption Differences 254

    Cross-Browser Behavioral Profiling 255

    Challenges in Dataset Merging 256

Conclusion 257

References 259

Chapter 10: Leveraging Machine Learning for Mobile Forensics 263

Machine Learning (ML) Techniques for Mobile Digital Investigations 265

    Supervised Learning 265

    Unsupervised Learning 267

    Semi-supervised Learning 268

    Reinforcement Learning 270

Data Preparation and Feature Engineering 271

    Data Collection 272

    Data Cleaning 273

    Feature Extraction 276

    Feature Selection 282

Machine Learning Models and Algorithms 285

    Overview of Common Models 285

    Model Selection Criteria 291

    Training Machine Learning Models 293

    Overfitting and Underfitting 295

    Fine-Tuning and Optimization 296

Evaluation and Validation of Machine Learning Models 302

Applications of Machine Learning in Mobile Digital Investigations 302

Integration of Machine Learning with Forensic Tools and Processes 303

    Workflow Integration 303

    Challenges and Limitations 304

Summary 306

References 307

Chapter 11: Applications of Machine Learning in Mobile Forensics 309

Machine Learning Models for Anomaly Detection in Mobile Phones 310

    Common Mobile Data Anomalies 311

    Supervised Learning for Anomaly Detection 312

    Unsupervised Learning for Anomaly Detection 313

    Semi-supervised Learning for Anomaly Detection 314

Machine Learning Approaches Predictive Analysis 315

    Use of Predictive Analysis in Mobile Forensics 315

    Appropriate Machine Learning Model for Predictive Analysis 318

Financial Fraud Detection 319

    Machine Learning Models for Fraud Detection 320

Social Media Forensics in Mobile Phones 323

    Machine Learning Methods for Analyzing Social Media Data 324

    Insights and Value from Social Media Analytics 326

Threat Intelligence in Mobile Phones 327

    Machine Learning Roles in Enhancing Threat Detection in Mobile Phones 329

Email Forensics in Mobile Phones 331

    Machine Learning Techniques for Emails Forensics 332

    Example of Email Classification in Forensic Investigations 335

Case Study 336

Summary 339

References 339

Chapter 12: Deep Learning Techniques for Mobile Forensics 341

Introduction to Deep Learning in Mobile Forensics 341

    Scope, Relevance, and Urgency of Mobile Forensics in the Digital Age 341

    How Deep Learning Surpasses Traditional Rule-Based Methods 342

    Use Cases Where Manual Forensic Approaches Fail or Scale Poorly 343

TABLE OF CONTENTS

    Digital Evidence Extraction Using Deep Learning 346
        Types of Digital Evidence on Mobile Devices 346
        Role of CNNs, RNNs, and Transformers in Evidence Parsing 347
        From Raw Data to Labeled Evidence: The Deep Learning Pipeline 348
    Deep Learning Models and Applications in Mobile Forensics 351
        Overview of Deep Learning Models in Mobile Forensics 351
        Mobile Malware Detection 352
        Malicious URL Detection 353
        User Behavior Analysis 355
        Mobile Network Traffic Classification 356
        Web/Browser Threat Identification 358
        Integration into Forensic Workflows 359
    Multimodal Data Analysis for Forensics 359
        The Need for Multimodal Analysis in Mobile Forensics 360
        Real-World Examples of Multimodal Analysis 360
        Deep Learning Models for Multimodal Fusion 361
        Emerging Models for Multimodal Mobile Forensics 363
        Challenges and Future Directions 363
        Conclusion 364
    Conclusion and Future Scope 365
        Superiority and Applicability in Mobile Forensics 366
        Legal and Ethical Challenges 366
        Future Scope: Emerging Trends and Innovations 368
        Closing Remarks 370
    References 371

Chapter 13: Privacy and Security Considerations in Mobile Forensics 373

    Privacy and Security in Mobile Forensics 373
    Security Concerns in Mobile Forensics 374
    Challenges in Mobile Forensics 375
    Chain of Custody and Data Integrity 376

- Chain of Custody in Forensic Investigations 377
- Data Integrity 379

Best Practices for Documenting and Preserving the Chain of Custody and Maintaining Data Integrity 381
- Good Practices on Recording and Maintaining Chain of Custody 381
- Good Practices for Maintaining Data Integrity 382
- Mobile Devices and Issues Within Chain of Custody and Integrity of Data 384

Role of Consent in Forensics 387
- Privacy Laws and Users' Rights 388
- Laws Affect Consent in Investigations 391
- Informed Consent 391

Forensic Tool Security and Reliability in Mobile Forensics 393
- Criteria for Selecting Secure and Reliable Forensic Tools 393
- Tool Reliability 395
- Testing and Certification 396
- Common Standards for Forensic Tool Certification and Reliability Assessment ... 398

Key Summary 399

Chapter 14: Future Trends and Emerging Technologies in Mobile Forensics 401

Introduction 401
- Importance of Adapting to Rapidly Evolving Mobile Technologies 401
- Limitations of Traditional Forensic Models 402
- Aim and Scope of This Chapter 403

5G/6G Network Forensics 405
- Technological Advancements and Forensic Potential 405
- Forensic Challenges in Next-Gen Networks 407

Augmented Reality (AR) and Virtual Reality (VR) Forensics 408
- Evidence Generation in Immersive Environments 409
- Analytical and Reconstruction Techniques 410

Cross-Domain Collaboration and Information Sharing 412
- Multi-stakeholder Forensic Workflows 413
- Privacy-Preserving and Auditable Forensics 414

Roadmap and Future Directions 417
    Timeline of Key Technological Integrations 417
    Research and Policy Recommendations 419
Conclusion 421
References 425

Chapter 15: Ethical and Legal Frameworks in Mobile Forensics 427

Introduction 427
    Ethical Foundations in Mobile Forensics 428
    Legal Mandates and Evidentiary Integrity 429
    Human Rights and Public Trust 430
    The Need for an Integrated Framework 430
Ethical Guidelines for Forensic Practitioners 432
    Professional Conduct and Data Integrity 432
    Privacy Respect and Minimization Practices 433
Legal and Regulatory Frameworks While Handling Evidence 435
    Jurisdictional Boundaries and Legal Access 436
    Chain of Custody and Evidence Integrity 437
Legal Standards, Continuous Monitoring, and Regulatory Compliance 439
    Compliance with Legal Frameworks and Industry Regulations 439
    Monitoring and Process Audits 442
Conclusion 444
References 445

Index 449

About the Authors

Dr. Ravi Sheth has more than 16 years of teaching experience. Currently Dr. Sheth is working as Assistant Professor in School of IT, AI and Cyber Security at Rashtriya Raksha University, Gandhinagar. His areas of interest are in multimedia security, machine learning, pattern recognition, VAPT, and Cyber Forensics. He has a conflate of combined experience in teaching, research, and training. Dr. Sheth had organized several high-quality training and extension programs for law professionals, teachers, and students from across India. He has published more than 50 research articles in international refereed journals on cyber security, cyber forensics, and artificial intelligence. Dr. Sheth has presented many research papers in international and national conferences and seminars. He has been invited as a resource person in many workshops, seminars, and training programs. He has also delivered many expert talks on various aspects of multimedia security, social media investigation, cyber forensics, pattern recognition, and darknet investigation for law professionals, teachers, and students from across India. He has filed a patent for a "Portable mobile forensic data acquisition tool." He is providing research guidance to students of Rashtriya Raksha University in the areas of cyber security, cyber forensics, machine learning, and pattern recognition.

Keshav Kaushik is a distinguished cybersecurity and digital forensics expert, currently serving as an Associate Professor at the Center for Cyber Security and Cryptology, Sharda School of Computer Science & Engineering, Sharda University, Greater Noida, India. A key member of the Cybersecurity Centre of Excellence, he has been instrumental in advancing research and education in cybersecurity, AI-driven security solutions, and digital forensics. Recognized among the World's Top 2% Scientists

ABOUT THE AUTHORS

by Stanford University and Elsevier (2024), he has made significant contributions to academia and research. His academic journey includes a prestigious faculty internship at IIT Ropar during the Summer Faculty Research Fellow Programme 2016, demonstrating his commitment to continuous learning. With over 200 publications, including 30+ peer-reviewed SCI/SCIE/Scopus-indexed journal articles and 80+ Scopus-indexed conference papers, he has established himself as a leading researcher. Additionally, he has authored and edited 40+ books and 30 book chapters, reinforcing his expertise in cybersecurity and digital forensics. An innovator in his field, Dr. Kaushik holds fifteen granted patents, six published patents, and five granted copyrights. His editorial leadership includes serving as Guest Editor for the *IEEE Journal of Biomedical and Health Informatics* (SCIE, IF: 7.7) and Associate Editor for journals such as *IECE Transactions on Emerging Trends in Network Systems, IECE Transactions on Sustainable Computing, International Journal of Sensors, Journal of Combinatorial Mathematics and Combinatorial Computing* (JCMCC) (Scopus-indexed), *Journal of Computer Science* (JCS) (Scopus-indexed), *Wireless Communications and Control* (Scopus-indexed, Bentham Science), and *International Journal of Information Security and Privacy* (IJISP, ESCI, Scopus-indexed). He is also an Editorial Board Member of Springer's International Cybersecurity Law Review. Dr. Kaushik is a Certified Ethical Hacker (CEH v11), CQI & IRCA Certified ISO/IEC 27001:2013 Lead Auditor, Quick Heal Academy Certified Cyber Security Professional (QCSP), and IBM Cybersecurity Analyst, further validating his expertise. He is a Vice Chairperson of the Meerut ACM Professional Chapter, Senior Member IEEE (SMIEEE), an appointed Bentham Ambassador by Bentham Science Publishers, and a member of the International Association of Engineers (IAENG). A dynamic speaker and mentor, he has delivered over 50 national and international talks on cybersecurity, AI security, and digital forensics. His mentorship was recognized in the Smart India Hackathon 2017, earning appreciation from AICTE, MHRD, and ISRO. He has also contributed to cybercrime investigation training and has been honored by the Uttarakhand Police for his efforts in cybersecurity education. A two-time GATE qualifier with a 96.07 percentile (2012 & 2016), he continues to drive impactful research, innovation, and education in cybersecurity, making significant contributions to academia and the industry.

ABOUT THE AUTHORS

Dr. Chandresh Parekha is an academician for more than 21 years. He completed B.E. (Electronics) in 1994 from BVM Engineering College, Vallabh Vidyanagar, and M.E. in Electronics Communication Systems from DDIT, Nadiad, in 2009. He completed Ph.D. in Electronics and Communication Engineering from Gujarat Technological University, Ahmedabad, in 2020. He started his professional career as an R&D Engineer at Crown Television Limited, Gandhinagar. He also worked as an R&D Engineer at Videocon International Limited, Gandhinagar. He started teaching in the year 2002 in the Electronics and Communication Department. He worked as an Assistant Professor in the Electronics and Communication Department of Sankalchand Patel College of Engineering, Visnagar, from July 24, 2002, to October 24, 2011. He has been working as an Assistant Professor—Telecommunications (Senior Grade) with Rashtriya Raksha University, Lavad, since October 25, 2011. He has been actively involved in academics by publishing more than 50 research papers in reputed journals/conferences, undertaking various funded research projects, delivering expert talks, and organizing technical events. He received the Geospatial Excellence Award for the research project "GIS based Crime Mapping of Ahmedabad City" in 2014.

Narendrakumar Chayal completed his BE in IT, MBA in Information Management, MTech in Cyber Security and is currently pursuing a PhD in Malware Analysis in the field of Cyber Security. Currently, he is working as a Malware Forensics expert and providing assistance and support to various law enforcement agencies of the Government of India. He worked as an Assistant Professor—Cyber Forensics in Raksha Shakti University, now Rashtriya Raksha University. He has more than seven years of experience in the field of cyber/digital forensics and cyber security. He has visited various crime scenes to collect digital evidences for incident response process and worked with various commercial mobile, memory, network, and malware forensics tools. As a malware forensics expert,

ABOUT THE AUTHORS

he was involved in the cases where national threat and APT cyberattack is involved. He was part of the core team of cyber security during G20 in 2023. He has also provided hands-on training to state and central police officers of various law enforcement agencies of the country. As a researcher, he is working on challenges in malware detection and analysis techniques.

About the Technical Reviewer

Bhargav Rathod works at Salesforce as a Security Analyst and is ignited by the thrill of uncovering hidden digital trails and solving complex cybersecurity puzzles. With an insatiable curiosity and a relentless pursuit of knowledge in Digital Forensics and Incident Response (DFIR) and Malware Analysis, he brings a unique perspective to the field, transforming challenges into opportunities for innovation.

He has a passion for mentoring young and aspiring cybersecurity professionals and fostering a culture of continuous and innovative learning. His interest areas are DFIR and Malware Analysis (iOS and macOS) and he holds certifications like GIAC iOS and macOS Examiner (GIME) and GIAC Reverse Engineering Malware Certification (GREM). Additionally, he serves as a member of the Organizing Committee for the DFRWS APAC and USA conferences.

He shares his work with the community at `https://www.malwr4n6.com/`.

CHAPTER 1

Introduction to Mobile Forensics

Mobile phones have become an integral part of contemporary life, holding large amounts of personal, work-related, and sensitive information. With this rise in usage, mobile devices have increasingly become vital sources of evidence in criminal investigations. This chapter introduces readers to the foundational concepts of mobile forensics, covering the evolution of the field, key mobile operating systems, the necessity of mobile forensic practices, and the methods used for data extraction and analysis. Additionally, it highlights official guidelines, challenges in the domain, and the role of mobile forensic processes in modern law enforcement.

Evolution of Mobile Forensics

Mobile forensics has evolved alongside mobile technology, adapting to increasingly complex operating systems, diverse data types, and stronger security features. Understanding this evolution provides the historical context necessary for grasping the current state and challenges of mobile forensic investigations. The following section highlights how mobile forensics has grown from basic call log retrieval to advanced cloud-based and encrypted data extraction techniques. Figure 1-1 describes the evolution of mobile forensics, starting from early mobile phones to latest iOS and Android phones.

CHAPTER 1   INTRODUCTION TO MOBILE FORENSICS

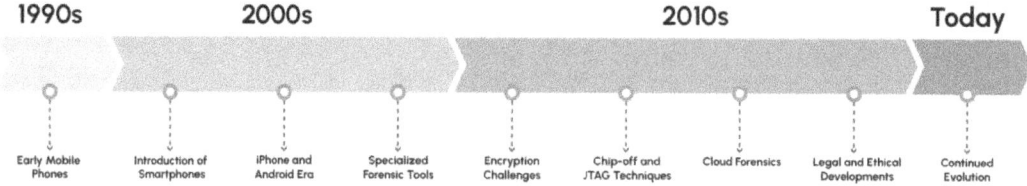

Figure 1-1. *Evolution of mobile forensics*

The evolution of mobile forensics went hand in hand with the development of mobile phone technology and the increasingly complex digital systems integrated into these devices. During the early 1990s, the focus of mobile forensic analysis was primarily on recovering basic data, such as call logs and text messages, because early mobile phones had very limited storage and functionalities. For example, many mobile phones at that time could store only a few dozen contacts and had storage capacities ranging from just a few kilobytes to around 1 MB. This is in stark contrast to modern smartphones, which now offer storage in the range of several gigabytes to terabytes, supporting a wide variety of complex applications and data types. With the development of feature phones and early smartphones in the 2000s, the area of forensic analysis expanded. The development of more feature-rich devices, such as Nokia feature phones and Symbian smartphones, encouraged the creation of specialized tools, like the first forensic tool from Oxygen Forensics in 2004 [2].

Then, the iPhone in 2007 and Androids flooding the market really changed the face of mobile phones—complex operating systems, app ecosystems, and a huge range of data types, such as emails, app-generated information, and geolocation data. Companies like Cellebrite and Oxygen Forensics quickly provided forensic tools to match these innovations, with more advanced extraction techniques. The beginning of the 2010s brought new challenges, with stronger encryption and heightened security in mobile operating systems. Default encryption and secure boot mechanisms demanded that forensic experts develop advanced techniques to access data without compromising its integrity.

Chip-off and JTAG methods became essential in this regard and thus provided direct access to either the memory chips or hardware debugging ports, respectively, to bypass the system security. Simultaneously, the increasing dependence on cloud services for storage grew the scope of forensic analysis to include cloud forensics, with tools like Oxygen Forensics pioneering the extraction of data from systems such as iCloud and Google Drive. These developments also brought to the fore ethical and legal considerations, in which investigators had to respect strict protocols for the admissibility of evidence while considering the laws of privacy.

Currently, the complexity of mobile devices with 5G connectivity, AI-driven applications, and integration with IoT is increasing by leaps and bounds in the 2020s. Modern smartphones now feature advanced AI-integrated operating systems, such as Apple Intelligence on iOS, Gemini on Google phones, and Galaxy AI on Samsung devices. These AI ecosystems are deeply embedded into the core functionalities of mobile devices, enabling real-time processing, predictive assistance, and intelligent personalization. This rapid evolution significantly increases the complexity of mobile forensics, not just in terms of data volume but also in dealing with dynamic AI-driven behaviors and cloud-synced interactions, both now and in the foreseeable future. Contemporary connected devices and decentralized platforms involve a host of progressive techniques, like AI-assisted analysis, memory forensics, and correlation of data from many sources, in forensic investigations. Various inbuilt security features, such as biometric authentication, secure enclaves, and full-disk encryption, continue to pose complex challenges and demand innovative approaches toward drawing out valuable information. Meanwhile, both cloud and hybrid forensic roles continue to evolve, involving such diversity and decentralization common in modern means of storage and synchronization. As mobile forensics continues to evolve, the field remains a key juncture of technology and law, weighing the demands of successful digital investigations against the ethical imperatives of privacy and security. In the next section, the fundamentals of mobile forensics methodology have been outlined to give an initial understanding of identifying, collecting, analyzing, and preserving data from mobile devices.

Introduction to Mobile Forensics

Mobile Forensics is a branch of digital forensics that deals with retrieving, analyzing, and preserving data from mobile devices, such as smartphones, tablets, and wearable technology, for either legal or investigative purposes. It typically involves extracting information, including call logs, messages, photos, videos, location information, and social media use, which may have been deleted or hidden. This is done by first acquiring the data physically, logically, or through the cloud, followed by analysis to comprehend the information for investigative relevance. Foremost is the assurance of integrity through forensic imaging and good traceability through its chain of custody. However, mobile forensics remains complex, given the numerous devices housing the information, matched by various operating systems and security measures like

CHAPTER 1   INTRODUCTION TO MOBILE FORENSICS

encryption. Despite these challenging factors, mobile forensics remains utterly relevant to law enforcement, corporate investigations, and cybersecurity, where mobile devices are usually the key repositories of information in today's digital age. In the following subsections, we explore mobile operating systems in detail to understand how they affect the forensic process [1].

Mobile Operating Systems

The operating system of a mobile phone is one of the most important components in data collection, examination, and analysis. Mobile operating systems have made great strides with a lot of functionality, from low-end feature phones to smartphones. Mobile operating systems have a direct impact on how people can use their devices. A deep knowledge of these mobile platforms aids the Forensic Examiner in making a solid forensic conclusion and can help in conducting a thorough investigation. While there is a huge variety of smart mobile devices available today, it is important to recognize that, following the decline of Blackberry, the market is now largely dominated by three main operating systems: Google's Android, Apple's iOS, and Microsoft's Windows (though Windows Mobile has significantly diminished in recent years). However, a considerable segment of the global population still relies on feature phones that operate on systems such as Java-based platforms, Symbian, and Nokia's Series 30+ (S30+), which typically do not support Internet connectivity or advanced applications. These phones, often preferred by users who are not comfortable with touchscreens or complex smartphones, remain widely used in many regions and present a unique dimension to mobile forensic investigations.

- **Android:** Android is a Linux-based operating system, developed by Google for mobile platforms and it is open source. Android has the largest market share in operating systems. Google developed Android as a free and open-source platform for hardware manufacturers. Android is considered the software of choice for companies because android is a low-cost, customizable, lightweight operating system that can be used on their smartphones without developing a new Operating System from scratch. Due to Android's open-source nature, developers can build their own User Interface skins and also develop a huge number of applications and upload them into the Play store. The Mobile phone users can download these applications through official app stores; however, especially on Android devices, users also have the option to sideload applications from third-party app stores or directly install APK files. Additionally, custom Android operating systems like VivoOS and ColorOS often come with their own dedicated app stores, offering alternative platforms for downloading and updating apps outside of the Google Play Store. All these features make Android a very user-friendly Operating System. In addition to Android operating system by Google, many smartphone manufacturers have developed their own variants of Android OS, with their own user interfaces, preloaded applications, and services not present in the original Android framework. For example, OnePlus offers OxygenOS, which is essentially a near-stock Android experience with some additional customization options. Vivo smartphones run on Funtouch OS or Origin OS, giving their systems a distinct difference in terms of visual design and native apps. ColorOS, for example, used by Oppo and Realme, brings system-level modifications and services. JioOS, designed for Jio-branded devices in India, is another custom Android platform with region-specific apps and services. Not only do these custom OSes alter the user experience but they also introduce additional layers of security, application stores, and system behavior that a mobile forensic investigator has to consider when analyzing data from such devices. The detailed architecture and forensic acquisition method is discussed in Chapter 2.

- **iOS:** iOS is a mobile operating system developed and provided exclusively only by Apple Inc. It was originally known as the iPhone OS. It is the world's second most popular mobile operating system after Android. It also serves as the foundation for Apple's iPadOS, tvOS, and watchOS operating systems. iOS is a Unix-like operating system that is derived from OS X, with which it shares the Darwin foundation. iOS takes care of the device's hardware and gives the tools needed to build native apps. iOS also includes a number of system programs, such as Mail and Safari, that offer users common system functions. Apple maintains rigorous rules, and the App Store is the main and most secure installation platform from which iOS users may download applications. Compared to Android, sideloading an app on iOS is very limited and generally remains unsupported unless one jailbreaks the device, which poses a considerable risk to security and warranty. As a result of this policy, almost every iOS application must be installed directly through the App Store, allowing for tighter control over app security, quality, and privacy. The detailed architecture and forensic acquisition method is discussed in Chapter 3.

Feature Phones (Symbian, Java OS, and Proprietary Systems): Although Android and iOS dominate most modern smartphones, a large percentage of the world still uses feature phones running on lightweight operating systems like Java platforms and company-owned platforms such as Nokia's S30+ (Series 30+) operating system. They are especially common among those looking to use the bare minimum in terms of functions such as voice calls, SMS, and simple apps like FM radio, calculators, and flashlights, compared to the advanced smartphone capabilities. They are predominant in rural regions and developing nations, as well as among people who prefer rather inexpensive, durable, and long-lasting phones without the complexities of Internet access or lower-resolution touchscreens. Some popular ones include Nokia HMD, Vtex, Kinbo, and Kechaoda phones, mainly produced in China and known for their long battery lifetime and durability. These devices may seem simple, but they are present in the current market and are unique in terms of mobile forensics. The data in feature phones is less standardized and can be harder to acquire forensically due to varied encryption and often non-smart file systems. Also, the limited connectivity and lack of app ecosystems provide fewer artifacts to analyze, though they are nonetheless of vital importance in voice data, SMS records, address books, and sometimes even multimedia files saved on external memory cards. In the next section, the actual need for mobile forensics has been described by considering the rapid growth in digital items.

Why Do We Need Mobile Forensics

As we know, data is the most important element in today's legal, corporate, and research investigations. Thanks to the great advancement in the field of mobile technology, most digital data can be found on mobile devices as most of us use a mobile device today, be it a smartphone, tablet, or even wearable device. Mobile devices contain a huge amount of qualitative and quantitative data that can be helpful for the investigation process. Implementing mobile forensics becomes easier with the assistance of proper mobile forensics tools. Such tools help in accessing, extracting, and analyzing data from mobile devices. These forensic tools are designed to empower the law enforcement departments which results in effective investigations and quality digital evidence. The advancement in mobile forensic tools helps investigators enormously by providing better identification of evidence and investigation of the crime.

Currently, desktop computers are being substituted by smartphones as they are more convenient in everyday life for a vast majority of tasks, ranging from personal use to organizational purposes, thus making smartphones the device that carries every single information regarding the user. With the help of mobile forensics, information such as recent chats, call logs, recent locations of the device, contents of the gallery, or the whole volatile memory of mobile device can be obtained. All this information can give an almost clear picture of the user which helps in identifying the person and gives a basic image of the person.

Another important aspect of mobile forensics is that time can be utilized better than traditional investigation techniques. Traditionally investigators used to gather information from relatives or someone who is close to the victim or suspect. But ever since technologies have emerged enormously, gathering information from smartphones is more than enough to get an idea about the people involved in a crime, and also with the help of the GPS location of the mobile phone, the suspect or the victim can be easily tracked down much faster.

Mobile devices help in the reconstruction of crime scenes to an extent. Crime scene investigators can depend on the mobile phones of victims or witnesses and could identify the unusual activities that took place and find out valuable clues and evidence. Also, there is always a risk of losing the evidence in every criminal case. A proper mobile forensic tool helps the investigators to retrieve, acquire, and preserve the evidence at the same time. Mobile devices hold a wide range of information that is stored in different formats or maybe even hidden. As the number of mobile devices is emerging day by day, advancement in the field of mobile forensics is also a must. The next part breaks down different mobile forensic techniques. It explains how experts get and examine information from cell phones. This depends on the kind of device, what system it runs, and its security features.

Mobile Forensics Approaches

This section reveals the various methods of extracting different types of data from mobile phones. Just like any Forensic investigation, there are several methods that can be used for the acquisition, examination, and analysis of data from a mobile phone/smartphone. The type and model of a mobile device, the operating system, and the security settings generally tell what kind of procedure an investigating officer should follow. Each investigative case is unique in its own way and with its circumstances. Due to this uniqueness, it is impossible to create a straight path of procedural approach for all cases. The following sections discuss the most used general approaches used to extract data from mobile devices.

The Mobile Forensic Data Extraction/Acquisition Methods

The acquisition and analysis of mobile phones require both manual work and the use of automated tools. For conducting mobile forensics, there are a number of tools available. It is critical to recognize that each and every tool has its own advantages and disadvantages and no tool is adequate for all tasks. As a result, forensic examiners must be familiar with a variety of mobile forensic technologies. A mobile device forensic tool classification system established by Sam Brothers comes in helpful for examiners when determining the proper tools for the forensic acquisition and examination of mobile phones. This mobile forensics extraction classification system was created to give the investigators a simplified overview of the tools that are available for evidence extraction. The difficulty for usage of these tools gains gradually in this classification. The goal of the mobile device forensic tool classification system is to allow examiners to classify forensic tools based on the tool's examination approach. Starting at the bottom and going up, the procedures and tools grow more technical and complicated and the analysis time also increases. There are advantages and disadvantages for each extraction method and the mobile forensic examiner should be aware of these concerns. If the offered method or tool is not used properly, the evidence can be completely deleted. As you progress up the pyramid, the level of risk and complexity in mobile forensic extraction increases significantly. Each existing mobile forensic tool typically falls under one or more of the five categories represented within this pyramid structure. Figure 1-2 significantly provide better clarity in visualizing this hierarchy, it includes a pictorial representation of the pyramid, illustrating the categories and the increasing level of complexity and risk as one moves from the base to the top. We will start from the bottom of the triangle with selective extraction and finally, end with micro read which is the topmost layer.

CHAPTER 1 INTRODUCTION TO MOBILE FORENSICS

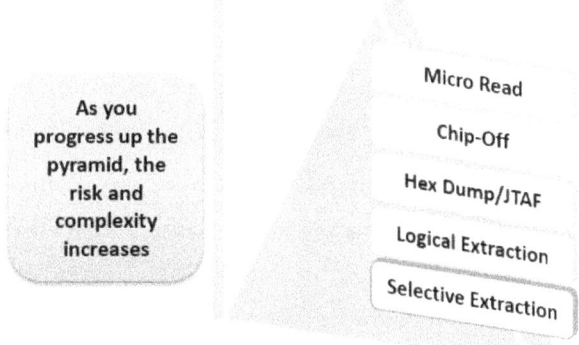

Figure 1-2. *Risk analysis in mobile forensic process*

As mentioned in Figure 1-2, there are basically five primary types of data extraction methods used in mobile device forensics, each differing in complexity, risk, and the depth of data they can retrieve:

- Selective extraction
- Logical extraction
- Hex Dump
- Chip-Off
- Micro Read

In the next session, we will study all the types of data extraction process in detail.
Data Extraction Types:

1. **Selective Extraction:** Selective extraction is by far the easiest forensic recovery tool because it simply involves an investigator scrolling through the contents. The data that is uncovered through this scrolling is then recorded through photography. This extraction process is quick and easy, and it works on practically all phones. This method can be easily mastered but it is prone to human error. For example, an Android user may miss out on valuable information present on an iOS device due to unfamiliarity with its interface and functionality, and vice versa. There's also a risk that data could be destroyed or altered after collecting, which could cause issues in legal processes. This

method can be quick, but it can also take a considerable amount of time if the information that is being looked at is hidden in a large amount of data. The examiner also needs the device in an unlocked state to see the contents.

2. **Logical Extraction:** In this extraction process, the mobile device needs to be connected to any appropriate forensic devices or a forensic workstation using a connection cable, USB cable, or a Bluetooth connection. When the workstation and the device are connected, the workstation transmits a command to the device, which is then processed by the device processor. Following that, the requested data is extracted from the device's memory and returned to the forensic workstation. The examiner can then look at the data. In this level of classification, commercial forensics tools currently available in the market, this method allows investigators plenty of time to go through the data and it is generally quick and easy. On the contrary, the operation can cause writing of data to the mobile device that is connected which in turn compromises the evidence's integrity, and also deleted data cannot be viewed from this method. The detailed Android and iOS logical acquisition methods and its tools have been discussed in Chapters 2 and 3, respectively.

3. **Hex Dump:** The Physical Extraction method is also known as Hex Dump, and it is quite different from Logical Extraction. Whereas Logical Extraction acquires accessible, active data via a standard system interface, Physical Extraction (also known as Hex Dump) is used to obtain a true bit-to-bit copy of the device's memory and can be used to recover deleted, hidden, and system-level files. Logical Extraction is shallow, secure, and fast, yet it has low depth; physical extraction has high depth, is riskier, and may require bypassing security systems. The decision about utilizing these particular methods relies on the aim of the investigation, the level of depth required for the data, and the risk to the integrity of the device. In this method the mobile device is connected to a forensic workstation and an unsigned code or a bootloader is unloaded into the phone. This code tells the phone to dump the

memory present in the phone to the workstation. The generated image data is in a raw format and it requires technical knowledge and expertise to examine. The procedure is also cheap and it also recovers the deleted files present on almost all mobile devices. The detailed Android and iOS logical acquisition methods and their tools have been discussed in Chapters 2 and 3, respectively.

4. **Chip-off:** In the chip-off method, the data is directly acquired from the device's memory chip. In the case of mobile devices, most of them use the NAND Flash memory. In this phase, the storage chip present in the device is physically removed from the handset, and the data stored on the chip is extracted using a chip reader or a second phone. Since different mobile devices use different chips for running, this method is more technically demanding. The process is expensive and the examiner requires hardware-level knowledge as it involves heating and desoldering of the chip. Specialized training is required to successfully perform this type of extraction. Errors in the process can cause damage to the chip and this in turn can make the data unrecoverable. In most scenarios, other methods of extraction are done first before the chip-off method is implemented. The reason is that the chip-off method is a destructive method. The data that we get from the memory is in an unprocessed/raw format which must be analyzed, decoded, and interpreted to get the data in readable format. In circumstances where it is very important to maintain the state of memory as it is present on the device, we use this extraction method. This method is very useful in conditions where a device is destroyed but the memory chip is still functional.

The Joint Test Action Group (JTAG) approach is frequently used to read the chips on the device. In JTAG approach, the suspected device's Test Access Ports are connected to, and the device's processor is directed to transfer the data contained on memory chips to, the forensic workstation. This approach is commonly used on devices that are operational but inaccessible with conventional extracting tools. Both of these methods operate even if the device's screen is locked.

5. **Micro Read:** Micro reading entails manually inspecting and analyzing data on a memory chip. The examiner uses an electron microscope to examine the physical gates on the chip and these gates are converted to 0s and 1s to ascertain the ASCII letters. The entire process is time-consuming and costly, and it necessitates substantial memory and filesystem expertise and training.
Micro read would only be tried for high-profile cases similar to a national security crisis and only as a last resort, after all other level extraction approaches have been explored, due to the significant difficulties involved. At this time, the procedure is rarely used and poorly described and furthermore there are currently no commercial micro read tools available.

In the next section, we walk through how mobile phone evidence is extracted and how digital investigation is done by following proper procedures to ensure everything is done correctly and in line with the law.

Mobile Phone Evidence Extraction and Digital Investigation Process

This section provides detailed step-by-step procedure for evidence extraction and investigation process. Evidence extraction and inspection vary depending on the type of mobile phone; there is no such uniform procedure in mobile forensics evidence extraction. As a result, it is critical to keep to a single method during the extraction and assessment of the evidence. In other words, consistency is as crucial as proof in this procedure. Documenting the actions properly and on time will also help to maintain the findings defensible and unchangeable. Data extraction from mobile devices necessitates testing, validation, and thorough documentation.

Step 1 Evidence Intake: In mobile forensics, irrespective of the method adopted by the investigator, in this phase, the evidence intake is considered the first and foremost step. In this phase, the investigator calls for the request forms and the documents that prove the ownership details of the device and the details related to mobile devices such as the type of mobile device, and the incident in which it was involved. The objectives of the further examination are framed on the basis of information collected in this step. Hence this phase becomes very important. The aim and the goals of the examiner regarding the evidence examination will be cleared so that the investigation and examination process can be systematically performed. In this critical phase, the investigator should make sure that the data in the device hasn't been subjected to any modification. They should also keep in mind that the evidence is collected for further examination, therefore, whatever opportunity that aids in the investigation positively should not be avoided. For example, the password of the device should be disabled if the device is in unlocked mode at the time of the seizure.

Step 2 Identification: In this phase, the examiner has to perform a detailed examination of the mobile device and the identification of the following details:

- **The legal authority:** In most cases, the legal authority or limitations on the media will be present there for the acquisition and the forensic examination of the mobile device. The examiner should be constrained strictly to these and should determine as well as document the details on this legal authority prior to investigation. For example, on the off chance that the versatile gadget is being looked compatible with a warrant, the analyst ought to be careful of limiting the look to the restrictions of the warrant.

- **The goals of an examination conducted:** The lawful authority decides what data should be collected to aid in the investigation. Accordingly, the forensic examiner has to examine the collected mobile devices. Prior to that they have to decide how deep the examination should be, what all the tools and techniques should they use, etc. This will help the investigation to be more efficient and make the goal of the examination crystal clear.

- **The features like brand, model, version, etc., and identifying details for the device:** The make, model, carrier, current phone number associated with the phone, etc., determines how the examination should be carried out and what specific tools and techniques have to be used during the examination of the mobile device. The identification and documentation of the same are done in a forensically sound manner.

- **Removable and external data storage:** A wide variety of mobile phones offer the option of expanding the memory with a removable storage device, for example, Trans Flash Micro SD memory expansion card. In those cases, in which a card is available on the

CHAPTER 1   INTRODUCTION TO MOBILE FORENSICS

acquired mobile device which is sent for testing, the card must be removed and processed by following the standard digital forensic procedures and techniques. Acquiring the card along with the mobile phone will ease the analysis part if the data found on the memory card acquired and the mobile device are linked to each other.

- **Other possible sources of potential evidence:** It is very important to wear gloves during the collection and handling of the device by forensic investigators because mobile phones are a potent source of other evidence such as fingerprints, biometrics, etc. Prior to the digital forensic examination of the device, this biological evidence should be collected in a forensically sound manner.

Step 3 Preparation: In this phase, after the identification of the specifications of the mobile device, detailed research with respect to the device that has to be examined and about the tools and techniques to be utilized during acquisition and the examination of the device is performed. These are strictly based on the make and model of the mobile device, its build operating system, current version, etc. The goals of examination, type, specifications of the device, presence of removable storage media, etc., determine the tool which has to be used for the examination.

Steo 4 Isolation: Since most of the devices have features like Bluetooth, Infra-Red, Wi-Fi, etc., and the ability to communicate over the cellular network, it is very easy to get remote access to the device through any of these facilities in order to tamper the evidence existing in the mobile device taken for the examination. Even the data will get added to the mobile device seized if it is connected to a network such as incoming calls, messages, notifications, etc. The data present in the device might get wiped off if the device comes into interaction with high-frequency IR rays or, in case of remote access, the attacker might use complete wipe-off commands to tamper with the evidence available. Isolation always remains an important part of mobile phone evidence collection because this step will prevent the interaction of the device with the network as well as remote access. Even a single call or SMS may overwrite or tamper with the existing data in the device. There are a few methods that can be followed to ensure that the device is isolated from these kinds of interactions.

One of the top methods is keeping or wrapping the device in a radio frequency shielding cloth and switching the mobile device to airplane mode (flight mode) to isolate it from the network available. This will disable the communication channels like cellular radio, Wi-Fi, Bluetooth, GPS, SMS services, hotspot, etc., of the device. But in most devices, switching the device to airplane mode is not possible if the device is in lock screen mode. So, this becomes a challenging phase for the investigator. Also, it is possible to connect to Wi-Fi even if the mobile phone is in flight mode so that the device will be able to communicate through that network. As a solution for this, we use Faraday Bags to store the mobile devices while collecting. Using jammers, if possible, will also help in isolating the device from the network.

The use of a Faraday bag is very important in cases in which the mobile phone is turned on. Faraday bag blocks all the radio signals incoming and outgoing from the device. These bags are made up of copper/silver/nickel with RoHS double-layer conductors which are capable of blocking the external static electrical fields, including radio waves. Some of the Faraday bags used in mobile forensics even have a charging cable with an emergency battery for keeping the device switched on till it is taken to the lab. The charging cord will also be isolated to ensure that it too won't receive any communications. This will completely eliminate any sort of communication to the mobile device so that wiping and tracking of the same can't be done. Even Faraday Tents and rooms are there to ensure the convenience of working with the seized device.

Step 5 Processing: This phase starts after the isolation of the evidence, that is the mobile device which is seized. The acquisition of the mobile device is done using a repeatable and forensically sound tested method.

During this stage, data are acquired physically and logically. Physical acquisition is more desirable as it acquires the full raw information on the mobile device, such as deleted data, system data, and unallocated space. This approach is more secure because the machine is usually turned off when the approach is being undertaken, and thus the chances of tampering with the original information are minimized. Nevertheless, marginal modification may still be witnessed in the device through the process of acquisition. Logical acquisition, on the other hand, is the extraction of available files and folders through conventional device protocols without extracting the entire memory content. It only recovers the file system and files that were actively in use. Deleted and hidden files are not accessible. Logical acquisition is typically undertaken when physical acquisition is not feasible and when there is a need to obtain a quick snapshot of user data. It is worth mentioning that logical acquisition mostly provides information in a parsed form, along with metadata, and may help investigators identify areas to focus on specifically when a physical image is acquired subsequently.

Table 1-1 below highlights the key differences and typical use cases for physical and logical acquisition methods:

Table 1-1. Difference Between Physical and Logical Acquisition

Feature	Physical acquisition	Logical acquisition
Data extracted	Full memory (raw data, deleted files, system files, unallocated space)	File system (active and accessible files only)
Device state	Typically powered off	Typically powered on
Reliability	High (closer to original data)	Moderate (parsed and may miss hidden data)
Risk of data alteration	Minimal but possible	Minimal
Time required	Longer	Faster
Typical use case	Detailed forensic investigation, legal cases	Quick assessment, when physical acquisition is not feasible

Step 6 Verification: The exactness of the information has to be verified to make sure that the data in the device haven't been subjected to any modification. There are a few methods to verify the extracted data.

- **Comparison of the extracted data to the available data in the handset:** Here we compare the extracted data and the data which is shown on the mobile device to see whether they match or not. The data is compared either directly with the device or else it is compared to the logical report as per the situation demands. It is better to compare with the logical image. If the original device itself is used, it should be used very carefully because there are chances that the data might get modified which results in tampering with the evidence.

- **Using different tools and comparing of results:** The comparison of data is done using different tools to check the accuracy.

- **Using hash values:** The hash values of all the image files have to be taken after the acquisition of the data from the mobile device so that the integrity of the information can be ensured. In cases of the file system extraction, the hash values of the extracted files are computed. Later the calculation of the hash values of individually extracted files is done and compared with the original hash values to make sure that the data hasn't been tampered with. There might be changes in hash values if the device is turned off and then powered on. These changes should be explainable.

Step 7 Documentation and Report: It is an inevitable step in the mobile forensics evidence examination process. The examiner has to document each and every step he has done during the acquisition and examination. After completion of the examination process, the report should be subjected to any kind of peer review, so that the integrity of the data is checked and the investigation is said to be completed. The examiner should note down the following details in their documentation:

- Date and time at which the examination of the mobile device started.
- Detailed note on physical condition of the mobile device.
- Photographs of the mobile phone, if any other components found along with it, take individual photographs of the same.
- Note down whether the mobile phone was switched on or off, that is, the status of the phone during receival.
- Details like make, model, year, etc., of the mobile phone.
- Notes on the tools used during acquisition.
- Notes on the tools used in examination.
- Detailed note about the data found on examination of the device.
- A detailed note based on the peer reviews.

Step 8 Presentation: After the completion of the investigation, presenting the findings of the examiner to any other examiner or to the court of law is very important. For that, the examiner should have to maintain a detailed and legible report of every step they have followed during the investigation. The data extracted from the mobile device at the time of acquisition and the analysis of the same should be documented along with the conclusion or findings of the examiner. The report may be in both paper and electronic format. The findings of the examiner should be clear, consistent, and repeatable. There are a variety of commercial mobile forensic tools, which offer features like timeline and link analysis which help to report and to describe the findings of the examiner across multiple mobile phones. It is possible for the examiner to compile the methods beyond the communication of several mobile devices.

Step 9 Archival of Evidence: As we know, the court cases may go on for years until the final judgment. So, during this time, the data extracted and documented from the evidence should remain intact, without undergoing any change, so that it can be used for future reference during the case period. The data should continue to have a usable format, the file should not undergo corruption. To the need for appeals, most courts will ask for the data to be retained for a long period. Nowadays, with the advancement of technology, it is possible to take out data from a raw, that is, a physical image, using modernized methods and then the examiner can come back and see the data of evidentiary value by taking out required copies from the archives.

The following section provides a clear guide for law enforcement agencies, showing the best approaches to examine digital evidence in forensics for use in legal proceedings.

Forensic Examination of Digital Evidence—A Guide for Law Enforcement

The Department of Justice (DOJ) of the United States has released important directions regarding the careful examination, interpretation, and discussion of results achieved in digital investigations. The Digital Evidence Policies and Procedures Manual, released by D. B. Muhlhausen (Director) [3] in 2020 as a product of the National Institute of Justice, US Department of Justice, is one of the most remarkable sources. This manual outlines detailed policies and procedure standards that govern the credibility and legal practicality of digital forensic practices. Besides the DOJ manual, there are other universally recognized best practices available in the Scientific Working Group on Digital Evidence (SWGDE) that highlight the necessity of maintaining the integrity, authenticity, and admissibility of digital evidence during investigations. In the next subsection, specific guidance and best practices recommended by both the US DOJ and the SWGDE are described. These principles empower forensic examiners to analyze and provide interpretations of digital evidence effectively, ensuring that the findings are not only forensically sound but also admissible in a court of law.

General guidelines

It is usually emphasized in general guidelines for digital evidence analysis that you must pay attention to documenting all evidence, follow a clear chain of custody, and observe set procedures. Following these guidelines guarantees that digital evidence is handled, examined, and protected for use as court evidence. When investigators follow these steps, they can ensure their results are reliable and credible, which supports the pursuit of justice. The following are the key parameters to be considered while conducting forensic examination.

Ownership and possession: At a specific date and time, pinpoint files of interest in non-default locations, recover passwords that indicate possession, and identify the contents of files that are specific to a user to determine who created, modified, or accessed a file, as well as the ownership and possession of questioned data.

Application and file analysis: Evaluate file content, correlate files to installed applications, establish relationships between files to find information relevant to the investigation (e.g., email files to email attachments, identify the importance of unknown file formats, examine the system to inspect file metadata, and change configuration settings providing information about the authorship of the work).

Time frame analysis: Review any logs and date/timestamps in the filesystem, such as the last modified time, to determine when activities on the system occurred to associate usage with an individual. In addition to call records, the date/time and content of texts and emails can be informative. Billing and subscriber records kept by the service provider can also be used to verify the information.

Data hiding analysis: By correlating file headers to file extensions to show intentional misdirection, gaining access to password-protected, encrypted, and compressed files and gaining access to steganographic information detected in images, you can recognize and recover hidden data that may indicate knowledge, ownership, or intent.

Evidence Guidelines

The information contained on a mobile phone is increasingly being used as significant evidence in courtrooms. A detailed understanding of the rules of evidence is required to present evidence in court. Because mobile forensics is a relatively new and dynamic field, laws controlling evidence admissibility are not universally recognized, and they differ by region. However, for evidence to be significant in digital forensics, there are five general criteria of evidence that must be observed. If you ignore these rules, your evidence will be ruled inadmissible, and your case will be dismissed. The five rules are as follows: acceptable, authentic, comprehensive, dependable, and credible.

Admissible: This is the most basic guideline and a metric for determining the validity and relevance of evidence. Evidence must be collected and stored in a form that allows it to be used in court or anywhere else. Many mistakes can lead to a court ruling that a piece of evidence is inadmissible. For example, evidence obtained through illicit means—such as unauthorized access methods like jailbreaking or rooting a device without proper legal approval—is frequently found to be inadmissible in court. Jailbreaking (for iOS devices) and rooting (for Android devices) refer to bypassing the manufacturer's security restrictions to gain deeper system access, which can compromise the integrity of the evidence and violate legal procedures if done without authorization.

Authentic: To prove something, the evidence must be meaningfully related to the incident. The forensic examiner must be held accountable for the origins of the evidence.

Complete: Evidence must be presented in a clear and thorough manner, and it must reflect the entire story. It's not enough to gather proof that only illustrates one side of the story. Presenting insufficient evidence is riskier than offering no evidence at all, as it may result in a different decision.

Reliable: The evidence obtained from the device must be trustworthy. This is contingent on the tools and methodologies employed. The methodologies utilized and the evidence gathered must not create doubt about the evidence's authenticity. Unless the examiner was directed to do so, the evidence is not included in the examination utilized techniques that cannot be replicated. This could include destructive techniques like chip-off extraction.

Believable: A forensic examiner must be able to describe the techniques they employed and how the evidence's integrity was protected in a clear and straightforward manner. The examiner's evidence must be straightforward, simple to comprehend, and believable to the jury.

Forensic Good Practices

The gathering and preservation of evidence follow good forensic procedures. Following proper forensic procedures guarantees that evidence is considered genuine and correct in court. Evidence tampering, whether deliberate or unintentional, can have an impact on the case. As a result, forensic examiners must be well-versed in the best procedures.

Secure the Evidence: With enhanced smartphone capabilities like Find My iPhone and remote wipes, protecting a mobile phone so that it cannot be remotely deleted is critical. In addition, when the phone is turned on and connected to the Internet, it is continually receiving fresh data. To protect the evidence, utilize the proper equipment and procedures to disconnect the phone from all networks. Isolation prevents the phone from receiving any new data that would cause active data to be erased. Depending on the circumstances, standard forensic methods such as fingerprinting or DNA testing may also be required to establish a link between a mobile device and its owner. If the gadget is not handled carefully, physical evidence may be tampered with and rendered useless. Gather all peripherals, linked media, cables, power adapters, and other devices that may be present. If the device is detected to be connected to a personal computer at the investigation site, the data flow will be interrupted. Instead, before removing the device, it is a good idea to take a snapshot of the computer's memory, since this contains crucial information in many cases.

Preserving the evidence: Evidence must be preserved in a court-acceptable state when it is gathered. It's possible that working directly with the original copies of evidence will cause them to alter. As soon as you recover a raw disc image or files, make a read-only master copy and duplicate it. There must be a mechanism to assure that the evidence provided is similar to the original gathered in order for it to be accepted. This is accomplished by producing a hash value for the image. By providing a cryptographically strong and non-reversible image/data value, a forensic hash is used to validate the integrity of an acquisition. Calculate and compare the hash values for the original and duplicate after copying the raw disc image or files to ensure evidence integrity. Any hash value changes should be documented and explained. Any further processing or examination should be done on the evidence copies. Any use of the phone has the potential to change the information stored on it. As a consequence, limit yourself to only doing what is absolutely necessary.

Documenting the evidence and changes: When possible, keep a record of every visible data. It's a good idea to take pictures of the phone as well as any other items you find, such as wires and accessories. This will come in handy if any environmental issues arise in the future. Do not touch or place your hands on a mobile device when taking photos. Make sure to keep a record of all the techniques and equipment you used to gather and retrieve evidence. Make your notes as thorough as possible so that they can be repeated by another examiner. A court may rule that your work is inadmissible if it cannot be replicated. It's crucial to keep track of the whole recovery process, including any changes made during acquisition and review. For instance, if the forensic tool that was used to extract the data diced up the disc image in order to preserve it, this must be documented. All changes to the mobile device, including power cycling and syncing, should be documented in your case notes.

Reporting: The process of creating a thorough summary of all the procedures done and conclusions reached throughout the examination is known as reporting. The examiner's report should include details about all of the significant steps they took, as well as the acquisition's outcomes and any conclusions reached from them. The majority of forensic tools include built-in reporting functions that will automatically create reports while also allowing for customization. In the following section, we look at the most important challenges in mobile forensics, focusing on problems related to technology, laws, and daily operations.

Challenges of Mobile Forensics

Mobile phone data can be stored, accessed, and synchronized across multiple devices. The data stored in volatile memory can be quickly transformed or deleted remotely, so more effort is required for the preservation of data. Mobile Forensics when compared with Computer forensics presents more serious and unique challenges for Forensic Investigators. Law and Forensic investigators often struggle to obtain evidence from Mobile phones.

Here are some reasons why evidence is hard to obtain:

- **Hardware Differences:** The mobile phone/smartphone market is very vast, with different manufacturers and their different models. Mobile forensic investigators may come across devices that differ in model, size, features, and hardware components. Also since mobile phones are depreciating assets, new models emerge very quickly in the market. Due to this reason, Mobile Forensic experts require different cables and tools to connect with different models.

- **Operating Systems:** In the market the two largest operating systems currently used are Apple's iOS and Google's Android. There are also other Operating Systems that are still being used in some phones like Blackberry OS, Microsoft Windows OS, and KaiOS. All of these Operating Systems have several versions which make the task of investigators even more difficult. Since android is an open-source operating system manufacturer, it changes the way it looks and feels according to its liking. This also confuses the investigators. It should be noted that iOS and Android frequently provide security fixes and software patches, which may dramatically affect the operation of mobile forensic tools. Such updates will regularly add new security layers, encryption schemes, or system constraints that will render previously developed forensic tools incompatible or, in certain instances, unusable. This could make it difficult for forensic analysts to extract data, or they could lose access to parts of the device that were previously usable. To remain compatible and maintain extraction opportunities, forensic tool vendors must constantly monitor such updates and release timely tool improvements or updates. When discrepancies involve updating forensic tools, a strategic issue may arise regarding lagging behind, where new and improved devices or operating systems are not compatible with outdated software, creating a critical gap in the investigation process. Therefore, it is advisable to stay up-to-date with patches and updates to devices and tools to guarantee successful acquisition and analysis in the forensic process.

- **In-built security features:** To safeguard user data and consequently their privacy, all the latest mobile operating systems, such as iOS and Android, come with highly sophisticated built-in security systems. Although these features are very helpful to users, they tend to be a great challenge for forensic investigators. Most new devices use multiple layers of security systems, such as fingerprint or face detection, in addition to numerical or alphanumeric passwords. Furthermore, there are file-level and full-disk encryption systems, which can be applied automatically at the device level or manually supported for specific files. What adds to the complexity of forensic acquisition is that iOS and Android phones are programmed to automatically reboot after a certain allocated amount of time of inactivity or after a certain number of unsuccessful unlock attempts. When the device is restarted, it usually requires a complete passcode re-entry, making the biometric unlock method inaccessible and requiring more effort to access in forensics. Additionally, some Android models come with Device Theft Protection functions, where the device can automatically lock itself or even wipe itself when it detects an active unlock followed by a violent removal from the user's hand. Although such evolving security measures are necessary to protect user privacy and device security, they severely impede data collection and examination in mobile forensic investigations. To preserve volatile evidence prior to the activation of such automated security reactions, investigators have little choice but to work quickly, carefully, and within tight legal parameters.

- **Data Modification:** The digital forensic expert has to make sure that the data on the device is not modified. When extracting data from a phone, the expert has to make sure that the data is not modified. This is practically impossible as even turning on the mobile phone can cause data change. There may be processes running on the mobile phone background when the phone is asleep and sometimes even when the phone is switched off. The best example of this is the alarm application.

- **Anti-Forensic Techniques:** Data hiding, data obfuscation, data forgery, and secure wiping are all anti-forensics techniques that are commonly used, that make investigations difficult.

- **Password Recovery:** Smartphones, if most people have some kind of screen lock, can be a four-digit number, an alpha-numeric password, pattern lock, fingerprint ID, or face ID. There are multiple ways to bypass the screen lock but there is a chance that it may not work on all models or software versions.

- **Lack of Resources:** With the growing number of mobile phone models, the need for tools and resources needed by Digital Forensic Investigators also increases. Experts need USB cables, batteries, write blockers, chargers, and above all, the latest and cutting-edge software that can analyze as well as crack the passwords of all the models of devices.

- **Dynamic Nature of Evidence:** Digital evidence can be altered either intentionally or unintentionally. Even browsing an application on the phone might alter the data stored by that application on the device.

- **Device alteration:** Sometimes, people intentionally try to hide or destroy evidence on their mobile devices. This can be done by moving app data, renaming files, or making deep changes to the device's operating system, like rooting or jailbreaking, which can seriously alter the device and affect the investigation. But it's not just digital tampering—physical damage is also a common tactic. For example, someone might break the charging port, smash the phone, drop it in water for a long time, or even try to burn it to prevent access. These actions make the forensic process much more difficult and time-consuming, often forcing analysts to spend extra effort just to recover even small amounts of usable data.

- **Communication Shielding:** Mobile devices communicate over cellular networks, Wi-Fi networks, Bluetooth, and infrared. As device communication might alter the device data, the possibility of further communication should be eliminated after seizing the device.

- **Lack of necessary tools:** There are so many different mobile phones out there, each with its own hardware and software. No single forensic tool can handle them all. Investigators usually need to rely on a mix of tools to get the job done, and figuring out which tool works best for a particular phone isn't always easy. On top of that, these forensic tools can be extremely expensive, often costing thousands of dollars each, with strict license limits. When several tools are needed, the costs can quickly pile up, making it a big challenge for labs and organizations to manage their budgets.

- **Malicious Programs:** Mobile Devices may contain malicious software or malware like trojans or viruses. This malware can spread and infect the device and can spread to the connected investigating devices.

- **Legal Issues:** The Digital Forensic Examiner should be aware of the nature of the crime and the regional laws.

Summary

The following chapter outlines the basic ideas and principles of mobile forensic investigation in modern times. This field puts much emphasis on the recovery, analysis, and preservation of data from mobile devices for legal and investigative purposes.

CHAPTER 1   INTRODUCTION TO MOBILE FORENSICS

Key areas to be discussed will include how mobile devices have evolved, the types of data retrievable, such as calls, messages, and applications, and the challenges posed by encryption and privacy concerns. Specialized tools and techniques, adherence to legal standards, and the integrity of evidence are emphasized. This chapter lays the foundation for understanding mobile forensics as a dynamic field, with constant adaptation to emerging technologies.

References

[1]   Practical Mobile Forensics Third Edition. (2018). Docslib. https://docslib.org/doc/2758664/practical-mobile-forensics-third-edition#google_vignette

[2]   https://www.oxygenforensics.com/en/resources/mobile-device-forensics/#history

[3]   Muhlhausen, D. B. (Director) (2020). Digital Evidence Policies and Procedures Manual. In National Institute of Justice. U.S. Department of Justice. https://www.ojp.gov/pdffiles1/nij/254661.pdf

CHAPTER 2

Android Forensics Fundamentals

Android forensics refers to the process of acquiring, examining, and reporting on digital evidence from Android devices, which is of paramount importance because the Android operating system is currently the most popular in the world. Investigations on Android-based devices involve understanding how the Android operating system is structured and how information is stored. The Android operating system's openness and its applicability on various hardware make forensic analysis very challenging but also beneficial in numerous cases, ranging from cybercrime to legal situations. What is the structure of the Android system?

It can be divided into several layers: Linux kernel and native libraries, the application framework, and user applications. All these layers contain different data that may aid investigations, such as user data, files, system logs, application data, etc. This chapter examines these components to provide knowledge of where possible evidence is located. Forensic investigators must also understand the different data acquisition methods: physical, logical, and cloud. Physical acquisition implies the creation of an exact duplicate of all the files on the device, including not only files that are easily visible and accessible but also those that are underlying, hidden files, deleted data, and even areas of storage no longer actively in use by the system. This backup is similar to doing a full backup of the phone, including items that most users may not be aware are included. Logical acquisition, on the other hand, will only retrieve files that are visible in the phone's operating system, such as photos, messages, contacts, etc.—data that can be easily accessed. In simple terms, physical acquisition operates at a much deeper level and is capable of retrieving lost or deleted data, whereas logical acquisition focuses on current data available to the user. Cloud acquisition pulls data that is synchronized

CHAPTER 2 ANDROID FORENSICS FUNDAMENTALS

with cloud services, such as Google Drive, iCloud, Dropbox, Outlook, and other online storage or backup platforms. The approach to choose mainly depends on the circumstances, device availability, and legislation.

After retrieving data, analytical tools are used to obtain useful information, such as communication records, multimedia data, a person's location record, or data that has been erased. Investigators are required to collect information and ensure evidence accuracy in such a manner that allows its use in court. The peculiarities of these processes will be described in this chapter and will therefore be helpful to individuals engaged in Android investigations. In the next session, the basic architecture of an Android device is explained, which helps the user understand the various types of information stored on Android-based mobile phones.

Android Device Architecture

The architecture of Android is composed of levels or layers, where each component has its own purpose and is independent from the rest. This layered system allows for enhanced performance and less rigidity in the system, since it works in conjunction with other layers without interfering with one another's processes. The organizational structure of Android systems is of great importance to forensic analysts, as it enables them to know where to search and what type of information to expect during an investigation. The Linux kernel is at the very bottom and handles the phone's hardware, keeping track of system logs and device drivers. Above that are the native libraries and Android Runtime (ART), where most of the phone's core functions are executed and may contain useful system files and cached data. The application framework layer enables developers to create apps that can communicate with the system; this is where app settings, app permissions, and service data are typically stored. The top layer is where all daily used apps reside, such as messaging, social media, and photo apps, and where much user-generated content is stored. A diagram illustrating these layers and the types of data they contain would be helpful, allowing readers to visualize exactly where specific pieces of information are stored. Figure 2-1 provides some insight into this, but additional information about the data in each layer would make it clearer. Understanding which information is stored in each layer, forensic examiners then attempt to identify where a particular piece of evidence might be stored. For example, user information would be found in the application layer; information about the system could reside in the kernel or framework layers. This knowledge is crucial for

CHAPTER 2 ANDROID FORENSICS FUNDAMENTALS

choosing the proper instrumentation and techniques for access and analysis during an Android forensic analysis. In the next section, the information contained in each layer is explained in detail.

Overview of Android System Components

Android architecture is basically a Linux kernel that provides the most basic system functions related to multitasking, memory management, and device control. Each layer in the Android architecture contributes uniquely to its functionality components, making it highly efficient and modular. Here's a closer technical explanation of each layer. Figure 2-1 precisely describes the basic overview of android system.

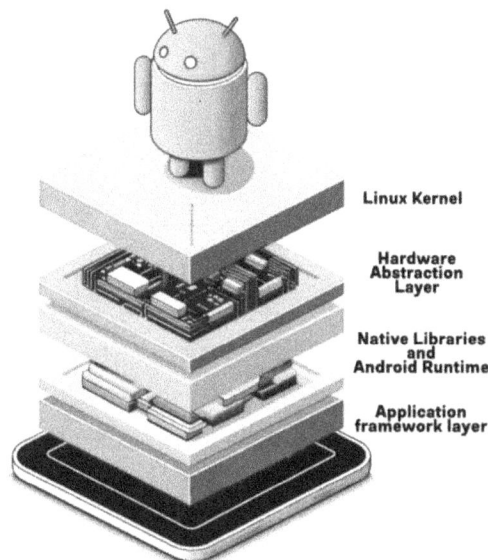

Figure 2-1. *Basic overview of Android system*

CHAPTER 2   ANDROID FORENSICS FUNDAMENTALS

Linux Kernel

The Linux kernel forms the core of the Android operating system, serving as a basic foundation that controls all essential low-level system operations.

- **Security:** As part of the Android platform, the Linux kernel has deep inner security features. One of the core features is SELinux, an abbreviation for Security-Enhanced Linux. Typically, SELinux implements MAC (Mandatory Access Control). This means that processes are tied to certain permissions through restricted settings based on previously defined security policies. This enhances security; it reduces the attack surface, thereby preventing unauthorized access to system resources, in addition to isolating apps, even if one is compromised. It utilizes process isolation and namespaces in the kernel; these establish secure isolated environments for both applications and system services, thus limiting their interference with each other.

- **Memory Management:** Android's memory management is a kernel-driven mechanism that employs the Virtual Memory Manager in the Linux kernel to execute most memory allocation, paging, and swapping operations. The kernel provides virtual memory, which allows each process in the system to have its own isolated memory space, so that it cannot leak memory and does not gain unauthorized access to other processes' memory. It also makes use of methods like caching and buffering to improve performance. The other significant mechanism is the Out-Of-Memory (OOM) Killer, which proactively kills processes when the system is under memory pressure, so the most important processes are not killed.

- **Process Scheduling:** The Linux kernel uses advanced algorithms for multitasking and efficient allocation of CPU resources. CFS, or Completely Fair Scheduler, is the default scheduler, which creates fair access to the CPU between processes according to priority and workload. In doing so, it balances the foreground processes of running applications with background tasks, such as system services, to ensure a high degree of real-time performance for active tasks by users. It also supports multi-threading; thus, Android can run several processes simultaneously, which aids in responsiveness and optimal resource utilization.

- **Hardware Drivers:** The Linux kernel interfaces directly with the device's hardware by means of drivers. These drivers, modules for the kernel, enable the kernel to interact with and control hardware components: networking (Wi-Fi, Bluetooth), storage (eMMC, SD cards), input/output devices (touchscreens, sensors), and power management systems. Android can run on various hardware platforms without requiring a different version for each hardware because it exploits the standardized interface provided by the kernel for these drivers.

- **Resource Control and Stability:** The kernel exercises control over how applications and system services access the device's resources, such as CPU, memory, and I/O. This type of control is implemented through mechanisms like cgroups (control groups), so that resource usage of all processes is well-contained and isolated from one another. Thus, one application cannot consume excessive system resources—a sure way to keep a working system stable and responsive even under substantial workloads. Thus, the kernel's ability to isolate and handle processes allows Android to utilize it with attributes such as kernel pre-emption and enables Android to perform multiple tasks efficiently without impairing system performance.

Fundamentally, the Linux kernel provides Android's foundational services in hardware abstraction, security, process scheduling, and resource allocation, thereby guaranteeing the smooth running of applications across a wide range of devices and use cases.

Hardware Abstraction Layer (HAL):

The Hardware Abstraction Layer (HAL) is the base abstraction layer between the hardware found in devices and the higher-level software of Android. It enables the system to communicate with and control hardware effectively without needing specific changes for each model or manufacturer.

Abstraction of Hardware Details: HAL abstracts device-specific hardware complexities with a standardized set of APIs exposed to higher levels of the Android framework. Each type of hardware (camera, GPS, audio, etc.) can be matched up with a matching HAL module that defines how the Android framework should talk to the underlying hardware without direct interaction with the kernel or low-level device drivers. This abstraction allows Android to run on a wide variety of hardware, from cell phones to wearables. The operating system does not have to change in order to support different hardware.

Modular Design: HAL comprises several modules, each representing a type of hardware component. Figure 2-2 gives basic understanding about HAL. Examples include different HAL modules for

- **Camera**: Regulates access to camera hardware for image and video capture
- **GPS**: Provides location-based services via GPS or other location-tracking hardware
- **Sensors**: Manages accelerometer, gyroscope, barometer, and other sensor data
- **Audio**: Controls input and output for audio playback and recording, including volume and streaming media
- **Bluetooth and Wi-Fi**: Manage wireless communication with other devices and networks

Every module in the HAL has an interface. The Android framework accesses the hardware module using function calls and callbacks to manage the hardware. The modularity of the HAL ensures that specific hardware components can be upgraded or replaced without affecting other aspects of the system.

CHAPTER 2 ANDROID FORENSICS FUNDAMENTALS

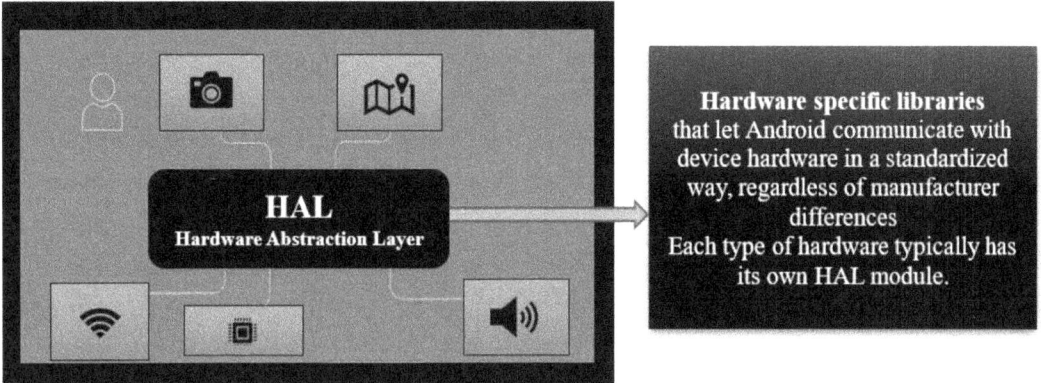

Figure 2-2. Basic understanding about HAL

Standardized API: The HAL defines a standard interface for every hardware component in the form of a C-based API. APIs are thus a kind of contract between the Android system and hardware, which allows the OS to consistently call their respective API calls while interacting with hardware, irrespective of the actual device or manufacturer. This is really important for hardware portability. For the lower layers, that is, device-specific hardware drivers, developers and manufacturers can maintain those. Provided they follow the HAL API specification, Android can interact with that hardware to any degree of detail without any problem from its higher layers. The API abstracts all that device-specific detail; consequently, if a higher layer—let's use the Android Framework—wants to access the camera or microphone, it simply makes function calls to the appropriately named HAL module. In return, the HAL module translates those higher-level function calls into lower-level instructions that actually involve the hardware.

Interfacing with Device Drivers: The HAL is essentially an intermediary between the Android framework and the Linux kernel, which has real hardware drivers. This is important because direct hardware access is contingent upon the services of these hardware drivers. The HAL won't communicate directly with hardware; rather, it communicates through either *syscalls* or *driver interfaces* provided by the kernel. Then, the details are encapsulated within the driver, and the HAL module takes away complexity from the higher-level Android framework. For example, if an application were to request camera permission, the Android framework would talk to the camera HAL, which would, in turn, talk to the underlying camera driver through the kernel. The driver will take care of lower-level operations, such as focusing, capturing, and saving images. Again, this is translated into higher-level functions that the Android framework will understand because of the HAL module.

33

Device Portability and Flexibility: One of the main advantages of the HAL is that it makes Android very portable across different devices and hardware platforms. Due to this, manufacturers can easily develop their own specific HAL modules corresponding to specific hardware; the main framework of the core Android code won't have to be touched as the HAL uses standardized APIs; the same version of the operating system can run easily across different devices. That's the kind of consistency an ecosystem needs. The HAL's modularity also ensures the possibility of integrating new hardware pieces without changing the full OS. Thus, for example, by writing a new HAL module, one can easily add a new sensor or communication protocol.

Android Runtime (ART):

Android Runtime, or ART, refers to the runtime environment in which an Android application runs. Compared with its predecessor, Dalvik Virtual Machine, it is a great leap forward in terms of execution performance and resource management. Here's a more technical description of how ART works and its salient features:

Ahead-of-Time (AOT) Compilation: One of the greatest advantages of ART over Dalvik is that it allows for AOT compilation. That is to say, while installing, ART compiles the application bytecode, which are DEX files, to native machine code. The output of compilation, of course, is stored as an OAT file, which means the application bypasses interpretation or runtime compilation when executed. The overhead of Dalvik's just-in-time (JIT) compilation is eliminated; a number of benefits accrue from this. In addition, execution speed is quite fast since the code is in machine language for execution, and the app performs well since CPU demand is lowered with no runtime processing. Apart from this, apps tend to launch faster because code precompilation often occurs. This is optimized, along with inlining, loop unrolling, and dead code elimination, through AOT processing, with the potential to make runtime performance even better.

Just-In-Time (JIT): ART also supports Just-In-Time (JIT) compilation, which compiles code dynamically while the app is running. This is useful for parts of the app that haven't been precompiled through Ahead-of-Time (AOT) compilation or when frequently used code paths need further optimization. JIT helps improve app startup times if the app wasn't AOT-compiled or if updates invalidate some precompiled code. Additionally, ART uses runtime data to profile and optimize frequently executed code paths by recompiling them in real time. This hybrid approach, combining both AOT and JIT, allows ART to balance between static performance gains from pre-compilation and dynamic optimizations during app execution.

Memory Management: ART uses advanced memory management in an application for efficient memory use and preventing crashes due to memory leaks. It performs memory management through heap allocation, which is divided into several sections. The "young generation" is the place to keep short-lived objects, whereas the "old generation" maintains objects that survive longer and exist through numerous garbage collections. Another thing is that ART adjusts memory allocation dynamically, which means applications get whatever memory is required without consuming too much of the operating system's resources. This smooths out operations and allows for better resource management.

Garbage Collector (GC): ART makes use of a garbage collector, which means memory is often cleaned up more efficiently, hopefully bringing down the performance impact of cleaning out unused objects. This platform has a huge feature—concurrent GC to run alongside the app, meaning it can clear out memory without having to wait until the end of an application run, thus avoiding those long pauses that cause an app to freeze or slow down. ART also uses the same low-latency GC approach to reduce interference and promptly remove unused objects so as not to interfere with app performance. This does not make an application feel laggy, especially during memory-intensive operations.

Application Lifecycle Management (ALM): Application Lifecycle Management in Android is handled by the Android Runtime (ART), which sees an application through all its phases: from launch until termination. The ART creates the process for the application, and once an application is launched, it loads its code into memory and initiates the process, where it starts executing at the entry point in the application. Other activities, such as main UI threads and background threads, are also controlled, thus ensuring effective scheduling and handling so as not to degrade the user experience with sluggish application processing. Another benefit is that ART ensures that all applications close if the app is terminated or closed. Thus, all resources get freed, and memory is reclaimed. Additionally, Android's task management allows ART to suspend and resume apps so that multitasking is optimal and resource usage fits well.

Memory Profiling and Optimization: ART gives developers the ability to profile memory, enabling them to track how their apps use memory and thus identify patterns, pinpoint inefficiencies, and better manage memory usage. Studying this data ensures that memory is allocated efficiently and reduces the possibility of memory leaks that could make an app unstable or slow over time. ART can also make real-time optimizations based on this profiling data by incorporating JIT optimizations. For example, when an application frequently creates many small objects, ART may optimize memory management by pooling objects or adjusting garbage collection settings, thereby preventing unnecessary slowdowns. These smart updates ensure that apps run faster, launch quicker, and consume less battery, providing a smoother, more responsive user experience.

Code Verification and Security: ART (Android Runtime) makes sure that the bytecode is checked before it even compiles or runs any code, so only safe and valid bytecode is allowed to run. This way some sort of security risk like buffer overflow or illegal memory access can't occur because the code is simply avoided. ART also introduces other key security measures. It utilizes memory protection to ensure that each application's memory is separate so that no other application can access its data. ART also uses code signing, which checks the integrity of apps using cryptographic signatures to ensure that only trusted and authorized code runs on the device. Combined, these precautions create a safe haven for Android apps to operate under ART.

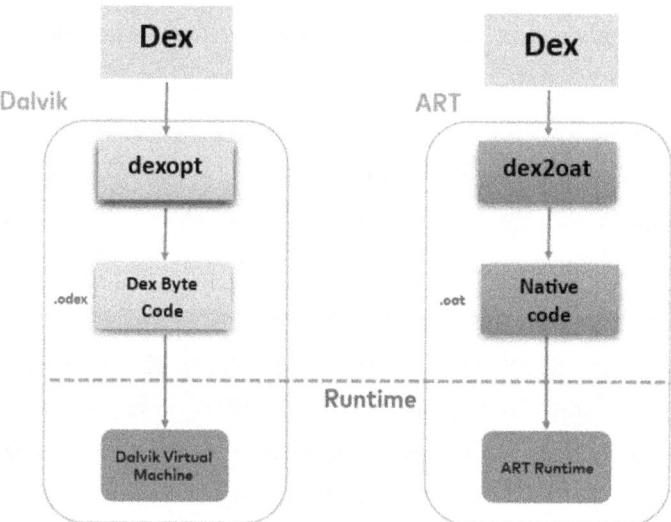

Figure 2-3. *Dalvik vs. ART basic difference*

In summary, the supremacy of ART owes to the dual combination of AOT with JIT compilation, efficient memory management, advanced garbage collection, and optimizations for modern hardware architectures. Thus, applications under Android are faster, use less memory, and feature a more refined user experience compared to older

Dalvik VMs. The basic difference between Dalvik and ART compiler has been described in Figure 2-3. Additionally, it is backward compatible and takes into consideration the vast Android ecosystem.

Application Framework layer

The Android application framework is a comprehensive set of core libraries and APIs that allow developers to simplify the development process on the Android platform. It abstracts the complexity of underlying operations of the system, providing intuitive tools to application developers so they can build richer and more responsive applications. At its core, the framework provides essential components for managing various aspects of an app's life cycle and interactions; some of them are described below.

- **Core Libraries:** Core libraries provide reusable code and functionality across different applications. This includes the User Interface (UI) framework, which provides a range of components, from buttons and text fields to layouts. The View System, part of the UI framework, handles how UI elements are drawn to the screen and how they process user input, such as touch gestures. It is class-based, and classes for this include View, ViewGroup, and particular widgets, such as TextView, Button, or RecyclerView. Then, for arranging these UI components on the screen, classes like LinearLayout, RelativeLayout, ConstraintLayout, and others can be used. Apart from this, the framework provides resource management, which handles localization, themes, and styles, and makes UI adaptation for an app on different screen sizes and orientations very easy. Another area where the Android framework is very powerful is graphics and animation. Developers have the ability to use libraries such as Canvas, Drawable, and the Animation APIs to create custom graphics, animations, and visual effects that provide a rich user experience. This combination of robust tools and libraries makes the Android framework a flexible and powerful foundation for creating applications.

- **Application Components:** The Android application framework bases itself on four primary application elements. These elements are the lower-level building blocks for an application and comprise *activities* which represent a user interface screen. An activity will then deal with interactions between the app and the user and control life cycle events, such as creation, pausing, resuming, or destroying, and might even launch other activities from within the same application. Another crucial part is *services*. These services run in the background to perform operations that do not consume time and do not require user intervention. Some examples of services are audio playing, data syncing, and downloading files. Services can be started and stopped independently of the user interface, making them ideal for long-running operations. Next important part is *broadcast receivers*. Broadcast Receivers listen for system-wide broadcast messages, such as changes in connectivity when the network is turned on or off, or incoming text messages. They let apps communicate with other apps as well as system-level events, meaning that if anything changes across the device, apps can react to that. Last but not least, *content providers* enable applications to manage and share structured data with other apps. They present a standardized interface for performing CRUD operations, such as Create, Read, Update and Delete, in a fashion that enables developers to share data across different apps securely. To sum up, all these components form the backbone of Android app architecture.

- **System services**: Android exposes access to various system services through the Context class, where developers can take advantage of specific device functionalities. Among these services is *location services*, which allows an application to access the device's location using GPS or network-based positioning; thus, it would be ideal for location-based applications. *Notification Services* enable applications to send and track notifications, thereby enabling an application to inform its users of important events or changes in status. The second service is *telephony services*, which give access to phone-related features, such as initiating a call or sending an SMS message, for communication features within applications. *Connectivity Services* enable applications to connect to a network in order to maintain various services connected to Wi-Fi, mobile data, or other networking capabilities, so that applications have easier online interaction. Last but not least, *power management* enables developers to control their application's battery usage by managing features like wake locks, thus allowing applications to manage their own battery usage efficiently and maximizing battery usage. These system services help create feature-rich Android applications that can interact easily with device hardware and system-level resources.

These services are accessed via the ***Service Locator Pattern***, allowing developers to retrieve service instances using the ***getSystemService*** () method.

- **Inter-Process Communication (IPC):** Android applications usually need to interact with other applications or services in the system; this interaction can be achieved using several Inter-Process Communication (IPC) mechanisms. One of the major IPC tools is ***Intents***, an abstract description of an operation that needs to be performed. Thus, Intents make it possible for interaction to occur among components inside a single application and across different applications. They can be explicit, targeting a particular component, or implicit, requesting an action without specific specification of the target, allowing the system to infer the appropriate component. Another crucial mechanism is ***AIDL (Android Interface Definition Language)***, which is used to define a programming interface for communication between services and clients in different applications. AIDL simplifies the sending of complex data types between apps and allows seamless IPC across app boundaries. ***Content providers*** are also useful in app inter-communication. They help applications share data safely and efficiently. They enable apps to perform read-and-write operations across different applications using a standardized interface to access and modify shared data. These IPC mechanisms are central to making interaction and data sharing within the Android ecosystem friendly and smooth.

- **Data storage:** It provides several data storage mechanisms within the Android framework that can assist developers in efficiently managing an app's data. Amongst the choices, one of the most basic is Shared Preferences, which is a lightweight, key-value store suitable for storing small amounts of data, like user preferences or the application's state. It provides an easy way to store and retrieve simple data, such as settings. For more structured, complex data handling, Android uses the **SQLite Database**, which is a more powerful relational database system that operates on SQL queries to manipulate data. The framework also possesses the SQLiteOpenHelper class, which simplifies database creation, version handling, and upgrading. Moreover, through File Storage, apps can read and write files to internal or external storage, where they can temporarily hold persistent data. The difference between the two types of storage is that internal storage contents are private to the specific app, while external storage contents may be shared among applications, users, or both. These differentiated storage options allow Android applications to perform well regarding the types of data each application requires.

- **Life cycle management:** To control the life cycle of these android-based applications and their components, the Android application framework manages them in an efficient manner. The nicely defined life cycle for both activities and fragments is controlled by the framework, giving developers excellent opportunities to override specific life cycle callbacks, such as ***onCreate(), onStart(), onResume(), onPause(), onStop(), and onDestroy()***, to specify precisely how the application should behave at different stages. These methods allow developers to initialize resources when needed, update the UI if the application comes into focus, and release resources if the application is paused or destroyed. Besides, configuration changes, such as screen rotations, language changes, and window resizing, are represented. Developers can address these changes through mechanisms like ***onSaveInstanceState()*** and ***onRestoreInstanceState()***. This helps avoid data loss when making changes through these mechanisms. This way, user information is retained even with application interruptions or configuration changes. Handling these life cycle events and configuration changes ensures optimal resource management and an elaborate user experience on Android.

- **Flexibility and Scalability:** The Android framework is designed to operate on a wide range of devices, from smartphones to tablets and TVs, ensuring that apps can run smoothly on many different hardware configurations. It abstracts device-specific details so that developers need not worry about apps being different on various systems. Moreover, the framework has modular and extensible properties. This inculcates reusability of libraries and components created to encourage scalability among developers. The Android Jetpack library can be used; it is a set of libraries developed to assist and simplify best practices and eliminate boilerplate code, with great concern for compatibility with older or new versions of Android devices.

- **Security and permissions:** The Android framework relies on a mature security model with the intent of safeguarding sensitive data and other system resources against unauthorized access. This approach is actually a permissions-based mechanism, whereby every permission that an application requires must be declared by the developer in the manifest file. The effect of these permissions is very much like "gates" for accessing features on a system—for instance, camera, microphone, location, or contacts. At the time of app installation by a user, or when an app tries to access a resource for the first time, Android forces the user to review the requested permissions. This ensures that the user knows what data or resources an app requires and also enables the user to control access on a granular level. Android provides a consent-based process, enhancing security and empowering users to protect their privacy by not allowing applications to auto-access sensitive resources without specific permission.

- **Testing and debugging:** The Android framework provides a comprehensive set of testing tools and libraries that allow developers to automate testing and ensure the reliability and stability of an app; this includes Espresso for UI testing and JUnit for unit testing. Additionally, it has the Android Debug Bridge, which provides a command-line interface for connecting to a device to debug, monitoring an app's performance, and capturing detailed logs. Together, these tools make the development process much easier, pointing out problems to be addressed at an early stage and helping improve the application's quality.

The Android framework provides developers with a robust and flexible platform on which to build applications, since it offers core libraries, standardized APIs, and all crucial components for managing user interaction, system resources, and data. Its modular design, life cycle management, security features, and support for testing applications on a vast scale create a stable platform through which developers can develop applications that vary widely across a range of Android devices.

Types of Data on Android Devices

The understanding of the three types of data on an Android device provides a good foundation for carrying out investigations, as different categories depict unique characteristics that differ in their forensic values. Generally, user data, system data, and app data well classify information on any Android device, with specific locations where each data type should be found. Below are the detailed breakdowns of each type of data and the locations of all key artifacts.

User Data

Android devices allow the creation of various user accounts, similar to user profiles on a computer. Each user can have their own apps, settings, and personal data that are separate from those of another user using the same device. This can be particularly helpful in shared devices, such as tablet computers used by family members or in an office environment. From a forensic perspective, it's essential to recognize that different user accounts store their data separately, so investigators should thoroughly examine every user account on the device to ensure that no relevant data is missed.

User data encompasses information directly generated or stored by the user, making it essential for forensic investigations. The following are key components of user data:

- **Contacts:** Contacts are usually kept in the device's Contacts app or synced with cloud services, such as Google Contacts. The contact information includes names, phone numbers, email addresses,

- and metadata, such as profile pictures or notes. Through forensic analysis, investigators can reveal the contacts, as well as when they were created and last modified; this provides some insight into the social network and communication patterns used by the user.

- **Messages**: Messages sent via SMS and MMS are stored in databases, such as the mmssms.db file, in the /data/data/com.android.providers.telephony/databases/ directory. They include text messages, multimedia files, timestamps, and sender/recipient information. In addition to traditional SMS/MMS, chat apps like WhatsApp store messages in their databases, often using SQLite, and may include encrypted chat histories, media files, and metadata about message delivery (read receipts, etc.).

- **Call logs**: Call logs maintain records of incoming and outgoing calls, along with missed calls. Call logs are available in a local database accessible to the call telephony provider. Analysis of call logs may provide some insights into patterns that the user uses to make calls, such as how many calls, what time of day, or any other information that might aid forensic analysis. Google's Phone app now comes with a smart feature that helps users avoid annoying spam calls. When a call is detected as spam, the app can automatically block it without even ringing the phone, saving users from unnecessary interruptions. It also clearly marks these calls as spam in the call history, so users can easily see which calls were blocked and decide if any action is needed. This not only makes life easier but can also be useful during investigations, as these spam flags and call records might provide valuable clues [1].

- **Pictures and Videos**: Most devices carry media files. Most media files are saved under /sdcard/DCIM/ or /sdcard/Pictures/. Most pictures have been taken using the camera, but some have been downloaded from the Internet. EXIF metadata is provided with each media file, including information about when and where it was taken, and by what device. This data will help set timelines and places in the user's activity.

- **Browser History**: The browser history is all about the sites a user has browsed, including cached pages, cookies, and bookmarks. Such information is stored in an SQLite database that varies from browser to browser. For example, in Chrome, the location is /data/data/com.android.chrome/app_chrome/Default/History. Browser history analysis provides information about a user's online behavior, interests, and possibly some criminal activity. In case of other Android devices, such as those from Vivo, Oppo, or Xiaomi, may come with their own custom browsers pre-installed as the default. These browsers often store their history, cache, downloads, and saved data in different storage locations specific to each brand or browser. Therefore, forensic investigators must be aware of the device manufacturer and the browser being used, as the location and structure of browsing data can vary across different Android versions and custom user interfaces.

System Data

System data comprises information that is generated and maintained by the Android operating system, which is essential for the functioning of the device. Key components include:

- **Log Files**: The Android has some log files, such as logcat, that record system events, application errors, and interactions. These logs carry a detailed history of what was happening on the device and what might have contributed to an application crash, as well as which other applications might have interacted with those causing the incident.

- **System Settings**: All settings governing device operation, including network settings, screen settings, and security settings (dubbed system settings), exist within the /data/system/ directory. This helps determine how the device was configured and potential changes made to security features.

- **Cached App Data**: Cached data is temporary data that applications maintain for better performance and faster loading times. Cached files appear in /data/cache/ directories; most contain images, web pages, and temporary files. Although generally ephemeral, cached data can occasionally contain valuable information, such as user sessions or recent activity.

- **Geolocation Data from System Services**: Android typically maintains location information through both GPS and network-based operations. This information is stored in databases like fused_location_provider and obtained through system services. Geolocation information, for example, informs us about user navigation behavior and visited places, proving crucial for investigations into alibis or locational evidence.

App Data

App data refers to information stored by third-party applications, which may include user-generated content and application-specific configurations. Important aspects include

- **Internal App Databases, such as Protocol Buffers (protobuf) and SQLite**: Many apps use SQLite databases to store structured data. These databases typically reside in an app's data directory, such as /data/data/com.example.app/databases/. Such databases may consist of various types of user-generated content (notes, contacts, media, etc.), preferences, and cached data. It's important to note that some modern apps also store data using Protocol Buffers (protobuf), a compact and efficient data format developed by Google. Protobuf is commonly used to store and transfer structured data in a faster and more space-efficient way compared to traditional formats. Forensic investigators should be aware of this, as identifying and correctly decoding protobuf files can uncover valuable app data that might otherwise be missed if focusing solely on SQLite databases. API Tokens: Most applications, by default, include authentication tokens that provide secure access to APIs. These tokens can be stored in shared preferences or databases. Such tokens play an extremely important role in how an app interacts with backend services. Their forensic recovery outlines the secure access points to external systems and services.

- **Authentication Credentials**: Apps may store users' authentication credentials, such as usernames and passwords, which are typically encrypted. These credentials are usually stored in Android's secure storage mechanisms, like SharedPreferences and KeyStore. Recovering these credentials might reveal user accounts and methods of unauthorized access to services.

- **User Preferences**: User preferences, such as application settings and configurations, are preserved in the shared preferences of the app's data directory. This data can help forensic analysts understand how the user customized their app experience and may indicate certain behavioral patterns or specific preferences during application use.

Locations of Key Artifacts

Understanding where to find the data is critical for forensic investigations. The following directories are key locations on Android devices:

- **/data**: This directory, which contains data for users as well as applications, has all its information considered sensitive, whether created by the user or stored by the application. Such directories may be accessed only with root permissions due to the intended protection of user data privacy.

- **/system**: This directory contains the Android OS and system-level settings and configurations. It encompasses various essential system files and applications, such as the Android framework and libraries. Thus, forensic analysts can analyze it to gain knowledge about system configurations and system integrity.

- **/cache**: Applications use this cache directory to store temporary files, which may contain data meant to accelerate application performance, as well as residual files useful during an investigation. Cached data can help analysts trace user activities that might not be saved permanently.

- **/sdcard**: Users can save media files, downloads, and app-specific information to the external storage directory, commonly known as SD cards. Since users typically have access to this directory without root privileges, all data stored here is user-generated content, making it another critical area for forensic data recovery.

All the data specific to Android devices hold importance in a forensic sense, along with implications for real-time investigations. There are various types of data and locations that help forensic analysts retrieve them for further analysis, providing details on user activities and behavior relevant to legal cases or security incidents.

Forensic Acquisition Methods

Obtaining data from an Android device requires specialized acquisition methods to retrieve relevant digital evidence without compromising the integrity of that data. Normally, the choice of method is based on whether it is physical or logical access and the kind of device being used—for example, if it has encryption and what version of the operating system is used. This section discusses different forensic acquisition methods and the problems arising during the acquisition process in depth.

Physical Acquisition

Physical acquisition involves creating a bit-by-bit copy of the entire data stored on an Android device, including user files, system data, and even deleted files. This is considered the most exhaustive means of data collection, as it captures all data held in the storage device, even files that are deleted and have not yet been overwritten. In forensic investigations, attention paid to deleted information is much higher because this information is more likely to contain critical evidence. Normally, physical acquisition occurs by directly accessing the storage media of the target device. This is made possible through the use of specifically designed hardware tools that may connect to the target or its peripherals, for instance, JTAG (Joint Test Action Group) and Chip-off methods. Basically, these techniques involve either getting a direct interface with the memory chip on the motherboard or accessing the internals of a device using the JTAG interface. However, acquisition is not easy in modern Android devices, as full-disk encryption is implemented in most of them to protect users' data. In that case, forensic analysts have to obtain the decryption key or unlock the device before acquiring it. Sometimes, forensic tools require a specific exploit or method to bypass the security features of a locked device.

Logical Acquisition

Logical acquisition process acquired files and data that could be pulled and accessed using the Android file system without entering into the raw storage. These are usually less invasive methods compared to physical acquisitions and include many important considerations. This process involves the extraction of user files, application data, and other system settings, which can be found through the hierarchy of a file system. This includes SMS messages, contact information, photographs, as well as application-related data. Since logical acquisition does not require accessing the raw storage of the device, it is generally less intrusive to the device's functioning. This procedure can usually be done without rooting the device or altering its state, which is why this is often the preferred choice. However, one of the main limitations of logical acquisition is that it usually doesn't recover deleted files or unallocated space. Due to this, if evidence critical to the investigation has been deleted, this method may miss capturing it, which may affect the forensic investigation. Common examples of logical acquisition methods include using ADB backup, which allows data extraction via Android's debugging bridge, and MTP (Media Transfer Protocol), which is typically used to access user-visible files like photos, videos, and documents. These methods provide practical ways to collect accessible data while minimizing impact on the device.

File System Acquisition

File system acquisition is a hybrid method that targets specific data within the file system. Therefore, it enables deeper analysis of the storage structure of the device by combining elements of both physical and logical acquisition. It is employed to help forensic analysts recover specific directories, files, or application data, maintaining the file-system structure so analysts can extract specific directories, files, or app data while preserving the file system structure, which in turn allows analysts to explore the data hierarchy and assess permissions and access rights. By focusing on the structure of the file system, it enables them to recognize the mechanisms by which applications may store data and inspect the contents of SQLite databases with a view to analyzing how files and folders inter-associate to derive further enlightenment on user actions and interaction with applications. On the other hand, acquiring file systems may, based on the tools chosen, require a device to be rooted or unlocked in order to have access to several directories. In many cases, when the device is unlocked, a forensic agent or specialized application is also deployed onto the device to facilitate deeper access to system files and directories that are otherwise restricted. This agent helps in securely extracting data while maintaining the integrity of the file system during the acquisition process. While some tools extract a file system directly from an Android device with a forensic tool without any root unlocking, this scope may affect data acquisition possibly.

Challenges in Acquisition

Although there have been several advances in forensic tools and methodologies, there are still several problems in acquiring Android devices. These may prove stumbling blocks for the forensic processes, leading to adverse changes in evidence.

- **Encryption**: One of the common security features of modern Android devices is full-disk encryption. This means that all data is encrypted, and access will be denied to the data if proper credentials are not entered. The data even stays encrypted when a physical image of the device has been obtained. To bypass this encryption, one may need to obtain a user's PIN, password, or pattern or use specialized tools that exploit an unpatched software or kernel vulnerability specific to the particular device model.

- **Rooting**: Rooting a device offers a pathway to system-level files and data that would otherwise be inaccessible. This allows access to system information to acquire data, but it poses some potential risks. Rooting the device alters its state, potentially compromising evidence and risking its admissibility in court. Moreover, the security aspect worsens since a rooted device is more susceptible to threats such as malware and unauthorized access, which may interfere with the investigation. Additionally, rooting typically voids the device's warranty and may affect future service support offered by the manufacturer. All these factors necessitate caution when rooting a device for forensic investigations. Android has implemented some security measures around rooting, like if the device is rooted, some native apps can detect it and not allow certain apps/services to run and may crash. This may affect the data acquisition since an examiner would not know if the complete data was acquired or not.

- **Anti-forensic techniques**: Some applications or malware employ anti-forensic techniques to prevent data acquisition. These methods include data obfuscation, which involves encrypting or hiding information to prevent unauthorized access or detection, often through complex storage formats. Additionally, data deletion is used, where files are overwritten multiple times, rendering them irretrievable. Furthermore, some applications control the file system, altering its characteristics to evade normal acquisition, thereby preventing access to stored data. Software-based obfuscation is not the only challenge that can make acquisition difficult; physical device damage can also be a significant obstacle. Even if the phone is severely damaged—whether it's water-damaged, has a smashed screen, is bent, or has been exposed to fire—the chances of recovering data from a dead phone are slim, as the internal parts and storage area are likely damaged, making it impossible to rescue any data on your own.

- **Device variability**: Because there are hundreds of thousands of different devices from several manufacturers, with various models and versions of Android, approaches to data acquisition become inconsistent. Sometimes, specific devices have their own unique security features or proprietary software, which make the analysis even harder. Therefore, forensic analysts constantly search for new and recent developments in Android security and forensic techniques, including emerging features such as Theft Protection Lock and advancements introduced in Android 16 like Advanced Protection and Inactivity Reboot, which can impact data access and analysis. In a nutshell, forensic investigation acquisition methods for Android devices entail selecting different techniques based on the investigation's context. The forensic analyst needs to be aware of the strengths and limitations of physical, logical, and file system acquisitions, while being aware of the challenges posed by encryption, rooting, anti-forensics, and device variability. Such considerations ensure the integrity of evidence, demanding a proper forensic investigation.

Forensic Tools and Techniques

Forensic tools and methods are needed for data extraction and analysis from Android devices. There are many types of tools, each focused with a specific purpose in mind, some for-data acquisition, while others are for forensic data analysis; investigators can use these tools to obtain evidence more effectively. Now, let's take a close look at some popular forensic tools, comparing their applications, data recovery techniques, and methods for dealing with encrypted data.

Overview of Popular Forensic Tools

Several tools have gained prominence in the field of Android forensics due to their effectiveness in data extraction and analysis:

- **Cellebrite UFED**, **Magnet AXIOM**, **Oxygen Forensic Detective** and **Belkasoft:** These are the most popularly known tools for physical and logical acquisition of data from Android devices. It is compatible with a huge variety of hardware and OS's and allows forensic experts to recover data including SMS messages, contacts, call history, and application data. It's an especially useful tool for law enforcement because it can break through security and get information from even locked devices.

- **Autopsy:** Autopsy is open source and cross platform and can analyze a huge number of data structures (it's especially good with SQLite databases—a lot of Android apps use SQLite), and it's all graphical, so it's easy to use. Autopsy allows investigators to view the file system and recover files for forensic analysis. Its architecture is extensible so it's possible to add more modules which provide more functionality such as different types of forensic analysis.

- **Android Debug Bridge (ADB):** ADB (Android Debug Bridge) is a command line tool that allows communication with an Android device. ADB requires a lot of manual intervention and setup, but it's very powerful and can be used for a lot of system-level analysis. Forensic analysts also use ADB to pull files and log and run shell commands on the device. It is portable and can talk directly to the android file system so it is a very valuable tool in many forensic cases.

CHAPTER 2   ANDROID FORENSICS FUNDAMENTALS

Steps for Logical, Physical, and File System Acquisition for Android Mobile

In the following sections, the logical, physical, and file system acquisition processes are discussed. Steps 1 to 2 and steps 7 to 10 remain the same across all three acquisition methods, while steps 3 to 6 differ, as outlined in Table 2-1 and Figure 2-4.

Step1—Device Preparation: Ensure the Android device is charged and connect it to a forensic workstation using a USB cable. Enable "USB debugging" on the Android device to allow the forensic tool to interface with it.

Step 2—Select Commercial tool: Choose one of the available forensic tools intended for logical acquisition of data from an Android device, such as Cellebrite UFED, Magnet AXIOM, Belkasoft, XRY or Oxygen Forensic Detective.

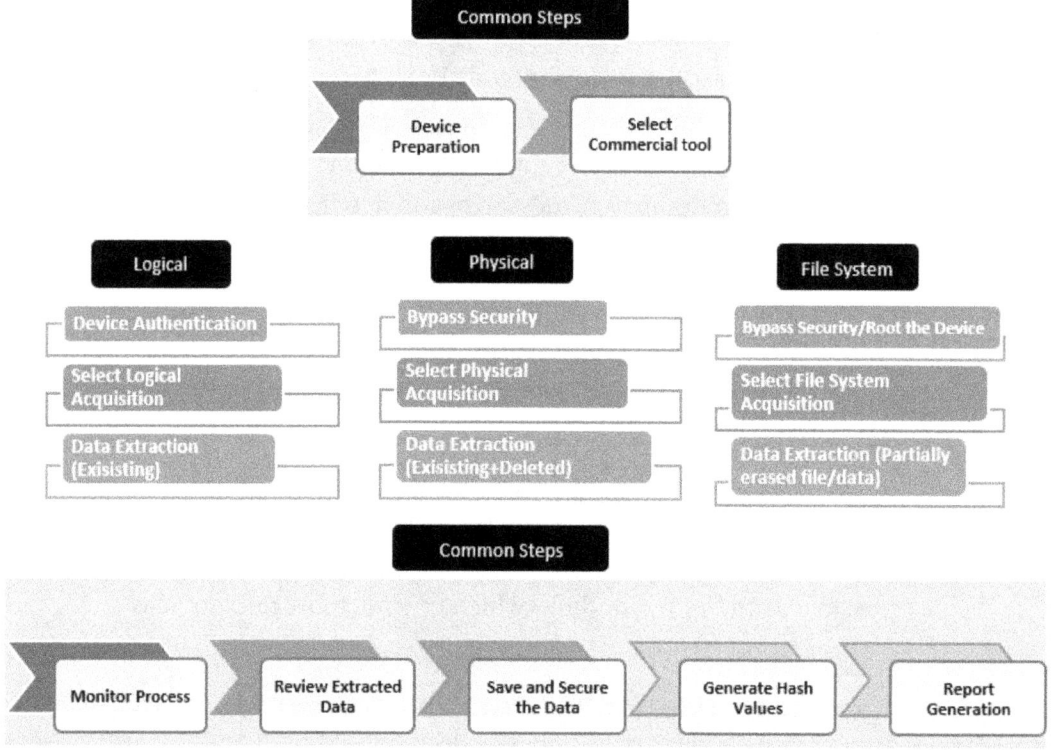

Figure 2-4. Basic process of logical, physical, and file system acquisition process

CHAPTER 2 ANDROID FORENSICS FUNDAMENTALS

Table 2-1. *Different Steps Involved in Logical, Physical, and File System Acquisition*

Logical Acquisition	Physical Acquisition	File System Acquisition
Logical acquisition involves extracting active files and data from the device's file system without accessing deleted or hidden partitions.	Physical acquisition creates a bit-for-bit image of the device's entire memory, including deleted data.	File System Acquisition provides access to the device's file structure, including system, application, and user data, without capturing a full physical image.
Step 3—Device Authentication: If the device is locked, authenticate using the passcode or the appropriate biometric authentication to unlock it. Verify any prompts for trust or permissions that may arise on the device to allow the forensic tool to access the data.	**Step 3—Bypass Security**: If the device is locked, bypass the screen lock using techniques such as passcode cracking or unlocking modes accepted by the forensic tool. Sometimes, rooting or other vulnerability exploitation may be necessary to access the device's entire file system.	**Step 3—Bypass Screen Lock or Root the Device**: If the device is locked, bypass the screen lock. Depending on the device and forensic tool, you may need to root the device to gain full access to the file system. Rooting grants elevated privileges, allowing deeper access to the Android file system.
Step 4—Select Logical Acquisition: Under the forensic tool's interface, select the logical acquisition option. This process extracts user-accessible data, including contacts, messages, call logs, photos, app data, and media files but does not access the full file system.	**Step 4—Select Physical Acquisition**: Under the forensic software interface, choose physical acquisition. This method helps extract data from hidden, system-level, and deleted files. Once physical acquisition is selected, examiners can also input specific keywords related to the investigation, enabling the tool to highlight or filter relevant data during the extraction process.	**Step 4—File System Acquisition**: Under the forensic tool, select File System Acquisition. This acquires system directories, application data, and metadata, providing access to files usually unavailable through logical acquisition, such as application databases, system logs, and user preferences.
Step 5—Data Extraction: Initiate the logical acquisition process. The application will start extracting available data from the Android device. Logical acquisitions typically include application data, address book, SMS, call logs, and multimedia data but do not extract deleted or hidden files.	**Step 5—Initiate Data Extraction**: By running the acquisition process, the forensic tool will then take an image of the device's storage, capturing all active and deleted data. This may include contacts, messages, media, app data, and numerous system files.	**Step 5—Extract File System**: Start the extraction process. This accesses the file system to extract information from relevant app folders and other files with system-level access, potentially retrieving partially erased files or data.

53

Step 6—Monitor Process: Logical acquisition can be relatively fast, depending on the target device's memory capacity. Monitor the process to ensure there are no errors, and the device remains online throughout. Physical acquisition can be relatively time-consuming, depending on the target device's memory capacity. Monitor the process to ensure there are no errors, and the device remains online throughout. File system acquisition takes time, depending on the device's storage size. Monitoring ensures the device remains connected, avoiding interference or incomplete extraction.

Step 7—Review Extracted Data: After extraction, review the data extracted from the system, including application data, system files, logs, and accessible databases. Ensure extraction captures the required data for the investigation.

Step 8—Save and Secure the Data: Save the extracted file system data securely in a predefined folder on the forensic workstation to store the data intact for further analysis.

Step 9—Generate Hash Values: Validate the data by generating an MD5 or SHA-256 hash to ensure the file system image has not been tampered with.

Step 10—Report Generation: Prepare a comprehensive report detailing the file system acquisition process, data obtained, and any problems encountered. This report may serve as evidence in legal or forensic proceedings.

Data Recovery Techniques

Data recovery techniques can greatly assist forensic investigations by uncovering, retrieving, and collecting deleted or hidden files on a computer:

- **Recovery of Deleted Files**: Forensic analysts employ file carving tools to recover deleted files. The file carving process examines the raw data of a storage device for file signatures, enabling the re-creation of deleted files even if file system entries have been deleted. This is often helpful in recovering deleted evidence that users may have intentionally deleted.

- **Accessing Backup Files:** Investigators can also find data by accessing backup files usually stored in cloud services or hidden directories on the device. Many Android users enable automatic backups to services like Google Drive, while some devices also store backups on vendor-specific cloud accounts such as Vivo Cloud, which contain valuable information, such as app data, contacts, and settings. Retrieval and analysis of these backup files can provide additional evidence that may not be readily obtainable directly from the device.

Handling Encrypted Data

Encryption is one of the major challenges that Android data access presents, although forensic tools have workarounds for most of these:

- **Accessing Encrypted Partitions**: Most modern Android devices use full-disk encryption (FDE) to secure user data. Although this encryption may prevent access to some data, forensic tools usually have ways to access encrypted partitions if the user's passcode, password, or fingerprint is known. Advanced forensic tools can sometimes exploit vulnerabilities or use other methods to bypass encryption, depending on the device model and Android version.

- **Forensic Strategy**: Here, analysts can strategize by obtaining cooperation from the device owner to access the device or maximize available forensic techniques to extract as much information as possible without compromising the chain of evidence. If encryption cannot be bypassed, analysis may focus on retrieving data from unencrypted regions or searching available backup files, which can also provide relevant information.

In summary, various forensic tools and techniques are applied to extract data from Android devices for thorough analysis. It is crucial that forensic analysts understand the capabilities and limitations of tools like Cellebrite, Autopsy, ADB, and Magnet AXIOM, among others. Additionally, employing data recovery techniques and developing strategies to handle encrypted data enhances the comprehensiveness of forensic investigations, ensuring that critical evidence is not overlooked.

Android File System and App Analysis

The Android file system is the most significant area of study in forensic examination, as it involves understanding how an application's data is stored and managed. All information regarding file systems, directory structures, APKs, and other relevant tools provides valuable insight into the forensic investigation process.

File Systems Used in Android

The EXT4 (Fourth Extended Filesystem) and F2FS (Flash-Friendly File System) have been incorporated in most Android devices. Their primary function consists of holding data, organizing it, and retrieving particular information when needed.

- **EXT4**: The EXT4 is currently the most utilized file system for Android devices. It is more advantageous than its predecessors because it is more stable and has the ability to support devices with larger memory capacity. It also includes journaling features that lessen impact in case of sudden shutdowns and aid in data recovery. The operating system uses block groups to distribute file storage, which enables easier data access and storage optimization.

- **F2FS**: The Flash Friendly File System is the most effective storage system for smartphones and tablets, which makes use of NAND flash memory. It has been built around specific features of flash memory, including wear leveling and read/write bandwidth. Nadri Expansion is an essential part of any marketing mix: F2FS enables organizing of files to reduce write amplification and increase device performance for the target market.

- **YAFFS**: YAFFS (Yet Another Flash File System) is also a flash memory file system that is generally used on older Android devices. Although it is becoming less predominant in contemporary devices, knowledge of YAFFS file system can be useful in forensic examination of older devices. The YAFFS design avoids complexity and achieves speedy access to small files, but it lacks some advanced capabilities found in EXT4 and F2FS.

Analyzing Android Package Files (APKs) and App-Specific Data

APKs are the files that are used in distributing and installing applications on Android devices, holding all the code, resources, and manifest information necessary for the app's rightful use.

CHAPTER 2   ANDROID FORENSICS FUNDAMENTALS

- **Decompiling the APKs**: Forensic analysis also uses the decompiling of the APKs to analyze the app's functionality and behavior. Investigation would derive characteristics of permissions the application requests, assess for any vulnerabilities, and check how the app handles user data. This process has also lent itself to the detection of malicious behavior such as the unauthorized accessing or sending of data to external servers.

- **App-specific Data**: Beyond the APK itself, app-specific data often contained in the /data directory could be used as crucial evidence. Forensic investigators need to analyze that data so that they can track interaction behaviors, analyze preferences, and recover deleted items from the app databases and can create a fuller picture of how the app is used.

Key Tools for File and App Analysis

Several tools are essential for analyzing the Android file system and APKs:

- **APKTool**: APKTool is a reverse-engineering decompiling tool. APKTool is powerful enough to decompose an Android application package, or APK file, into its most basic elements: resources, manifest files, bytecode, and much more. APKTool decodes an APK. Now, the app structure is rebuilt so that it can be read in human language. This makes it easier to learn a lot about the app. For example, the manifest file contains all the important metadata of an application: its permissions, activities, and services. Such information is crucial for understanding how an app behaves with the operating system and what data it can access. Another feature of APKTool is the ability to edit the decompiled APK, which allows analysts to make changes and recompile the application for further testing. This feature would be especially useful for examining app behavior under changed conditions or researching vulnerabilities, as analysts simulate different attack vectors or check how the app reacts to unanticipated inputs. APKTool helps forensic analysts completely understand the structure and functionality of an application while ascertaining any potential security weaknesses within it.

CHAPTER 2   ANDROID FORENSICS FUNDAMENTALS

- **JD-GUI:** JD-GUI is a tool with a graphical interface that helps analyze Java bytecode in APK files. It plays a key role for forensic analysts who need to examine decompiled Java code. The tool gives a clear organized view of how an app works inside. JD-GUI lets analysts look through class files from the APK showing a structured layout of methods and fields in the code. This ability to trace method calls helps analysts understand how the app runs and how its parts work together. By looking at the decompiled code, analysts can see how the app handles sensitive data, like how it stores and sends user info. This helps them find weak spots in security. Also, JD-GUI helps spot possible harmful actions hidden in the app such as trying to access system resources without permission or hidden features that could cause problems. In short, JD-GUI gives forensic analysts what they need to take a deep look at Android apps. It shows important details about how the apps behave and how secure they are.

- **Frida:** Frida is a dynamic instrumentation toolkit that allows forensic investigators to inject their own code into running programs on android devices. It's just that nothing else can compare to the level of understanding about how an application works that it provides. Forensic experts can interact with the app in real time using Frida and can change its functionality while playing with data and watching what happens so they can see how the app reacts to certain inputs. Which is especially helpful in exposing characteristics or weaknesses that may not be evident from static analysis. Frida supports JavaScript-based scripting, enabling automation of various tasks and development of scripts tailored to specific applications or security scenarios. This malleability lets forensic analysts write scripts on the spot, changing the way the program operates and discovering security holes, for instance, lack of data validation, or insecure API calls. Frida is a must for analyzing application behavior, on the fly evaluation of possible security threats, and detecting any malignant activities that may jeopardize user data or system integrity.

Understanding the Android file system, which comprises multiple file systems, directories, APK analysis, and specialized tools, is crucial for analysis. This knowledge enables forensic analysts to properly retrieve data from an Android device, supporting investigations and uncovering vital evidence.

Summary

Android forensics is a complex but crucial area in current digital investigations because Android devices carry highly valuable data to support cases in court. An in-depth knowledge of Android device architecture, varied file systems, and professional forensic tools can ease investigators' tasks in collecting, analyzing, and interpreting evidence. The mastery of the diversity in devices and constant updating of knowledge due to frequent updates would allow investigators to overcome such challenges, facilitate the uncovering of hidden information, ensure the integrity of evidence collected, and negate sophisticated anti-forensic measures that could otherwise compromise critical data acquisition.

Reference

[1] Hanif, H., & Anuar, N. B. (2018). Thesis for: Master of Computer Science (Research): Opinion Spam Detection using Supervised Boosting Models. In ResearchGate. 10.13140/RG.2.2.20701.87524/1

CHAPTER 3

iOS Forensics Fundamentals

iOS forensics is defined as the discipline concerning the systematic retrieval, examination, and preservation of digital evidence from Apple devices that run on iOS, primarily iPhones. While iPads now run a separate but closely related OS called iPadOS, many forensic principles overlap, given their shared architecture. Slight differences may appear at the file system level in behavior and supported features, thus potentially affecting the forensic process. iOS and iPadOS are highly secure and give a hard time to investigators. The enforcement of advanced security considerations thus places them under certain operational constraints that forbid unauthorized access. In a nutshell, these operating systems have an XNU kernel basis and layered architecture that separates system services, user interfaces, and application frameworks to provide strict security boundaries. This chapter discusses acquisition methods, device security models, categories of obtainable data in forensic investigations, and available tools for iOS and iPadOS analysis. Data on iOS devices may include user data, such as contacts, messages, and photos, as well as system data, which include logs and settings; app data comprises preferences and internal databases. Forensic investigators use acquisition methods to gain access to this data; they can acquire it logically, where the contents are accessible to extract files, or physically, which provides a bit-for-bit copy of the device's storage. Several commercial forensic tools are in use, such as Cellebrite UFED, Belkasoft, Magnet AXIOM, oxygen forensics and elcomsoft and open-source options like Libimobiledevice, in an attempt to find ways through the iOS security scheme to find critical evidence without compromising data integrity at any stage of the forensic process.

CHAPTER 3   IOS FORENSICS FUNDAMENTALS

iOS Device Architecture

Understanding how iOS devices are architected is very important in the realm of forensics, since the architecture fundamentally dictates how data are stored, accessed, and protected. The robust security features built into iOS are meant to protect the user's data; however, they also make it uniquely difficult for forensic investigators to extract and analyze the relevant information.

Introduction to iOS System Components

iOS is a closed-source system. It runs on the Darwin operating system. This system is based on UNIX. The architecture of this system contains layers with specific, different functions; these functions affect forensic analysis.

- **Core OS:** It provides the bottom foundational layer composed of the XNU Kernel; it manages hardware interaction, file systems, and security in terms of devices. It provides the basic services around memory management, process scheduling, and power management. These all are very important to understand how data is dealt with on iOS devices.

- **Core Services:** This is the layer that comprises basic services provided by the system, such as networking, file access, and data storage. It provides interfaces for applications to interact with other system resources, including APIs that allow developers to utilize the capability of the device.

- **Media Layer:** This layer is responsible for managing media-related tasks and handles functions such as image and video playback, audio management, and other multimedia processing. Therefore, analyzing this layer will help a forensic investigator understand how media files are processed and stored within the device.

- **Cocoa Touch:** This topmost layer provides the frameworks that develop user interfaces; it supports multi-touch interaction. It includes libraries that are used in managing graphical user interfaces and, therefore, it is of great importance while describing how user data is presented and interacted with in the application.

Secure Enclave and Encryption

The Secure Enclave is a segregated, robust security coprocessor on an iOS device that secures sensitive user data. It also securely processes biometric data, such as Touch ID fingerprint data and Face ID facial recognition data (the feature that allows iPhone X and later models to unlock with facial scans), encryption keys, and many other critical security functions. On account of its isolation from the main processor, in case the main operating system is compromised, the Secure Enclave ensures that the protected data cannot be accessed. Think of it as an ultra-secure vault in your device, complete with its own security system. The advanced hardware-based encryption, combined with software keys safely stored inside the vault, keeps user information resistant to attacks. If someone wants to access this information without the correct password, fingerprint, or face scan, or without legal permission, it is really tough. Due to the massive security provided to devices, old-fashioned methods to retrieve data may not succeed

Runtime on iOS and App Sandboxing

In iOS, , app sandboxing is a characteristic feature that is an important security feature holding each application in a separate isolated environment. Mechanisms like user permissions, file system isolation, and system-level security policies enforce this sandboxing. When you install an app from the App Store, iOS automatically places it in its own dedicated folder—usually found at /var/mobile/Containers/Data/Application/<UUID>. This is part of Apple's built-in sandboxing system, which keeps each app isolated to protect your data and the device. However, if a user installs apps from outside the App Store (known as sideloading), they may not follow the same strict sandboxing rules, which can affect how securely they behave—and how easily their data can be accessed during a forensic investigation. Such information ranges from preferences to caches and even content created directly by users—once again, these are impossible for apps and the system to access without permission. What this means is that there is tough isolation that makes the nefarious activities of a malicious app impossible in terms of accessing other sensitive information from other apps, hence enhancing user security. Inter-app communication is also tightly controlled through a set of APIs and frameworks, such as URL schemes and app extensions. For example, if one app wishes to interact with another, it must use an API that the target app has exposed beforehand. This makes forensic data access much more complicated because only apps designed to communicate can, indeed, communicate. Forensic analysts also face challenges in retrieving evidence from applications because established methods of data extraction do not necessarily work. In fact, they may need to use tools meant for exploiting vulnerabilities and, in certain situations, debugging to bypass sandboxing restrictions; they usually resort to jailbreaking the device to access app data at deeper levels.

System Security and Permissions

The iOS operating system uses an overall security model that includes strict access controls and permission management. Before an app can access sensitive user data—such as contacts, photos, camera, microphone, or location—Apple's Transparency, Consent, and Control (TCC) framework prompts users with dialog boxes the first time the app attempts to access such data. This permission model gives end users control over their personal data to prevent unauthorized access.

This model is extremely challenging from a forensic perspective. If a user denies permission to some apps, the forensic investigator will face difficulties trying to access pertinent data because that data is stored in a location inaccessible to apps without user permission. Moreover, additional layers of security encrypt the data, making it very difficult to retrieve without proper credentials. Therefore, forensic analysts utilize tools that facilitate both logical and physical acquisitions in order to bypass such restrictions. Logical acquisition acquires data through an application's APIs, while physical acquisition aims to make a bit-for-bit copy of the whole device storage, regardless of areas inaccessible due to sandboxing. These methods often rely on sophisticated techniques to maintain data integrity and maximize the extractability of relevant information by analysts. For instance, forensic software packages can include the ability to bypass security features, allowing analysis of user activity and application behaviors even under high security control.

Types of Data on iOS Devices

iOS devices are full of digital evidence and are also full of data, which can be highly useful for analyzing in forensic investigations. Usually, data on iOS devices can be divided into three primary types: user data, system data, and app data, with features of each type and their corresponding implications for the analysis process.

User Data

User-generated data are considered the most important type of data for forensic analysis, as it gives insights into their real-life behaviors, interactions, and other occurrences. Key components of user data include

- **Messages**: This type of message covers iMessages, SMS, MMS, and information obtained from frequently used extensive messaging services like WhatsApp, Signal, Telegram and many more. Native app messages may be integrated into message threads, which provide valuable contextual information such as date and time, sender and recipient details, as well as any attached images, videos, or documents. This may also reveal relationships, interaction profiles, compromising discussions, or anything in between—including recently deleted messages, which can be especially valuable during forensic investigations. Call Logs: Call logs offer a complete list of all calls made and received, including completed and missed calls. Conventionally, each entry may contain the phone number, the duration of each call, dates, and times when the calls were made. Call logs—including those from native and third-party apps like FaceTime, WhatsApp, Instagram, and Snapchat—can be used to build a user's social map and deduce interaction patterns and the roles of contacts from an investigator's perspective, offering deeper insight into the user's communication behavior. Photos and Videos: These provide a photographic record and history of the user, comprising photographs and videos taken and stored using the device. In fact, each media file has the possibility of containing additional metadata, for example, EXIF data, where available, contains geolocation coordinates and timestamps of the photo or video. This can help prove various aspects related to location tracking or pinpointing someone to a particular place at a given time in cases involving alibis, for instance.

- **Notes and Voice Memos**: User's personal notes and audio recordings saved as voice memos offer a window into their thoughts, plans, and intentions. These can be rich in sensitive information, personal reflections, or even discussions about questionable activities, making them valuable pieces of evidence in various investigations.

- **Location Data**: Many apps, especially those related to maps or location services, collect GPS data to track user's movements. This location data can be cross-referenced with other evidence—such as photo timestamps, social media check-ins, data from the Health app, and Apple's Significant Locations feature—to construct a detailed timeline of the user's movements and potential interactions. Significant Locations, in particular, can reveal where the user spent the most time during specific parts of the day, offering valuable insights during investigations. **3.2 System Data**

System data includes logs and settings that the iOS operating system maintains to ensure proper device functionality and user experience. The key components include

- **Log Files**: System activity logs are very important to a device's operational history. Some things recorded include app use, errors experienced during system operation, and even security incidents. Log files are therefore examined by forensic analysts for reconstruction purposes, as well as to establish timelines of actions taken on the device, with the help of such logs to determine user behavior or identify malicious activities.

- **Cache Files**: Cache files are temporary files created by the system and applications with views to speeding up processes and thereby enhancing performance. Often ignored, such files might contain fractions of what the user has done, previously accessed content, and even web browsing history. Recovered cache files could unveil useful evidence not found elsewhere, especially when the user's data has been deleted.

- **Settings and Configurations**: As a matter of fact, it contains critical data, such as network configurations, Wi-Fi password, and system preferences. These settings can be analyzed for inferences related to the user's connectivity and environment that may be relevant in ascertaining user behavior or intention.

Places of Key Artifacts in iOS

iOS devices systematically organize the crucial forensic structures in directories; each directory will store dissimilar forms of data necessary for forensic analysis. The following are some very important locations:

- **/var/mobile/Media**: This directory serves as the central location of a user's personal files, such as images and videos, and even high-quality audio recordings. It is filled with digital evidence that a forensic investigator looks for in both visual and audio forms.

- **/var/mobile/Library**: This is a vital directory that contains different system databases, storing users' information, such as SMS and call logs and data from other installed apps. This directory should be accessed when there is a need for much information about user activities and communication; it contains raw user data.

- **/private/var/log**: This directory contains system logs that hold the record of activities happening with devices and systems and interactions among different users. Logs also provide information on device performance issues, security intrusions, and other overall important timing details for events under investigation.

- **/private/var/mobile/Containers/Data/Application**: This path is mandatory for acquiring data pertaining to a specific application. Every application has one, and it stores user documents, preferences, caches, and other pertinent files. Forensic analysts typically explore these directories to understand how an application interacts with user data and to obtain necessary data that can be tricky to obtain otherwise.

It is always important to know the categories and paths of iOS information. Each type of data and location has its role in the reconstruction of what a user did and can contain important information in any investigation. Proper data analysis, including data extraction, will help to achieve better results in legal proceedings, as well as whenever this information is necessary. This is why iOS forensics should be an essential part of an expert's work.

CHAPTER 3 IOS FORENSICS FUNDAMENTALS

Forensic Acquisition Methods

Forensic acquisition is the systematic extraction of iOS data for the purpose of investigation. It often also depends on the type of data to be looked for, the security features on that particular device, and the tools an investigator has at his fingertips. Techniques range from logical acquisitions to complete storage images of a device through a process called physical acquisition, which captures deleted data.

The selection of the acquisition method depends on the required depth and accuracy of data collection. Forensic analysts must choose appropriate tools to ensure reliable evidence gathering. This section outlines four primary methods—logical acquisition, file system acquisition, and physical acquisition—each varying in complexity, access level, and data completeness.

Physical Acquisition Method

Physical Acquisition is a forensic process that acquires a bit-for-bit copy of all the contents of the flash memory of an iOS device. The procedure is essential since, through this method, forensic analysts are given access to every kind of data found in the device—from personal user files to system data or even deleted files that have yet to be overwritten. Through a complete physical image, investigators can perform intensive analysis and recover crucial evidence that may not be available through other acquisition methods. However, the practice of physical acquisition on iOS devices is usually complicated by several factors, especially the robust security architecture on such devices. Most current iOS devices provide full-disk encryption, where data at the hardware layer is encrypted using the Advanced Encryption Standard (AES) with a key derived from the user's passcode. In this way, analysts face difficulties in pulling the decryption key when not using the correct passcode that would help them access the data in the physical image. Also, the Secure Enclave—a coprocessor assigned purely for iOS devices—handles sensitive data, such as cryptographic keys and biometric information, separately from the primary processor without accessing it. Thus, acquisition becomes harder physically because accessing some key materials is impossible. Additional hardware-based security measures include the T2 Security Chip in newer versions, which adds another layer of protection to hardware security and complicates acquisition by requiring sophisticated techniques to extract data directly from memory chips. Physical iOS acquisition using forensic software tools is usually done using professional and specialized tools that bypass or overcome one or more of these security measures to achieve successful physical acquisition. Tools include Cellebrite UFED, MSAB XRY, and open-source options like Autopsy, all of which are widely used and effective in mobile forensic investigations. However, most require unlocking the device before uploading an image or may include exploits that bypass protection features. Thus, physical acquisition provides an overview of all existing data on an iOS device, albeit a laborious process due to high-end encryption, secure hardware components, and advanced security features integrated into the iOS architecture. Careful navigation through these obstacles is necessary for forensic investigators to gather a valid and workable data set for analysis.

Logical Acquisition

Forensic logical acquisition is a type of acquisition that involves obtaining accessible, active data from an iOS device using its file system without accessing the raw storage. This way, forensic analysts can obtain direct user data extraction from the operating system interface, including contacts, messages, call logs, photos, and application data. The logical acquisition method employs the built-in tools and APIs of iOS, such as the Apple File Conduit or Mobile Device Management (MDM) frameworks. The access to device data structures with very pragmatic attitudes toward integrity of the system is adopted.

This is, perhaps, one of the greatest advantages of logical acquisition—that it is non-intrusive and does not even require physical access to the device's raw memory and is also generally less intrusive to the operation of the device, while allowing the full functionality of the device during the process. This is particularly very useful in cases where the functionality of the device has to be maintained, or the owner of the device is around. However, logical acquisition is pretty limited; it would retrieve only active data. It will not capture deleted files, unallocated space, or any other data found in an encrypted partition that needs decryption keys to be accessed. Furthermore, the approach is also bound by the limitations that the iOS security model brings about; it enforces particular application-specific permissions for certain types of data access. For example, an application cannot gain access to personal information, like location data or contact information, without the user's permission, which may pose an obstacle in the forensic investigation. It has proven to be very efficient; however, it may be influenced by the security settings of the device applied by the user. Logical acquisition with Cellebrite UFED or Oxygen Forensics is one of the common tools because it might, in most cases, have features that can be used to ease data extraction by beating several iOS security permissions and limitations. Hence, although logical acquisition might be a good way to obtain relevant user information, it has not reached the comprehensive scope of physical acquisition and needs to be taken with great attention to the security environment of the device and other constraints in accessing the data.

File System Acquisition

File System Acquisition is such a deep forensic technique that it captures access and extraction of data from particular directories of the iOS file system, giving investigators more room for analysis compared to standard logical acquisition methods. This type of approach basically attacks the logical structure of the file system and therefore allows access not only to active user data but also to system files, application data, and even deleted files or at least parts thereof, depending on whether those blocks of data have been written over or not. File system acquisition will engage tools to interact with the iOS OS to look for the file system hierarchy, thereby accessing several key directories, including /var/mobile/Library, which normally contains application data, or /var/mobile/Media, where user media files are stored. Tools engaged in file system acquisition could employ methodologies that either work around APFS (Apple File System) snapshots or use HFS+ (Hierarchical File System Plus) structures to access required data. Understanding these file systems aids forensic analysts in systematically extracting files related to other databases and logs of both user-installed applications and native iOS services. One of the key advantages of file system acquisition is its ability to recover deleted files by accessing unallocated space and metadata structures, which are often not available through simpler acquisition methods. In iOS, generally, the operating system marks the space occupied by the deleted file as available, as opposed to overwriting it immediately when a user removes a file. As a result, since the deleted file area on storage is not overwritten with new data, a forensic analyst can recover files during acquisition. However, it requires high-end equipment that can identify and reconstruct files based on metadata of the file system and allocation tables for storage. Although file system acquisition offers several benefits, it also presents some challenges. This can mean a level of access to the device higher than standard logical acquisition, often requiring a jailbreak or unlock of the device, which changes the device's state and can thus impact its integrity. Also, the process might be curtailed by device encryption protocols because encrypted partitions cannot be accessed without proper decryption keys. For example, some forensic tools, like Cellebrite UFED or Oxygen Forensics, usually incorporate file system acquisition capability, enabling analysts to achieve targeted extraction while maintaining custody and integrity of evidence. Overall, file system acquisition is a really fundamental method in a forensic toolkit since it offers more detailed information regarding what is done with an iOS device. Thus, it is an approach through which an investigator can comprehensively analyze any case in question involving an iOS product.

Cloud-Based Acquisition

Cloud-based Acquisition remains one of the significant concepts in iOS forensics since it takes advantage of the synchronization between Apple iOS devices and iCloud. This method enables forensic investigators to extract a sundry of data backed up in iCloud, such as, but not limited to, photos, messages, friends' contacts, settings, and apps, among others. In light of the trend of data storage to the cloud, it is obvious that effectively investigating and conducting cloud-based acquisitions is critical to efficient forensic investigations. Using iCloud, iOS users have the choice to automatically back up their device data. This backup also includes such things as the data of installed applications, device configuration settings, as well as any content created by a device user, for example, photos and messages. Forensic analysts can obtain this as additional information to physical or logical images of the entire device. However, cloud-based acquisition poses its own challenges and needs. In order to begin a cloud-based acquisition, officers and investigators require the victim's login information to the iCloud account—Apple ID and password. Often, this requirement can be troublesome; where the user is uncooperative or the credentials cannot be identified. In such circumstances, the police and a court order from a local magistrate may be required to force Apple to open the iCloud to the account's contents. The legal requirements must be met to ensure that the acquisition is done legally and to keep the data clean.

Thus, if the investigator obtains authorized access to the iCloud account, they can use APIs or other relevant forensic tools to interact with iCloud and recover lost or deleted information. It may require downloading certain files or datasets that have been mirrored on cloud storage. For example, forensic investigators can easily extract photos backed and synced in iCloud Photos, messages backed and synced in iCloud Messages, and application-contained data that may be backed and synced in iCloud Drive. Furthermore, backups created on iCloud are sometimes encrypted; this complicates acquisition work. In cases where a backup is encrypted, investigators spend more time decrypting it, which may require the user's password or a decryption key. Without such information, it becomes very difficult to gain access to the backup's contents, hence a hindrance to the effectiveness of cloud-based acquisition of assets. Also, some data stored in iCloud synchronizes with the device and may represent a predisposed status of the gadget. Hence, it is important for investigators to be conscious of the fact that the cloud's contents might not be the latest information available on the device, especially when the last update has not been synced to the cloud. Tools used to carry out cloud acquisition include Cellebrite UFED Cloud, Oxygen Forensics Cloud Extractor, and ElcomSoft Cloud Explorer. These tools are designed to interface with iCloud, automate the extraction of relevant data, and provide analysts with a structured way to analyze the information obtained from cloud sources. Cloud-based acquisition is one of the important tools for iOS forensic investigations that permits significant user data acquisition, not available or directly via a device. Utilizing iCloud backups, forensic investigators can gain a broader view of a user's digital activity, often extending beyond a single device. Since iCloud may store data synced across multiple devices using the same Apple ID, it provides a more comprehensive and consolidated source of digital evidence in iOS-related investigations.

Acquisition Challenges in iOS Devices

Acquisition of iOS data is very problematic in the context of encryption, potential loss of data, and possible anti-forensic measures. Advanced techniques and tools have to be put into place with careful planning and execution in order to overcome such challenges when acquiring data from iOS devices, ensuring that evidence collected is not only reliable but also complete and admissible.

- **Encryption:** The encryption that comes by default in iOS devices creates a very strong barrier for the acquisition process. Full-disk encryption has been provided on all iOS devices since iOS 8. By default, all files, applications, and the operating system are encrypted, which means nothing can be accessed without the decryption key. In fact, encrypted data enjoys protection via a passcode or other forms of biometric authentication, such as Touch ID and Face ID. If the device is locked and its passcode is unknown, forensic investigators face severe barriers in accessing its data. Generally speaking, without proper authorization, forensic tools cannot bypass this encryption layer, which prevents finding the evidence in criminal investigations or other forensic contexts. Furthermore, from the point of view of user privacy and security, Apple does not provide law enforcement agencies with any backdoors; that hinders the process of obtaining access to more data.

- **Data loss:** Poor forensic techniques can result in the loss of data due to corruption during the acquisition process, and this is usually most costly when acquisition is involved. The volatile memory in iOS devices is used to temporarily store active data such as running processes, encryption keys, cache files, and session information. If someone agitates the device during the extraction process by restarting it or letting the battery discharge, this volatile data would be lost, resulting in an incomplete or partially corrupted data set that can obstruct forensic analysis. For example, if an investigator attempts to acquire data on a device that has been powered off, then any unsaved changes have a chance of being lost permanently. In addition, if the device is reset or updated during acquisition, it may overwrite critical data or lose evidence altogether. Forensic professionals should ensure that proper procedures are followed, including bit-by-bit imaging of the device's storage and other write-blocking techniques, to ensure that no changes occur during data acquisition. Any carelessness in these procedures can result in compromised evidence integrity, which can influence the outcome of investigatory or legal cases.

- **Anti-forensic Measures:** Some users and applications may use anti-forensic techniques to safeguard their data from easy extraction in case an investigation is carried out. Such measures can include data obfuscation, encryption, and automatic deletion of data. For example, some applications might use proprietary encryption algorithms to secure their data; such data will be very difficult for forensic tools to interpret without access to these keys. More importantly, some applications may include features that erase sensitive information based on specific time intervals or frequent incorrect login attempts. These functions can be very problematic in crime investigations, as key evidence may be erased before investigators have a chance to access the evidence. Furthermore, users can use data obfuscation techniques to hide the contents of their data or make data structures less easily understandable, further complicating analysis for forensic experts. Finding and mitigating these anti-forensic techniques requires advanced skills and specific tools, which further challenges the process of acquiring evidence during forensic acquisition.

Forensic Tools and Techniques

Various tools have been built for harvesting and analyzing data from iOS systems. The few tools have been described as mentioned below:

Cellebrite UFED is one of the most popular iOS forensic data extraction tools which has comprehensive capabilities for logical, physical, and file system extractions. Cellebrite UFED (Universal Forensic Extraction Device) is an industry-leading mobile forensics tool that provides high-level data extraction functionality from multiple types of mobile devices, including iOS. It is especially useful to forensic investigators, because it covers three main types of acquisitions, logical acquisitions, physical acquisitions, and file system acquisitions, with each one addressing a different investigation purpose.

Magnet AXIOM is a complete digital forensics tool used to acquire, analyze, and report on evidence from iOS or Android smartphones, computers, or cloud services. It supports various data acquisition types (logical, physical, and file system extraction), virtually eliminating limits to accessing digital evidence for forensic investigators. With its simplicity and intuitive visualization capabilities, Magnet AXIOM creates reports that provide very good assistance to investigators. It is a useful solution for law enforcement and forensic experts, offering advanced features such as messaging history analysis (which includes numerous apps) and location analysis, helping discover vital evidence and understand user behavior.

Belkasoft is a well-established digital forensics software firm and it is very much popular in the digital forensic field for its tool named Belkasoft X Forensic. This tool allows the forensic investigator to acquire, analyze, and report on computer, mobile device, and cloud service data. iOS, Android, Windows, macOS, and Linux are some of the platforms that can be analyzed using this tool. Logical, physical, and file system acquisitions, together with in-depth analyses of such artifacts as messages, call logs, photos, and app data, can be supported with Belkasoft. It is a very convenient tool for the work of the police, cybersecurity specialists, and forensic investigators due to its ease of use and comprehensive forensic functionality.

Oxygen Forensic Detective is a very powerful mobile forensic software, carrying out extraction, analysis, and reporting of data from different smartphone and tablet operating systems: iOS, Android, and Windows Phone. It follows logical, physical, and cloud data extraction methodologies. Information from messages, call logs, photos, and different applications can be retrieved by forensic analysts using the said tool. Oxygen Forensic Detective makes it pretty easy for an investigator to visualize and interpret the data through its very user-friendly interface. It can also become much more sophisticated in parsing databases, salving deleted files, and generating reports, all of which are very important in law enforcement and among digital forensic experts.

Process to Be Followed for the Data Acquisition

Phase 1: Preparation: Start the acquisition process with Cellebrite UFED, Magnet AXIOM, Belkasoft Evidence Center, or Oxygen Forensic Detective (start the application on your forensic workstation first). When this tool is opened, you will be prompted to create a new case; click the "Create New Case" option and fill out relevant information, such as the case name, case number, and investigator(s), to ensure correct documentation and organization. Now that the case has been created, the next step is to plug in the iOS device you want to examine. Plug the device into your workstation using the appropriate cable—typically a Lightning cable for devices below iPhone 15, and a USB Type-C cable for iPhone 15 and later models.

When the device is connected, it may prompt the user with a "Trust This Computer" dialog. Confirming this prompt initiates a trust relationship between the iPhone and the computer by exchanging cryptographic keys. This process is essential, as it grants the workstation authorized access to the device, enabling data extraction necessary for forensic analysis. The established trust relationship allows for secure communication between the workstation and iOS device, enabling data acquisition.

Phase 2: Data Acquisition: Mobile data acquisition is a defined process that involves identifying the forensic workstation, connecting a mobile device to it, determining an appropriate acquisition method (logical, physical, or file system), authenticating the device, and extracting data from it, followed by securely storing a copy for analysis. This ensures that forensic investigations and court proceedings maintain no degradation in

CHAPTER 3   IOS FORENSICS FUNDAMENTALS

data integrity. It is well known that the next significant phase, following the preparation phase, is data acquisition, in which several acquisition techniques are implemented to gather artifacts from iOS devices.

Logical acquisition Process

Step 1: Connect Device: The logical acquisition can be performed by first physically connecting the iOS device to the forensic workstation using an appropriate cable. Depending on the model of the iOS device, it may require a Lightning-to-USB connector cable for newer devices or a 30-pin connector for older models. It is important to use a cable that supports both charging and data transfer. After connecting, you need to ensure the device trusts your computer. If it is the first time you are connecting, the device will always prompt you to confirm whether it should trust this computer, which you must allow for future access. Confirming this trust creates a secure pairing between the device and the forensic workstation, enabling tools like Cellebrite UFED, Magnet AXIOM, Belkasoft Evidence Center, or Oxygen Forensic Detective to communicate with the device for data extraction. Without this trusted connection, the acquisition process is not possible.

Step 2 A Select Acquisition Type: After connecting the devices, the next step would be to start either Cellebrite UFED, Magnet AXIOM, Belkasoft Evidence Center, or Oxygen Forensic Detective and choose an acquisition type. Upon opening any of the aforementioned software, several acquisition methods will become available to the examiner, such as logical and physical data extraction and file system extraction. In this case, the logical option would be selected. This option enables the extraction of user-level data, accessible through the device's operating system APIs, like contacts, call logs, messages, and media files. Quite often, logical acquisition is preferred when trying to acquire only user-generated data without accessing encrypted partitions or system-level information. This process is simple and non-intrusive, relying on native iOS APIs to access the data in question.

Step 3: Authenticate Device: After the acquisition type has been selected, authentication of the device is required to access its data. Most iOS devices are usually unlocked with a passcode, Touch ID, or Face ID. In other words, without these authentication methods, the data on the device can hardly be accessed. If the device is passcode-protected, the examiner must provide the correct passcode. Alternatively, it must be unlocked using Touch ID or Face ID to bypass biometric security. Additionally, the trust relationship between the iOS device and the forensic workstation may need to be confirmed on the device itself by selecting "Trust This Computer." This is an important step, since if a device remains locked, most user data will remain inaccessible.

Only after successfully unlocking the device and establishing the necessary trust relationship is it possible to begin the actual process of data extraction. The Cellebrite UFED, Magnet AXIOM, Belkasoft Evidence Center, or Oxygen Forensic Detective software then starts interfacing with the device through publicly available APIs to access typically available data: contacts, call logs, messages, and media, such as photos and videos. Logical acquisition works by pulling data available through standard system interfaces without accessing deeper, encrypted areas of the device's file system. This method is efficient and non-intrusive but, at the same time, is limited to data that iOS allows public access to, which may exclude encrypted messages, application data, or system files. This process may take some time, depending on the amount of data stored on the device and the connection speed.

Step 4: Perform Logical Extraction: After unlocking and authenticating the device correctly, the forensic tool is ready to start the logical acquisition process. At this point, the investigator begins the extraction using specific software (Cellebrite UFED, Magnet AXIOM, Belkasoft, or Oxygen Forensic Detective). The tool interacts with the iOS device via trusted and publicly known APIs to retrieve user data that is accessible.

In the course of logical extraction, the software collects data categories consisting of various user data, such as contacts, call history, SMS/iMessage, media files, notes, calendars, and certain applications that iOS allows access to through standard interfaces. The examiner has the option to select specific data types or extract all available logical data.

The tool then reads and copies this data from the device without altering the original contents, thus making the acquisition non-intrusive. During the entire procedure, progress indicators in the tool display the extraction status, and logs are automatically generated to keep a record of every action taken. The length of time for this step varies according to the amount of data, the device model, the iOS version, and the transfer rate. After the extraction is finished, the tool organizes the acquired data into a structured format, which is then ready for review in the next step.

Step 5: Review and Save Data: After the completion of the extraction process, the examiner must review all extracted data to ensure that every bit of available information has been recovered. All the tools provide a clear view of extracted data, organized into sections, including messages, contacts, photos, and call logs. The examiner can browse and filter data to ensure everything is complete and highlight some of the most important artifacts for further investigation. Following data analysis, the exported data needs to be transferred to a secure location. The extracted data can be saved in various formats, such as HTML, PDF, or CSV, depending on the investigation's needs. Hash

CHAPTER 3 IOS FORENSICS FUNDAMENTALS

values are imperative throughout this process to ensure data integrity and confirm that no tampering has occurred. All actions should be documented to establish an unbroken chain of custody.

Physical acquisition Process

Step 1: Connect Device: The same procedure as explained in the previous section

Step 2: Select Physical Acquisition: Once an iOS mobile is properly connected to Cellebrite UFED, Magnet AXIOM, Belkasoft Evidence Center, or Oxygen Forensic, the next step would be to choose the Physical Acquisition method in the software. This type of extraction is the deepest because it attempts to capture a complete image of the device's storage, including both hidden and visible files, deleted data, and even some types of encrypted data. At this point, the software interface will ask the examiner to choose the acquisition type; for iOS devices, physical acquisition offers the best possible chance for data retrieval beyond logical extraction. This is an important step in every forensic investigation involving the recovery of deeply buried information or deleted data, as physical acquisition creates a complete bit-for-bit image of the device's storage.

Step 3: Authenticate Device: The same procedure as explained in the previous section

Step 4: Create Bit-for-Bit Image: The physical acquisition does not start until the device is unlocked and authentication is bypassed. A bit-for-bit image of the entire storage is then created. In this process, therefore, the resulting image captures all active data in high detail, including user files, messages, and application data, as well as remnants of deleted files that may still remain in unallocated space on the device storage. This process aims to replicate the device's storage in its current state without any data loss during extraction. This is crucial because it allows forensic investigators to recover deleted or hidden data that users no longer have direct access to. The created bit-for-bit image is an exact replica, and this raw data can later be analyzed to recover both visible and hidden information that may become crucial in investigations involving potentially deleted or tampered evidence.

Step 5: Address Encryption Challenges: These complications can become very crucial during a physical acquisition. Full-disk encryption is used in many newer iOS devices to protect data within the device from unauthorized access. The forensic examiner should go the extra mile to ensure that proper decryption protocols are followed. This may include acquiring a user's passcode or leveraging exploits to bypass encryption protections. Tools like Cellebrite UFED have capabilities to address out-of-the-box encryption types. A common example is that UFED can extract unencrypted parts of the data if the device has been unlocked. In other cases, iCloud backups or iOS security vulnerabilities might offer an alternative entry point for retrieving encrypted data. Correct handling of encryption is vital for thorough extraction, as poor decryption may leave valuable information inaccessible.

Step 6: Review and Save Data: The same procedure as explained in the previous section

File System Acquisition

Step 1: Connect Device: The same procedure as explained in the previous section

Step 2: Start File System Acquisition: Once the devices are properly and safely connected, the next step is to start acquiring the file system. This means selecting the "File System Acquisition" option on the respective forensic software in use. This acquisition method focuses on capturing a complete view of the device's file system, thereby enabling investigators to retrieve not only user-accessible data but also the underlying system files and directories. Unlike logical acquisition, which usually pulls data accessible through standard user interfaces, file system acquisition delves deeper into the device's storage to provide access to hidden and deleted files. By selecting this option, the examiner prepares to capture a wide range of data, including application data, configuration files, and potentially recoverable deleted files, which could all prove valuable evidence in forensic investigations.

Step 3: Specify Directories: At this step, the examiner might need to define exact directories or areas within the file system that they want to process. Options are typically available within the forensic software to target certain points-of-interest on the device's storage, such as directories linked to vital applications, system caches, and user data folders. Designating directories aids in streamlining the acquisition process and ensures that only the most important data is extracted. The examiner may decide to be more targeted and look in directories pertaining to messaging applications, the images directory, and app-specific important data, etc., which may hold vital evidential details for the examination. This helps make the acquisition faster, as the examiner can limit the scope and identify areas with more evidentiary value, also cutting down extraction time by excluding irrelevant data.

Step 4: Begin Extraction: Once the necessary directories have been set, extraction should be initiated. This phase is where the forensic software communicates with the iOS device to extract data from the predefined sections within the file system. Extraction captures active files and all recoverable deleted files that may still be available in storage. Because these forensic tools access system-level files and metadata, they can compile a more comprehensive dataset than may be available through other acquisition methods. This process may take quite some time, depending on the volume of data to be extracted and the file system complexity. During extraction, the software may display progress, and it is during this stage that the examiner must be vigilant to ensure it completes without hitches.

Step 5: Review and Save Data: The same procedure as explained in the previous section

Phase 3: Data Analysis: Data analysis is a crucial phase that follows acquisition using any of the above-mentioned tools in forensic investigation. The analyst starts by using onboard tools to view and interpret the data extracted from the iOS device. These tools provide means to organize data into understandable forms, which forensic experts can prepare into detailed reports outlining what was found during extraction. Messages, call logs, location data, and app-specific information, among others, need to be reviewed during this phase, as they may hold vital evidentiary value. In fact, critical artifacts can provide insights into how users interacted with the device or behaved. Additionally, extracted data may be correlated with other evidence pieces to provide a fuller picture of user behavior or specific incidents. The general approach bridges different data points, enhancing the overall forensic narrative and subsequent legal proceedings that may follow.

Phase 4: Reporting: Documentation and reporting is a very important phase of forensic investigations, as it legally binds the process to maintain consistent records of all protocols followed, findings discovered, for future reference as well as in case of scrutiny. In this phase, creating an accurate log of acquisition must be done, with each step timestamped, including actions used, solutions applied, and any problems that occurred along the way. Complete documentation provides transparency and ensures the chain-of-custody for collected items is well-documented. A final report is then composed, detailing results, methods, and analysis findings after the documentation step of the acquisition process. This report should be clear and precise enough to command the attention of stakeholders, including law enforcement, legal teams, and other investigating parties. It frequently contains sections on the scope of analysis, tools and methods used, important findings, as well as suggestions for further investigation. This methodical process not only supports the legal process but also fortifies the forensic investigation's veracity and reliability.

iCloud Data Acquisition

Acquisition of iCloud data plays a vital role in mobile forensic investigations. Investigators retrieve data from iCloud, a cloud-based service provided by Apple. iCloud stores various types of data transmitted from Apple devices, including iPhones, iPads, Macs, Apple Watch, and Apple TV. In addition to syncing device data, it also functions as cloud storage where users can manually upload and store files, photos, documents, and other personal content. Thus, it proves very useful for forensic purposes. Here is a step-by-step comprehensive guide to acquiring iCloud data. To access iCloud data, specific credentials or particular access are necessary to execute a successful extraction of information stored in the cloud. The easiest method to access iCloud is to obtain the target account's Apple ID and password. An Apple ID serves as the identifier to log into iCloud, while the password unlocks the stored data. Without these credentials, directly accessing the iCloud service and extracting data is difficult. However, most iCloud accounts have Two-Factor Authentication (2FA), which requires an additional verification form beyond the Apple ID and password. Commonly, 2FA sends a verification code to a trusted device or phone number associated with the account. To complete acquisition, investigators require this trusted device (usually an iPhone, iPad, or Mac) or the phone number to receive the code and complete the login process. Without this second factor, access to the iCloud account is effectively prevented.

Some forensic tools can use iCloud authentication tokens when the Apple ID and password are not available and investigators cannot bypass 2FA. Such a token essentially serves as a temporary credential stored locally on devices that have previously logged into a target iCloud account. These tokens can be extracted from the target device using forensic tools, such as Elcomsoft Phone Breaker or Cellebrite UFED, granting the ability to bypass the Apple ID password and sometimes even the 2FA step. This allows for access to and downloading of iCloud data without full login credentials. Although this approach is very effective, it assumes physical or remote access to a user's device.

Access iCloud Account

Once the required credentials or authentication tokens are available, the next step is accessing the iCloud account. It may be performed manually via a browser or with the use of specific forensic tools, each having particular advantages and limitations. Let's look at both methods in a little more detail.

1. ***Manual Access via Browser:***

    You can directly access iCloud.com with the user's Apple ID and password using a standard web browser. This is going to be a basic technique that will give you simple access to some iCloud-stored data. After logging in, you will have access to the following kind of data:

    - **iCloud Drive files:** These may be documents, spreadsheets, and other files that the user has manually or automatically stored in iCloud Drive.

    - **Photos and videos:** These are all the media that have been synced with iCloud Photos.

    - **Notes, reminders, and calendar entries:** These serve as great insight into the user's personal or work life, future events, and scheduled activities.

2. ***Forensic Tool Access***

    Forensic tools are a desirable methodology for iCloud data extraction, especially when in-depth extraction is necessary. These tools automate the acquisition process by pulling a wide range of data types, including backups and application-specific data. Here is how to proceed with using forensic tools:

    - **Install the Tool:** First, install and set up the desired forensic software on a forensic workstation. Common tools utilized for this task include Cellebrite UFED, Belkasoft, Elcomsoft Phone Breaker, Magnet AXIOM, and Oxygen Forensic Detective. These tools are designed to work specifically with cloud services like iCloud to ensure forensic-grade data acquisition.

    - **Login Credentials:** After installation, fill in the Apple ID and password used for the iCloud account being examined. If 2FA is enabled, you will also need the verification code sent to the trusted device or phone number. Some available tools automatically handle such verification through an interface once you enter the appropriate credentials.

- **Token-Based Access:** Some tools bypass the need for an Apple ID password by using iCloud authentication tokens. These tokens are temporary credentials cached on devices that have previously authenticated to the iCloud account. These tokens can be extracted from the device using tools, such as Elcomsoft Phone Breaker, which then allows investigators to access iCloud without needing the password or dealing with 2FA. This technique is most effective in cases where access to the device is possible but credentials are not.

- **Data Type Selection:** Once authenticated, the type of data to be downloaded needs to be selected. These tools offer granular-level selection, allowing you to choose only those categories most relevant to your investigation. The most common types of data that can normally be extracted include

  - **iCloud backup**: Full-device backups provide the most comprehensive dataset, including app data, system settings, and user preferences.

  - **Messages and SMS**: All messages are included, encompassing those sent through Apple's iMessage service and regular SMS text messages. This may also include deleted messages, provided they were included in the most recent backup.

  - **Contacts**: All contacts stored within the iCloud account are accessible, including name, phone number, email address, and physical address.

  - **Safari browser history and bookmarks**: The user's browsing history, saved bookmarks, and metadata (such as last visited date) are available.

  - **Photos and videos**: All photos and videos within iCloud Photos are accessible, along with metadata (timestamp and geolocation information).

- **Calendars, reminders, and notes**: Information on the user's schedule, personal reminders, and written notes provides contextual clues for reconstructing the timeline.

- **Application data**: Most applications share data in iCloud, enabling backups from apps like WhatsApp, Facebook, or Instagram for further analysis of user activities, chats, and file exchanges

- **Start Data Download**: After selecting the type of data one intends to get, it's time for the download to start. The forensic tool will start fetching the data directly from Apple's servers. Depending on how much data is being acquired and the speed of the Internet connection, this may take some time. Large iCloud backups or media-rich accounts with thousands of photos and videos result in longer download times. Tools will often provide logs or progress indicators that allow investigators to monitor the acquisition process.

With the help of forensic tools, you can get much more data than you would have manually with a browser, backups, data from service synchronizations, and content for some apps only. This is crucial in deep forensic investigations that require thorough data mining or recovery of deleted content. However, it is very important to ensure that authorized and verified methods are employed at all times to ensure the integrity of the data, with adherence to legal frameworks right from the start.

Analyzing iCloud/iOS Backups

iOS Backup Analysis is one of the cardinal processes in forensic investigations when direct access to the physical device is not possible. Backups made through iTunes or iCloud contain a vast amount of information that may prove to be a perfect source of evidence. The iTunes backup locally stores a wide range of data on a user's computer, including messages, call logs, various application data, photos, and system settings. By default, an encrypted backup will also include sensitive data, such as saved passwords and health data. iCloud backups, stored on Apple servers, will again contain similar data, but perhaps not updated as frequently as iTunes backups. Investigators often use specialized forensic tools, such as Elcomsoft Phone Breaker, to extract and decrypt this data. These utilities can even handle encrypted backups, including iCloud data, provided the credentials are available. Upon analysis, these backups may yield very important artifacts for an investigator, such as messages, multimedia, browser history, and application-specific data that could help reconstruct an accurate picture of user actions and behavior.

Data Recovery and Decryption Techniques

- **Jailbreaking:** Jailbreaking is a technique to bypass safety restrictions imposed by Apple, providing forensic investigators with root access to the iOS device. This removes inherent restrictions, allowing investigators to delve deeper into the device's file system for data and system files otherwise inaccessible. This technique is helpful in retrieving app-specific data, system logs, and other important evidence possibly lying in restricted directories. However, this modification affects the integrity of the evidence, as jailbreaking modifies system files. Additionally, Apple further reinforces its security measures and hardware defenses, diligently rendering jailbreaks more and more challenging—particularly on newer iPhones with state-of-the-art chipsets. Although some jailbreak techniques might still prove viable on older devices even when these are running the most current editions of iOS, newer models tend to be released without jailbreak support. Forensic examiners need to take cognizance of both the version of iOS as well as the particular iPhone or iPad model in question since compatibility is highly inconsistent. Moreover, equipment using beta releases of iOS or iPadOS tends to be even more problematic, as much of the forensic hardware and software tools and procedures might not work because of insufficient support or unsteadiness in the beta releases. Keeping abreast of the newest breakthroughs is key to successfully extracting and analyzing data.

- **Passcode Bypass**: Passcode bypass techniques are essential when the iOS device is in a locked state and the passcode is unknown. Some forensic tools exploit vulnerabilities within the iOS system to bypass the passcode, especially in older iOS versions. In more complicated cases, brute force attacks involve tools systematically attempting combinations to determine the correct passcode. Advanced tools are designed with mechanisms that avoid triggering security features like data erasure after multiple incorrect attempts. However, bypassing has become increasingly difficult as iOS security advances with Face ID, Touch ID, and longer passcodes. Once bypassed, investigators have complete access to the host device, enabling extraction of key data, including encrypted or deleted information.

Common Steps for Both Jailbreaking and Passcode Bypass

Step 1 Device Preparation: Preparation of the iOS device is the preliminary step in jailbreaking or bypassing a passcode. This is an initial preparatory phase in which the device has to be put into an optimal state for acquisition. Investigators have to make sure the device is fully charged to prevent interruptions during the procedure and that it is connected with the forensic workstation using a secure USB connection. A good connection is very important in facilitating data transfer and communication between the device and forensic software. Besides, it is essential to make sure that the respective forensic software or tool used acknowledges the device. This confirmation will be helpful in noticing any potential problems with connectivity or compatibility issues prior to proceeding with the next steps.

Step 2 Backup of the Original Device: It is very important to make a complete backup before attempting jailbreaking or passcode bypassing. This backup is useful in two ways: it preserves the original state of the device and guarantees data integrity, especially in dangerous procedures like jailbreaking or passcode bypassing. If something were to go wrong, such as data loss or corruption, the investigator can restore the device to its previous state. This step demonstrates that data preservation in forensic investigation creates a safety net, limiting risks when manipulating the device's security features.

Step 3 Tool Selection: The correct forensic tool selection is a very critical stage in the process. Investigators must decide on a tool that exclusively supports either jailbreaking or bypass code functionalities. Certain established forensic tools, such as Cellebrite, Belkasoft, and Oxygen Forensic Detective, support different iOS versions and device security features. The choice of tool may seriously affect the jailbreaking or bypassing outcome because not every tool is compatible with any model or iOS version of the device. It is therefore necessary for the investigator to study the complete specifications of the tool to ensure it meets the requirements for the device they are working with.

Step 4 Assess iOS Version and Device Compatibility: It is important before executing any jailbreaking or bypassing of passcodes that the iOS version and device compatibility be considered. In addition, a variety of different tools and methods may only work on specific versions of iOS or particular device models, which makes the verification an important part of this process. By cross-checking the iOS version and model against the capabilities of the selected forensic tool, an investigator will avoid possible issues that might arise because of using incompatible methods. Such an assessment allows for smoother acquisition and reduces the possibility of failure or complications in executing the selected methodology.

Step 5 Apply the Technique: Once the preparations and evaluations are complete, the next step would be the implementation of the selected methodology: jailbreaking or bypassing the passcode. Jailbreaking entails executing the forensic tool in order to remove the security restrictions imposed by Apple on its device; it provides root access to the file system. It involves the exploitation of some bugs in the operating system, iOS, that may allow an investigator to evade standard security protocols. On the other hand, if bypassing the passcode is in order, the investigator launches the process of cracking or brute forcing. This step needs to be closely monitored, since it may take time to unlock the device, especially if multiple passcode attempts are needed or specific vulnerabilities are being targeted.

Step 6 Monitor Progress: It is also very important to know the progress that's being made throughout the jailbreaking or bypassing of the passcode. The investigator must look into the interface of the tool for any type of updates or prompts that give indications about the current status of the operation. Some procedures can also be very time-consuming, especially when performing an exploit or brute force algorithm in guessing the passcode. It is essential that the computer remain coupled at this stage and not be disturbed; any disturbance may affect this process negatively and may even result in partial access or loss of information. Continuous monitoring will enable the investigators to deal with all unexpected issues immediately.

Step 7 Validate Access: Once the jail breaking or passcode bypass is done, the investigator has to verify that they have attained the desired level of access to the device. On jailbreaking, one checks whether files and directories at the system-level could be accessed, signifying that the root access had been granted. For passcode bypass, the investigator verifies that the device is unlocked and they can navigate through the interface of the device. Validating access is essential; this is what shows whether the work done for jailbreak or bypass is correct, or not, and the investigator can extract data from it.

Step 8 Data Extraction: The final step is to utilize the forensic tool by extracting the necessary data after bypassing the security of an iOS device. Once access of the device has been granted, investigators can perform various acquisitions—such as logical, physical, or file system extraction—depending on the investigation's goals. This enables the retrieval of valuable information, including application data, system logs, contacts, messages, and other relevant files that can serve as evidence during a forensic investigation. Therefore, complete extraction of information covering all data should be ensured, allowing investigators to base their analysis on comprehensive data and guaranteeing that evidence extracted from the device remains untampered.

iOS File System and App Analysis

The iOS filesystem and its structure are central to any forensic investigation process. It enables the investigator to systematically find, recover, and interpret data of interest. This is a hierarchical structure that further enables targeted investigations in areas of interest such as system logs, application data, and media files. This section gives an overview of some of the key components and techniques that form part of iOS-filesystem and application analysis.

iOS File Systems: APFS and HFS+

Modern iOS devices use, by default, the APFS, or Apple File System, which was introduced in iOS 10.3 for flash and SSD drives to offer better performance and security. APFS provides very effective storage management, such as copy-on-write and space sharing. Copy-on-write is a design where modifications can be safely made to a file without touching the original data until the write has been committed, thereby allowing enhancements in data integrity and preventing corruption during writes. Along with efficient storage facilities, APFS is designed with strong encryption; each file can be encrypted with its own key for fine-grained data security. This complicates things for forensic investigators, who need to access encrypted partitions without valid credentials or passcodes. The snapshots in APFS allow the operating system to refer to the specific state of the file system at any given point in time. This feature is especially useful in forensic analysis, as it provides a chance for investigators to recover deleted or modified data that might otherwise have been lost. Before APFS was implemented, older iOS versions used HFS+, or Hierarchical File System Plus. While HFS+ allowed for a decent file system, it had limitations in handling large volumes of data and lacked the advanced encryption features introduced with APFS. This migration to APFS constitutes a giant leap forward in security and efficiency and underlines Apple's drive to secure user data, while presenting forensic investigators with fresh obstacles to overcome.

Analyzing App Data and iOS Apps (IPA Files)

IPA files, or iOS Application Archive files, are crucial for understanding how an application behaves on iOS. These files wrap up all the executable code, resources, and metadata required to make an app work. In analyzing IPA files, forensic investigators glean valuable information about an application's behavior, permissions, and data handling practices. This is extremely important in investigations related to data breaches, unauthorized access, or malicious behavior. The IPA files are usually reverse-engineered for investigation purposes by forensic analysts. Hopper Disassembler and Ghidra are some of the commonly used tools. Using such reverse engineering tools, app binaries are disassembled, and investigators can review the structure of the code, use of API calls, and identification of security vulnerabilities. Through analysis of the app's logic, hidden functionalities can be discovered, such as unauthorized data transmission or data storage methods not compliant with regulations concerning privacy. Moreover, the permissions an application requests can serve as a source of discrepancies between user consent and actual data access. This becomes valuable during investigations involving apps that solicit more permissions than they need to function, making unauthorized data collection possible. Such detailed analysis within IPA files helps forensic analysts gain a complete understanding of the app's behavior and interaction with user data.

Tools for iOS File and App Analysis

The following special tools are developed to support forensic analysts in iOS file system navigation and data extraction of relevant information for applications:

- **iExplorer:** The integrated solution designed for forensic examiners to browse and extract data efficiently from the iOS device file system. This is a wide-ranging tool in investigations for the extraction of key data types such as messages, call logs, voicemails, and multimedia files. It provides the user-friendly interface which navigates the file system with great ease to assist forensic professionals in pinpointing their vital evidence for investigations.

- **iMazing:** More than a powerful device management tool, iMazing boasts robust forensic capabilities for data extraction and exploration from backups and live file systems. This tool supports extraction of specific data kinds, including application data, contacts, messages, and system logs. iMazing's backup management capabilities are particularly useful in cases where the physical device is inaccessible, allowing forensic analysts to recover data efficiently from iCloud or iTunes backups.

- **DB Browser for SQLite**: Since many iOS applications operate using local SQLite databases, this tool is very useful during forensic analysis of application databases. Using this tool, investigators can open, query, and modify SQLite databases, performing deep structure and content analysis. This may be essential in recovering deleted records or evidence of user activity within the application. By querying the database, forensic experts can extract relevant information related to an investigation, such as user events, transaction records, or logs related to sensitive data access.

Together, these utilities arm the forensic analyst with the necessary depth for an iOS device investigation, making the task of extracting and considering crucial evidence much less daunting due to the intricacies of the iOS file system and app architecture.

Summary

This chapter provides an overview of the principles and practices concerning forensic investigations on iOS devices. It covers, among other things, the different types of data stored on iOS devices, such as user data and system files, and describes the various acquisition methods available for iOS devices, namely, logical, physical, and file system extraction. The chapter also focuses on device preparation, data integrity, and proper tool selection, emphasizing the importance of jailbreaking and passcode bypass techniques. Additionally, it addresses legal and ethical considerations regarding best practices for forensic practitioners and preserving the chain of custody throughout the investigation process.

CHAPTER 4

Leveraging Blockchain for Mobile Forensics

This chapter explores the intersection of blockchain and mobile forensics, delving into the challenges, opportunities, and methodologies involved in extracting and analyzing blockchain data from mobile devices. It also covers the cryptocurrency transactions to decentralized applications and their investigation.

Introduction to Blockchain

Blockchain is developed using a cryptography and it is a method of securing information from unauthorized source by applying a mathematical algorithm and complex calculation to generate a combination of cryptographic keys to perform encryption and decryption. Researchers have developed a blockchain in pursuit of a secure encrypted decentralized distributed transactions systems (digital ledger) to record a digital transaction all over the world. This ledger ensures confidentiality, security, transparency and immutableness by applying a complex mathematical algorithm all over the digital network worldwide. Blockchains consist of a block which contains a list of new information like transactions, cryptographic hash of previous and current block, and timestamp of transaction. These blocks of data are continuous and consistent to each other by linking into a chain by which each block refers to the one before. All transactions have been verified in a distributed and decentralized ledger which updates a transactions record to all nodes of peer-to-peer network. Authentication and privacy are the major concerns during development of the blockchain to achieve unaltered transactions. It is a technology on which all cryptocurrencies developed. Bitcoin and blockchain are different. All cryptocurrency, likewise Bitcoin, is built on blockchain technology.

Blockchain contains various applications as follows:

1. Cryptocurrencies
2. Smart contracts
3. Supply chain management
4. Electronic voting system
5. NFTs (Non-Fungible Tokens)
6. Blockchain-based Healthcare systems
7. Identity management
8. Government and public services, etc.

Blockchain Applications for Mobile Device

Mobile Application developers have been using blockchain application to achieve confidentiality and integrity of data. Cryptocurrency wallets in mobile device provides a feature to securely store, send, and receive the cryptocurrency. Some the famous crypto wallets are trust wallets, Binance, Coindcx, etc., which support multiple cryptocurrencies. These applications are decentralized (dAPPS), run on peer-to-peer network or servers instead of single server controlled by single authority, and provide secure key storage, lower transaction fees, cross-border transactions,

Instead of being controlled by a single company or entity, the dApp operates on a distributed network of nodes. "Nodes" are simply computing devices of users that are interconnected to store copies of the blockchain. Therefore, the dApp works and is controlled by its users, not a large company like Meta or Google.

DApps can take several forms, from games and social networks to financial applications. Some of the dAPPs are as mentioned below:

Axie Infinity

Axie Infinity is one of the most popular play-to-earn games based on the Ethereum blockchain. It involves collecting, raising, and battling monsters known as Axies. Each Axie can have different classes and features.

PancakeSwap

PancakeSwap is a decentralized exchange built on the BNB chain. It employs smart contracts to allow swaps between BEP-20 standard tokens. The platform also offers several different avenues for users to earn from the platform. These include yield farming, staking, lotteries, and even NFT collectables.

Uniswap

Similar to PancakeSwap, Uniswap is a decentralized exchange (DEX). But Uniswap is built on the Ethereum blockchain and allows buyers and sellers to trade without an intermediary. It has a larger token base as it supports all ERC-20 tokens, which is an Ethereum standard.

Compound

Compound is a DeFi protocol that allows functions like borrowing and lending in a decentralized ecosystem. Built on Ethereum, Compound allows its users to earn from providing liquidity to borrowers on the platform.

Using an Automated Market Maker (AMM), the protocol automatically matches lenders and borrowers to facilitate a simple and seamless function. Owning the COMP token also enables its users to vote on key issues like treasury decisions and future updates of the protocol.

OpenSea

OpenSea is the most popular marketplace for buying and selling NFTs and other digital assets. It supports 8 blockchains, including Ethereum, Polygon, Solana, and BNB Chain. It provides the tools to create vibrant new economies that thrive on digital platforms.

Blockchain-Related Cybercrimes

Case 1: The DPRK's DMM Bitcoin Exploit

Attackers utilize sophisticated malware and social engineering techniques to infiltrate a company's network and IT assets. Attackers have utilized false identities, mediators, and hiring firms to be a part of a company to access the internal network by using remote work opportunities. North Korea-affiliated hack in 2024 involved Japanese cryptocurrency exchange, DMM Bitcoin, which suffered a security breach resulting in the loss of approximately 4,502.9 Bitcoin, valued at $305 million at the time. The attackers targeted vulnerabilities in infrastructure used by DMM, leading to unauthorized withdrawals.

Case 2: Gainbitcoin Cryptocurrency Scam—Multilevel Marketing Scheme

GainBitcoin was an alleged Ponzi scheme launched in 2015 by owner and their network of agents. The scheme operated through multiple websites such as `www.gainbitcoin.com`, etc., under the facade of a company named Variabletech Pte. Ltd.

The fraudulent scheme lured investors by promising lucrative returns of 10% monthly in Bitcoin for 18 months. Investors were encouraged to purchase Bitcoin from exchanges and invest them with GainBitcoin through "cloud mining" contracts. The model followed a multilevel marketing (MLM) structure, commonly associated with pyramid structured Ponzi schemes, where payouts were dependent on bringing in new investors.

Initially, investors received payouts in Bitcoin. However, as the influx of new investments dwindled by 2017, the scheme began to collapse. In an attempt to cover up the losses, GainBitcoin unilaterally switched payouts to their alleged in-house cryptocurrency called MCAP, which had significantly less value than Bitcoin, further misleading investors.

Forensic Analysis of Blockchain Data on Mobile Devices

Cryptoanalysis is an emerging field in digital forensics that focuses on extracting and analyzing data related to blockchain transactions and activity from mobile devices. With the growing use of cryptocurrencies and blockchain-based applications, mobile devices have become key platforms for managing and interacting with blockchain networks. This creates new challenges and opportunities for forensic experts investigating blockchain-related incidents or crimes.

Key Concepts in Blockchain Forensic Analysis on Mobile Devices

This section contains some important concepts to understand the blockchain related artifacts for performing forensic analysis.

1. **Blockchain Data**: Blockchain data refers to transaction records stored in a distributed ledger. These records can be publicly accessible (in the case of cryptocurrencies like Bitcoin or Ethereum) or private (in permissioned blockchains). Forensic analysis focuses on tracking transaction history, wallet addresses, and identifying suspicious or illicit activities.

2. **Mobile Devices**: With increasing mobile usage for blockchain-related activities (cryptocurrency wallets, decentralized finance (DeFi), non-fungible tokens (NFTs), etc.), smartphones have become critical devices for storing wallet keys, managing crypto funds, and interacting with blockchain applications. Investigating blockchain activity on mobile devices requires extracting relevant data and analyzing it to trace cryptocurrency transactions or identify fraudulent activity.

3. **Cryptobased application analysis**

    Cyber criminals are using fake cryptocurrency application and spreading these applications using social media platforms. Criminals are using social engineering techniques and platforms like Whatsapp, Telegram, YouTube to share the fake application and initially ask for small amounts for investment. Criminals form a chat group and showcase a fake profit via fake group members to gain trust of group members. They display fake screenshots of application to lure target group members. They give the profit initially, but the next time, they convince the investor to invest a larger amount to get good profit. Criminals lure victims by displaying fake profit in crypto applications and insist the victims invest more and fleece them. Application analysis reveals actual behavior of application, legitimacy, command and control details, Email ID, phone no., etc.

Steps for Forensic Analysis of Blockchain Data on Mobile Devices

1. **Data Acquisition**: The first step in blockchain forensic analysis on mobile devices is **mobile data acquisition**. This involves extracting data from the mobile device in a forensically sound manner to ensure that no evidence is altered during the process. Chain of custody should be maintained by investigating officer and forensics expert by making chain of custody documents. Some modern crypto-related applications often decrypt the data and load it in temporary memory like cache or RAM, which might contain private keys, etc. Commercial Forensic tool like MSAB XRY assists forensics experts to reveal some interesting artefacts which might support the investigation. But, in some cases, to capture the RAM dump, it is required to root android device or jailbreak IOS device, so forensics experts differs it.

    - **Mobile data acquisition process:** Create a bit-by-bit copy (image) of the mobile device's storage. This can be done through physical or logical acquisition methods, depending on the device's security settings.

- **Physical acquisitions process**: The technique of data acquisition is contingent upon the mobile device type (e.g., iOS, Android) and the forensic instrument employed. Physical acquisition retrieves all data from the device's storage, including system files, deleted data, application data, and concealed data. The iOS device is incompatible with physical extraction.

- **Logical Acquisition:** The logical acquisition of data from a mobile device involves extracting available information, such as files, application data, contacts, messages, etc., directly from the device's operating system using conventional techniques like APIs and file system access. In contrast to physical acquisition, which entails accessing the device's raw memory, logical acquisition concentrates on extracting data permitted by the operating system for user or tool access. Logical acquisition is generally more expedient and less invasive than physical acquisition, making it the preferred method for forensic experts and security professionals during digital investigations. Here is a comprehensive analysis of the logical acquisition process:

- **Data Sources**: Relevant data sources include mobile wallet apps (e.g., MetaMask, Trust Wallet, Coinomi), browser history (for web-based wallets or DeFi), messaging apps (for communications related to cryptocurrency transactions), and system logs. Some of the browser applications provide plugins and inbuilt features to manage crypto transactions: for example, Brave Browser. This browser maintains the database at the application level.

2. **Identifying Blockchain-Related Applications:**

    - **Wallet Apps**: Identify mobile applications that manage cryptocurrency wallets. Extract relevant files, such as wallet.dat, keystore files, or app-specific data (e.g., private keys, recovery phrases, transaction history).

- **DeFi and NFT Apps**: Applications related to decentralized finance (e.g., Uniswap) or NFTs (e.g., OpenSea) also store relevant blockchain data. Forensics investigators should look for transaction histories or wallet interactions with decentralized platforms.

3. **Data Parsing**: After acquiring the data from the mobile device, the next step is to parse the blockchain-related data and identify key pieces of evidence: Forensics tools like UFED and Oxygen contain features to identify the crypto transactions. These tools classify crypto-related transactions and activities automatically during parsing the raw data.

   - **Transaction Histories**: Extract transaction details, including wallet addresses, timestamps, transaction IDs, amounts, and blockchain network used (Bitcoin, Ethereum, etc.).

   - **Wallet Information**: Analyze stored wallet data (private keys, seed phrases) that may allow investigators to trace wallet addresses and transactions back to individuals.

   - **Communication Analysis**: Analyze messages, emails, or social media conversations related to cryptocurrency transactions (e.g., addresses sent, received, or offered).

4. **Blockchain Transaction Analysis**: Once blockchain data is extracted, forensic investigators can use blockchain explorers or any commercial tools like CipherTrace or Chainanalysis to trace transactions on the blockchain.
   Key steps in transaction analysis include

   - **Wallet Address Identification**: Trace wallet addresses identified on the mobile device to determine ownership and link them to known entities, if possible.

   - **Transaction Flow Analysis**: Analyze transaction flows to identify the origin and destination of cryptocurrency movements. Investigators should look for patterns that suggest suspicious activity (e.g., mixing services, rapid transfers).

- **Smart Contract Interactions**: If the mobile device is interacting with smart contracts (e.g., on Ethereum), investigators need to analyze contract interactions and identify the function calls, participants, and outcomes.

5. **Data Correlation and Reporting:**

   - **Linking Physical and Digital Evidence**: Investigators need to correlate physical evidence (e.g., device metadata, communication logs) with digital evidence (e.g., wallet activity) to build a timeline of blockchain-related actions.

   - **Report Findings**: Document the findings in a detailed, organized report that includes transaction details, associated wallet addresses, timestamps, and relevant context for each piece of evidence.

Tools for Blockchain Forensic Analysis on Mobile Devices

1. **Cellebrite UFED**: Cellebrite is a popular tool for mobile forensics that can extract data from mobile devices. It has capabilities to retrieve data from cryptocurrency wallet apps and social media platforms that may have blockchain-related activity.

2. **Magnet AXIOM**: Magnet AXIOM allows forensic examiners to analyze mobile devices and extract blockchain-related data from wallet apps, messaging apps, and other sources. It also includes features to parse and analyze cryptocurrency transaction data.

3. **Blockchain Explorers**: Blockchain explorers like **Blockchain.info** (for Bitcoin) or **Etherscan.io** (for Ethereum) are essential tools for analyzing blockchain transactions by providing detailed public data about transactions, wallet addresses, and smart contract interactions.

4. **ElcomSoft iOS Forensic Toolkit**: This tool allows forensic investigators to extract and analyze data from iOS devices, including cryptocurrency wallet information stored on mobile apps.

5. **XRY by MSAB**: XRY is another mobile forensic tool that can be used to extract data from mobile devices and analyze app data related to cryptocurrency transactions, wallet apps, and blockchain activities.

Key Blockchain-Related Artifacts on Mobile Devices

1. **Wallet Database Files:**
   - These files contain private keys, transaction history, and addresses associated with cryptocurrency wallets. They may be stored as wallet.dat or keystore files on mobile devices. Sometime private keys could be captured from memory if the device is already rooted or jailbroken.

2. **Backup Files:**
   - Many mobile wallet apps allow users to back up their wallet to cloud storage. These backups may contain critical data such as wallet keys or seed phrases that are essential for forensic analysis.

3. **Browser Data:**
   - Blockchain-related browsing activity, such as visiting decentralized exchanges (DEXs) or interacting with NFT platforms, may leave traces in the browser history or cookies on the mobile device.

4. **App Logs:**
   - Cryptocurrency wallet apps often generate logs that can provide information about user interactions with blockchain networks, such as transaction requests, transfers, and wallet synchronization.

Challenges in Blockchain Forensic Analysis on Mobile Devices

1. **Encrypted Data**: Many mobile apps, including cryptocurrency wallets, store private keys and transaction data in an encrypted format. Extracting and decrypting this data can be challenging without proper access of credentials (e.g., PIN, password, or biometric authentication).

2. **Distributed Nature of Blockchain**: Blockchain transactions are spread across multiple nodes (computers) in the network, making it harder to pinpoint specific devices or sources of transactions directly. Analyzing a mobile device alone may only reveal partial information.

3. **Lack of Standardized Forensic Tools**: Blockchain forensic tools are still developing, especially those designed specifically for mobile devices. Forensic investigators may face limitations in using existing tools for blockchain-related data extraction from mobile apps or wallets.

4. **Anonymity and Pseudonymity**: Many blockchain networks (such as Bitcoin) provide pseudonymous identities, where transaction details are tied to wallet addresses instead of real-world identities. Analyzing these transactions requires techniques to link wallet addresses to individuals or entities.

5. **Lack of Access to Private Keys**: Since private keys are essential for performing transactions on blockchain networks, gaining access to a mobile device with an active wallet often depends on the user's authentication (password or biometric). If investigators cannot unlock the device, they may not be able to access crucial information.

Decentralized Application and Investigation

Decentralized applications (dApps) have gained significant popularity, especially in the blockchain space, due to their ability to operate without a central authority. Unlike traditional applications, dApps leverage blockchain technology to offer transparency, immutability, and decentralized control. However, this decentralization presents unique challenges and opportunities in digital investigations. Forensic investigators and law enforcement agencies may need to adapt to these challenges when attempting to investigate crimes or incidents involving dApps.

What Are Decentralized Applications (dApps)?

Decentralized applications (dApps) are software applications that run on a decentralized network, typically powered by blockchain technology. Unlike centralized applications (like traditional banking apps or social media), dApps have no central server. Instead, they operate on distributed ledger technology, such as blockchain or peer-to-peer networks, and use smart contracts to handle the logic of the application.

Key features of dApps:

- **Decentralized**: They operate without a central authority.
- **Blockchain-Based**: They use blockchain to store data and manage transactions.
- **Smart Contracts**: They leverage self-executing contracts with predefined rules encoded into the blockchain.
- **Transparency**: Transactions and data are typically publicly accessible and immutable.

Examples of dApps:

- **DeFi platforms** (e.g., Uniswap, Compound)
- **NFT marketplaces** (e.g., OpenSea, Rarible)
- **Decentralized social media** (e.g., Steemit)
- **Gaming platforms** (e.g., Axie Infinity)

Investigating Crimes Involving dApps

Investigations involving dApps present several unique challenges, especially since the decentralized nature of blockchain removes a central authority that investigators can easily reach for data access or intervention. However, it is still possible to conduct investigations using a combination of digital forensics techniques, blockchain analysis, and traditional investigative methods.

Here are the main aspects of conducting **investigations** into decentralized applications:

1. **Data Acquisition**

    Unlike centralized applications where data is stored in a central server (which can be subpoenaed or accessed via traditional means), dApps store their data across a decentralized network. Investigators need to focus on acquiring **blockchain transaction data** and **smart contract interactions**, which are often publicly available.

    Data Acquisition:

    - **Blockchain Data**: Transaction details, wallet addresses, smart contract logs, and decentralized exchange (DEX) activity are recorded on public blockchains (e.g., Ethereum, Binance Smart Chain). Investigators can use blockchain explorers like **Etherscan** or **BscScan** to view this data.

    - **Off-Chain Data**: Some dApps store additional information off-chain (e.g., metadata for NFTs). Investigators need to explore decentralized file storage networks like **IPFS (InterPlanetary File System)** or **Arweave** where such data may reside.

    - **Wallet Data**: Access to user wallets and transaction history is vital. Investigators can extract wallet data (public keys and addresses) from dApp interaction logs and transaction history on the blockchain.

Tools for Data Acquisition:

- **Blockchain Explorers** (Etherscan, BscScan, etc.) for tracing transactions
- **IPFS clients** to access data stored off-chain
- **Smart contract analysis tools** (e.g., **MyEtherWallet**, **Remix IDE**) to examine contract interactions and source code

2. **Data Analysis**

Once data is acquired, investigators need to analyze the data to identify connections between transactions, wallet addresses, and potential criminal activity.

Key Analysis Steps:

- **Transaction Flow Analysis**: Investigators can track the movement of assets (cryptocurrency, NFTs, etc.) across different wallet addresses. This can help identify patterns such as illicit money transfers, layering (money laundering), or wallet clusters associated with known criminal entities.
- **Smart Contract Interaction**: By analyzing smart contracts, investigators can review code execution, see who interacted with it, and identify whether any malicious code was involved (e.g., hacks or rug pulls).
- **Wallet Address Linking**: Investigators can link wallet addresses to known individuals, services, or exchanges by tracing patterns or using **Know Your Customer (KYC)**/AML data from centralized exchanges (CEXs), if available.

Tools for Data Analysis:

- **Chainalysis**: A blockchain analytics platform that helps trace cryptocurrency transactions and identify patterns or illicit activities.
- **Elliptic**: A blockchain analysis tool that helps identify suspicious transactions, monitor wallets, and track crypto transactions.

- **CipherTrace**: Provides blockchain forensics tools for tracing transactions and analyzing wallets, which can help investigators identify illicit activities.

- **Maltego**: It displays the relation between different entities like wallets, transactions, users, meta data, etc. It is integrated with different crypto analytics platforms like Ciphertrace, Wallet explorer, etc.

- **Token Terminal/DeFiLlama**: DeFi analytics tools to track decentralized finance transactions and investments.

3. **Attribution of Crimes in dApp Ecosystems**

Attributing criminal activity within a decentralized system is often difficult due to the pseudonymous nature of blockchain transactions. However, investigators can use several techniques to enhance attribution:

Techniques for Attribution:

- **Wallet Address Mapping**: Investigators can map wallet addresses to known individuals or entities by cross-referencing with public data (e.g., KYC data from exchanges or social media posts).

- **Link Analysis**: Look for patterns of interaction between various addresses or dApp users, potentially identifying criminal networks.

- **Forensic Tracing**: Forensic tools can help link a particular transaction or address to a known illicit activity (e.g., darknet markets, stolen funds, or ransomware payments).

4. **Privacy and Anonymity Issues**

Privacy-focused cryptocurrencies and dApps can make it more challenging for investigators. Some platforms and dApps use privacy protocols like **Monero** or **Zcash**, which obfuscate transaction details, including wallet addresses and amounts.

Privacy Challenges:

- **Zero-Knowledge Proofs (ZKPs)**: Protocols like Zcash use ZKPs to hide transaction details, making it difficult to trace funds.

- **Mixing Services**: Cryptocurrency mixers can obfuscate the source and destination of funds, which can hinder tracing efforts.

Overcoming Privacy Challenges:

- **Cooperation with Exchanges**: Investigators can seek information from exchanges that have KYC/AML compliance to identify users behind specific wallet addresses.

- **Advanced Forensic Techniques**: Use heuristic analysis and other techniques to try to link transaction patterns despite obfuscation (e.g., **Elliptic** and **Chainalysis** provide tools for partially de-anonymizing privacy coins).

5. **Legal and Jurisdictional Issues**

The decentralized nature of dApps means there is no central authority to request data from. Many users operate anonymously or pseudonymously, which creates jurisdictional and legal hurdles.

Key Legal Considerations:

- **Lack of Centralized Authority**: There may be no clear entity to subpoena or request data from (as there would be with centralized applications).

- **Cross-Border Issues**: Blockchain transactions can easily move across borders, complicating jurisdictional authority in international investigations.

Legal Approaches:

- **Working with International Partners**: Investigators may need to collaborate with international law enforcement agencies to trace cross-border blockchain activity.

- **Cooperation with Centralized Exchanges**: Though decentralized in nature, many users of dApps ultimately move funds through centralized exchanges, which are more likely to comply with legal requests.

- **Requesting Data from Decentralized Protocols**: Some decentralized protocols may have on-chain governance or logs that could be useful in an investigation.

Cryptocurrency and Wallet Investigations

Cryptocurrency investigations have become a critical part of modern digital forensics, as cryptocurrencies like Bitcoin, Ethereum, and others are increasingly used for legitimate transactions and illicit activities alike. Wallet investigations, in particular, are crucial because they serve as the gateway to accessing and moving cryptocurrency on the blockchain. These investigations often focus on identifying illicit activity such as money laundering, fraud, ransomware payments, or terrorist financing.

Below is a detailed breakdown of how cryptocurrency and wallet investigations are conducted, key tools and techniques used, and the challenges investigators face.

Understanding Cryptocurrency Wallets

Before diving into the investigation, it's important to understand the two main types of cryptocurrency wallets are involved in investigations:

Types of Wallets

1. **Hot Wallets:** These wallets are connected to the Internet and are typically used for frequent transactions. They include exchange wallets, mobile wallets, and desktop wallets.

2. **Cold Wallets**: These are offline wallets, such as hardware wallets (e.g., Ledger, Trezor) and paper wallets. Cold wallets are considered more secure but may be harder to track if the owner has taken steps to remain anonymous.

How to Identify the Wallet

A wallet is identified by its public address (a string of characters) that allows others to send cryptocurrency to it. A wallet also has a private key that grants access to the cryptocurrency stored inside. Investigations generally focus on the public address and transaction history, but sometimes private keys may be crucial if the wallet's contents need to be accessed from seized devices.

Investigating Cryptocurrency Transactions

Cryptocurrency transactions are recorded on the blockchain, a decentralized ledger, which means they are permanent and transparent. Blockchain explorers allow investigators to trace transactions from wallet to wallet, offering valuable clues for investigations.

Transaction Analysis: Start by identifying suspicious transactions, such as large amounts of cryptocurrency, transfers to/from known criminal addresses, or transactions with privacy coins. Tools like Ciphertrace, Blockchain explorers, etc. (e.g., Etherscan for Ethereum, Blockchain.com for Bitcoin) help investigators see transaction history and associated wallet addresses.

Investigators trace funds from one wallet address to another by examining transaction paths. This can help uncover if the funds are being funneled through mixing services (e.g., CoinJoin) or tumbling services that anonymize transactions. Using wallet address analysis, investigators can uncover linked addresses that share common traits, such as using the same exchange or wallet provider. This could help uncover the identities behind multiple addresses.

Some cryptocurrency-related crimes occur through decentralized applications (dApps) and smart contracts. Investigating smart contracts can help identify fraudulent activities (e.g., scams or rug pulls in the DeFi space).

Tools like MyCrypto or Etherscan can be used to analyze the smart contract interactions associated with cryptocurrency transactions.

Wallet Analysis

Investigating cryptocurrency wallets involves accessing the wallet's transaction history and tracking the movement of funds. For mobile or desktop wallets, investigators may need to extract wallet data from the device itself, which might require technical expertise.

For Mobile or desktop forensics, investigators might need to extract wallet files (e.g., the wallet.dat file for Bitcoin or the keystore file for Ethereum) to analyze the private and public keys stored on the device. Many wallets encrypt their private keys for security. If investigators can access the device and obtain user credentials (e.g., PIN, password, fingerprint), they can decrypt the wallet and gain access to the funds. Once the wallet file is obtained, investigators can view its transaction history. This can include incoming and outgoing transfers, addresses interacted with, and the amounts transferred.

Techniques for Linking Wallets to Individuals

One of the most difficult aspects of cryptocurrency investigations is linking anonymous wallet addresses to real-world individuals. While blockchain provides transparency, it does not automatically reveal who controls a wallet address. However, certain techniques can help in this process.

Techniques for Linking Wallets:

Cross-Referencing with Centralized Exchanges (CEX): If the suspect has used a centralized exchange (CEX) such as Binance, Coinbase, or Kraken, investigators may obtain KYC (Know Your Customer) data for the wallet address, especially if they were involved in the exchange's services (depositing or withdrawing). CEXs have a strong regulatory requirement for KYC compliance, but investigators still need to request this information through legal channels. If an individual shares their wallet address on social media, blogs, or other public platforms, it could be linked to their identity. Investigators can use social media scraping or open-source intelligence (OSINT) tools to link wallet addresses to known public figures or entities. OSINT tools like Maltego or Hunchly can help investigators gather publicly available information that can link wallet addresses to individuals.

Challenges in Blockchain Mobile Forensics

Pseudonymity and Anonymity

Blockchain transactions are often associated with pseudonymous addresses rather than real-world identities. For instance, a user's activity is linked to their wallet address, not their name or other personal identifiers. Tracing the origin of funds or identifying individuals behind addresses is challenging without additional information (e.g., KYC data from exchanges).

Investigators must rely on blockchain analysis tools, external data sources (like centralized exchanges), and transaction patterns to piece together identities. Advancement in these tools creates opportunities to investigate such cases.

Decentralized Nature of Blockchain

Blockchains are decentralized, meaning there is no central authority or server where data can be easily retrieved. Data is distributed across nodes, which makes it difficult to gather comprehensive evidence from a single source. Investigators may struggle to acquire all relevant data, as blockchain data is not hosted in a central location and can be fragmented across multiple participants.

Blockchain explorers and distributed ledger technologies (DLTs) can provide public access to blockchain data, but investigators must use advanced methods to reconstruct complete transaction histories.

Data Encryption and Privacy Features

Many mobile apps, including cryptocurrency wallets and decentralized finance (DeFi) applications, use end-to-end encryption to secure private keys, seed phrases, and transaction data. If investigators do not have the encryption keys or passwords, they may not be able to access crucial evidence on mobile devices.

Forensic tools can be used to extract encrypted data, but success is contingent upon obtaining user authentication credentials (e.g., PIN, password, fingerprint).

Volume and Complexity of Data

Blockchain networks generate massive amounts of data (e.g., transactions, blocks, contract interactions). In mobile forensics, this data can be overwhelming to process, especially when dealing with decentralized applications and multiple blockchain platforms. Sifting through large datasets for relevant evidence can be time-consuming and may require advanced data analytics techniques.

Using blockchain analytics tools that can filter, correlate, and visualize blockchain data helps investigators manage the sheer volume of information and focus on critical evidence.

Cross-Border and Jurisdictional Issues

Blockchain transactions often occur across borders, making it difficult for investigators to determine which jurisdiction should handle the case. Investigating blockchain-related crimes may require international cooperation, especially when funds have moved through multiple countries and blockchain networks. Collaborating with international law enforcement agencies (e.g., INTERPOL, Europol) and using tools that track cryptocurrency movement across borders can help resolve jurisdictional issues.

Opportunities in Blockchain Mobile Forensics
Transparent and Immutable Data

Blockchain's core feature of immutability means that once data is recorded, it cannot be altered or deleted. This offers a clear and permanent record of transactions. Investigators can access unalterable evidence, such as transaction histories, timestamps, and wallet addresses, which can be useful in tracing illicit activities like fraud, money laundering, and ransomware payments. If a suspect moves stolen cryptocurrency through various wallets, investigators can trace the flow of funds using a blockchain explorer.

Blockchain explorers like Etherscan and BscScan provide public access to blockchain data, making it easier for investigators to trace transactions, view wallet balances, and identify contract interactions. Investigators can use these tools to publicly view transaction details, wallet addresses, and interactions with decentralized applications (dApps) on mobile devices. By analyzing a wallet address, investigators can identify its transaction history, check for suspicious activity, and even track transactions back to known exchanges.

Integration with Mobile Forensic Tools

Integration of blockchain forensics with mobile forensic tools is improving. As blockchain-based applications on mobile devices (e.g., cryptocurrency wallets, dApps) generate increasingly critical evidence, mobile forensic tools are evolving to support blockchain data extraction. Forensic investigators can now use mobile forensics tools (like Cellebrite and Magnet AXIOM) that integrate blockchain analysis features, enabling a more comprehensive examination of blockchain data and mobile device data together. Forensic tools can now extract wallet files from mobile devices and simultaneously analyze transaction data on blockchain explorers to identify suspicious activities and track transactions.

Smart Contract and dApp Investigation

With the rise of decentralized finance (DeFi) and NFTs, investigating smart contracts and dApps provides a way to access valuable data stored on public blockchains. By examining smart contract interactions and transaction logs, investigators can trace funds and identify vulnerabilities. Investigating smart contracts and dApp logs can provide insights into how certain contracts were executed and whether they were involved in fraudulent activities, hacks, or rug pulls. If a user interacts with a vulnerable DeFi contract, investigators can analyze the contract's code, review transaction history, and identify malicious actions.

Real-Time Monitoring of Blockchain Transactions

Blockchain forensic tools can offer real-time monitoring of transactions, enabling investigators to track illicit activities as they occur. Investigators can set alerts for specific wallet addresses or suspicious activities (e.g., large transfers) and respond in real time, reducing the time to detect and act upon crimes.

Real-time monitoring systems like CipherTrace can alert investigators when funds are moved from a wallet associated with known criminal activity, enabling prompt action.

Opportunity: Increased awareness and collaboration between blockchain analytics companies, mobile forensics providers, and law enforcement agencies have led to better tools and protocols for investigating blockchain-related crimes.

Law enforcement agencies are improving their blockchain-related expertise, which enhances the effectiveness of forensic investigations in areas like fraud, money laundering, and cybercrime.

Agencies like the FBI and Europol have partnered with blockchain forensic firms to track cryptocurrency movement across networks and arrest individuals involved in cybercrimes.

Summary

This chapter explores the blockchain technology and related cybercrime. Further attackers or cybercriminals uses blockchain to perform the digital transactions. Further, this chapter explores the tools and techniques used by forensics experts to retrieve the blockchain related artifacts from mobile device, wallet application, etc. This chapter also elaborates the challenges, opportunities, and methodologies involved in extracting and analyzing blockchain data from mobile devices and also covers the cryptocurrency transactions to decentralized applications and its investigation.

CHAPTER 5

Investigating Mobile Banking and Financial Applications

Mobile banking is employing application-based technology on mobile devices to conduct banking transactions. The smartphone revolution has generated chances for the financial sector to offer online banking services 24/7, 365 days a year through net banking and mobile banking. Users are utilizing smartphones, tablets, and laptops to do banking operations instead of visiting physical banks. Nearly all financial organizations, including banks, stock markets, insurance sectors, and capital markets, offer the capability to conduct transactions using mobile banking applications. Earlier, financial organizations utilized SMS for restricted banking services, allowing users to check their balance by sending a text to the bank's designated number. In early 2010, the advancement of mobile operating systems such as Android and iOS brought transformative changes to the mobile device business. The technological improvements in mobile application development demonstrated the diversity of programs that support both Android and iOS devices. Google's Android Play Store and iOS's App Store serve as the primary platforms for users to obtain and install applications on mobile devices. All banking and financial applications must be registered on these application stores to be available for download on mobile devices, where the application store conducts a thorough inspection of the source code and program features. This chapter includes diverse characteristics of mobile and financial applications; their security features and how cyber criminals are misusing financial applications to commit cyber fraud. Cybercriminals are generally using UPI as a medium to receive the fraudulent amount by sending QR code, UPI Id, etc. They imitate legitimate banking websites and applications via sending a phishing email, sms, social media messages, or via phone calls.

CHAPTER 5 INVESTIGATING MOBILE BANKING AND FINANCIAL APPLICATIONS

This chapter also explains about important artifacts, mobile forensic acquisition process, challenges and timeline analysis. Further it elaborates forensics analysis of mobile malware or malicious applications, tools, and techniques.

The Evolution of Mobile-Based Applications

In the early 2000s, individuals utilized simple feature phones that supported just voice calls and SMS; the concept of mobile banking applications did not exist at the time. In this era, Internet connectivity was slow, and the development of wireless protocols like Wi-Fi was inadequate. In 2007, Apple introduced the first iOS device, the iPhone, followed by HTC's debut of the inaugural Android smartphone in 2008. Blackberry and Nokia were also competitors, utilizing the Blackberry and Symbian operating systems. The enhancement of global Internet accessibility following the introduction of 3G services allows customers to utilize greater amounts of Internet data at a reasonable cost. During this period, information technology experienced significant growth as application developers created a diverse array of programs, including social media, gaming, utilities, taxi booking, online shopping, and banking, which are then registered on the Google Play Store and iOS App Store.

Local application developers initiated the creation of diverse news applications featuring innovative concepts such as online commerce, food delivery, health and fitness, and radio, but consumers exhibited limited awareness of cybercrime. In early 2016, the National Payment Corporation of India developed the Unified Payment Interface (UPI), which revolutionized the fintech business in India. During the same decade, India experienced a data revolution as Internet services became available at cheap prices. The proliferation of highly affordable mobile data plans has led to a significant increase in Internet users in India, resulting in a peak in mobile application usage. Firms have introduced financial technology applications like Paytm, PhonePe, and GPay, preferring these platforms for transactions over cash payments. Consequently, these revolutions facilitate opportunities for cybercriminals to exploit FinTech platforms for digital crimes.

Security Attributes of FinTech and Banking Applications

There are two main categories for financial applications:

1. **Financial technology applications**

   Fintech applications encompass UPI, insurance platforms, cryptocurrency wallets, merchant services such as PayPal and Razorpay, loan applications, stock market applications, and investing platforms.

2. **Banking applications**

   All scheduled and non-scheduled commercial banking apps that offer banking services to their clients. Application security is a significant difficulty in mobile applications because of the handling of sensitive data and information, including transaction details, personal data, and transaction passwords. Financial applications must guarantee the protection of user data both at rest and in transit over networks from cyber attackers and cybercriminals.

Security Features of Banking Application

Below are the security features used by banking applications

1. **Cryptography**

   Banking apps employ cryptographic techniques, specifically SSL/TLS protocols, to secure Internet connections and web servers, thereby ensuring data confidentiality and integrity. Application developers are employing SSL pinning to authenticate and validate the SSL certificate on the application server. SSL pinning entails the use of a special server certificate, created by the developer with a customized key size and public key, to mitigate the risk of man-in-the-middle attacks.

2. **Session administration and expiration**

   Session management and timeout are essential elements of mobile application security; they ensure effective handling of user sessions to avert unauthorized access or data breaches. Some characteristics are listed below:

a. **Session management in mobile applications** Session management pertains to the manner in which an application administers user authentication, preserves user state, and guarantees that users remain logged in alone when suitable. Inadequate session management may result in security vulnerabilities such as unauthorized access.

b. **Optimal Strategies for Session Management**

1. **Session Tokens**: Employ secure session tokens (e.g., JWT-JSON web tokens) for user authentication. Guarantee that these tokens are authenticated and encrypted to avert tampering.

2. **Brief Session Duration**: Restrict the lifespan of session tokens. Automatically invalidate the session following a period of inactivity or after a specified time frame.

3. **Session Storage**: Securely retain session information. For mobile applications, refrain from keeping tokens in vulnerable locations, such as plaintext in shared preferences. Utilize secure storage solutions such as Keychain for iOS or Keystore for Android.

4. **Session Revocation**: Establish protocols for user logout and guarantee the revocation of tokens on both the client and server sides to avert exploitation in the event of a compromised session.

5. **Multi-Factor Authentication (MFA)**: Whenever feasible, implement MFA to enhance security in session management. Banking apps employ multifactor authentication, incorporating one-time passwords (OTPs), biometric verification, and hardware-dependent authentication via registering mobile devices with specialized authenticator programs, such as Google Authenticator.

6. **Session Expiration**: Session timeouts prevent an idle session from remaining open indefinitely, thereby mitigating the danger of illegal access via abandoned devices or displays.

3. **Authentication and authorization of transactions**

   Transaction Authentication: Transaction authentication confirms the identity of the user requesting a banking transaction and ensures that it is launched by an authorized individual and device. This procedure often entails authenticating user credentials and verifying the authenticity of the transaction request.

   Transaction Authorization: Authorization denotes the procedure of approving or rejecting a transaction subsequent to its authentication. It ensures that users can carry out the requested action according to their account privileges, balance, transaction limitations, and other regulations.
   Banking application often checks the transactions associated with registered mobile device which are listed below:

   - **Behavioral Analytics**: Banking application's development institutions employ algorithms to identify anomalous transaction patterns, like abrupt substantial withdrawals or transfers initiated from an unfamiliar device. Financial institutions identify suspicious transactions, which may require manual examination before approval.

   - **Instant Notifications:** Application alerts users for each transaction, enabling prompt detection of fraudulent actions.

   - **Identification of IP addresses and devices**: Banks monitor users' IP addresses, geolocation, and device fingerprints to identify anomalies. Should a transaction request originate from an unrecognized device or location, supplementary verification measures may be necessary.

- For example, if a user's account is accessed from a country different from their typical location, the system may necessitate supplementary verification to verify its legitimacy.

4. **A protected application programming interface**

    a. Utilize HTTPS (TLS/SSL) for Secure Communication

    - **Data in Transit Encryption**: Consistently employ HTTPS (utilizing TLS) for a secure connection between the mobile application and the backend server. This guarantees that all sensitive data transmitted, such as login credentials, personal information, and transaction details, is encrypted and safeguarded against man-in-the-middle (MITM) attacks.

    - SSL Pinning: Integrate SSL pinning within your mobile application. SSL pinning guarantees that the application exclusively connects to a trusted server by verifying the host's certificate against a pre-stored, recognized certificate or public key; hence, it mitigates risks such as certificate spoofing.

    - **Code Obfuscation**: Obfuscate the mobile banking application's code to complicate reverse engineering efforts by attackers and obscure potential vulnerabilities. This safeguards important application logic and cryptographic keys from extraction.

    b. API Authentication and Authorization

    Utilize OAuth 2.0 for secure and standardized token-based authentication. OAuth 2.0 enables the application to acquire a bearer token (access token) after user authentication, which the application subsequently utilizes to securely access APIs without repeatedly transmitting passwords with each request.

    - **Authorization Code Flow**: Employ this method for enhanced security in authentication, wherein the mobile application swaps a transient code for a token.

- **Implicit Flow (for public clients)**: In specific scenarios, such as with single-page apps, the implicit flow can be utilized to directly issue access tokens; however, this method is comparatively less safe.

5. **Security at the device level**

   Encryption of Devices

   - **Full Disk Encryption (FDE)**: This algorithm ensures the entire device undergoes encryption to protect all data. Contemporary mobile operating systems such as iOS and Android offer integrated full disk encryption. This ensures data security even in the event of device loss or theft.

   - **File-Based Encryption (FBE)**: Android utilizes file-based encryption, which individually encrypts files based on the application's security requirements. This allows for more precise control over sensitive data and ensures that unauthorized users cannot access specific application data, even with the device unlocked.

   - **Minimize Permissions**: Adhere to the idea of least privilege when soliciting application permissions. Solicit solely the permissions essential for the app's operation (e.g., access to camera, location, storage, etc.) and verify that these rights are warranted by the app's functionality.

   - **In-App Security Features**: Integrate in-app security mechanisms like PIN or pattern locks to prevent unauthorized access within the application post-device unlocking.

   Root/Jailbreak Identification

   - **Identify Jailbroken or Rooted Devices**: Jailbreaking (iOS) or rooting (Android) eliminates inherent security constraints on the device, thereby increasing its susceptibility to malware and unauthorized access. Integrate root/jailbreak detection in the application to inhibit its operation on compromised devices.

- **Restrict App Functionality on Compromised Devices**: If the app identifies that it is operating on a rooted or jailbroken device, limit access to sensitive functions (such as financial transactions) or prohibit access to the app altogether.

- **Developer mode**: Banking application has permission to read the device configuration to check either developer mode is enabled or not. If developer mode is enabled, application restricts itself to function. Generally, developer mode is used by developer or attacker to install third-party application.

6. **Geolocation and IP-based identification**

Location-based detection employs the geographical position of a user's device to evaluate the legitimacy of a transaction, login, or request. Real-time location analysis enables prompt action in response to substantial discrepancies or potential fraud threats.

Methods for Geolocation Detection:

- **GPS-Based Geolocation Monitoring**:

    Most smartphones possess integrated GPS, enabling the acquisition of the user's location during app interaction. In mobile banking applications, GPS can verify that the user is engaging in activities at a place that aligns with their typical activity. If a person usually accesses their account from one region and a transaction request arises from another country, it may prompt a review flag.

    Geofencing entails establishing a virtual perimeter around a certain geographical region. If a user tries to access the app or start a transaction after this time limit has been set, the system may either refuse access, mark the behavior as suspicious, or ask for more proof (for example, multi-factor authentication).

Detection of Anomalies using Locational Patterns:

- Over time, one can develop a profile of standard sites for each user. A user's sudden login from an unfamiliar or atypical location (e.g., a different country, a new city, or a distinct region within the same country) may be deemed suspicious.

- Example: If a user consistently signs in from Chennai but attempts to access their account from a different place (e.g., Srinagar), the application may initiate an alert and necessitate further verification to proceed when the journey duration is impossible within a specified time frame.

Immediate Location Verification:

- Subsequent to determining a user's position, juxtapose it with the IP-based geolocation data. A significant disparity between GPS and IP locations may suggest dubious activity or the use of a proxy or VPN.

- **VPN/Proxy/TOR Identification:**

    Attackers frequently employ Virtual Private Networks (VPNs)/ TOR or proxy servers to conceal their actual IP address. By comparing the IP against a recognized database of VPN or proxy IP addresses, one might ascertain whether the user is attempting to conceal their actual location. If a banking application or net banking session detects a logon from a VPN or Tor network it restricts users to login unless it is whitelisted VPN application by banking department.

Mobile Application security Guidelines by Cert-In

Computer Emergency Response Team of India has issued a guidelines and best practices for secure application development life cycle. These guidelines contain 4 phases like context of the security in designing the application, secure development practice, audit of the application and secure deployment of the applications.

Phase 1: Establish the Context of the Security in Designing of Application.

- **Security by Design Approach**: It includes the integration of security controls, methodology, best practice and mechanism from the early stages of the implementation.

- Adoption of Secure Software Development Life Cycle (SDLC): Secure Software Development Life Cycle (SDLC) encompasses various models and frameworks:

    a) "Microsoft Secure Development Lifecycle (SDL)" is a widely known and adopted SDLC framework with seven phases.

    b) "Open Web Application Security Project (OWASP) Software Assurance Maturity Model (SAMM)" helps build mature software security programs with four levels and multiple security practices.

    c) "Agile Secure Development Lifecycle" integrates security practices within agile methodologies, including security grooming, security testing, continuous integration and deployment, security feedback loop.

    d) "NIST Secure Software Development Framework (SSDF)" is a comprehensive guide for developing secure software.

- Engagement of security trained designers and developers in the application development: Developers and designers must have understanding of cyber security fundamentals.

Phase 2: Implement and Ensure Secure Development Practices

- Authentication, Authorization, and Session Management
- Cryptographic Practices
- Version Control and Change management
- Ensure Secure Coding
- File and Memory Management
- Optimize to Manage Complexity

- Develop Software Technology Specific Security Checklist
- Security Test Driven Development (STDD)
- Implementation of Threat Modeling in Application Development
- Build Secure Environment for Application Development
- Secure Use of Environment Variables
- Stored Procedures Over SQL Statements
- Handle Error Messages, Commented Code and Exceptions
- Linear Data Structure and Multiple Inheritances
- Third Party and Open-Source Libraries, Components, and APIs
- Build Trust Boundaries
- Principle of Least Privileges
- Enhancing Maturity of Software Security

Phase 3: Guidelines for Audit of Applications

- **Source Code Review**: It examines an application's source code to identify security flaws or vulnerabilities. It ensures adherence to coding standards, uncovers bugs, and identifies potential threats.
- **Conduct Security Vulnerability Assessment**: Periodic vulnerability assessment should be conducted by referring the baseline documents and framework like OWASP Application Security Verification Standard (ASVS), OWASP Web Security Testing Guide, Mobile Application Security Verification Standard (MASVS), Mobile Application Security Testing Guide (MASTG), OWASP Application Programming Interface (API), etc.
- **Timeline for Completion of Audit**: Audit report should mention appropriate timelines for closure of vulnerabilities according to severity.
- **Penetration Testing**: Penetration Testing enables organizations to prioritize and fix vulnerabilities before they are exploited by malicious actors. It is advisable to conduct penetration testing

iteratively with each release or modification of the source code to continually identify and address potential security threats in the application.

- Logging and Audit Trails

Phase 4: Ensure Secure Application Deployment and Operations

- **Secure Deployment and Configuration**:
- Provision for Patch and Update
- Secure Development of Update, Patch, and Release to Mitigate Against Supply Chain Risk from Developers.

Detailed document: ref(https://www.cert-in.org.in/PDF/Application_Security_Guidelines.pdf)

Cyber Crimes and Fraud Associated with Mobile Banking

Mobile banking has emerged as a prevalent and handy method for individuals in India to execute financial transactions. Nonetheless, the surge in mobile banking utilization has concurrently resulted in an escalation of mobile banking-related crimes and frauds. Cybercriminals continually discover novel methods to attack weaknesses in mobile banking systems, resulting in substantial financial losses for both customers and institutions.

The following are prevalent mobile banking crimes and frauds in India:

- **Phishing** is among the most common techniques employed by cybercriminals to get personal and financial data. In a standard phishing attempt, perpetrators disseminate deceptive messages, emails, or phone calls that seem to originate from authentic financial institutions or mobile banking applications. These messages deceive users into disclosing important information, including usernames, passwords, and OTPs (One-Time Passwords).

- **SMS Phishing (Smishing)**: Users often receive deceptive SMS messages that include counterfeit links to fraudulent financial websites. These websites may require users to provide their account credentials, personal identification numbers, or one-time passwords.

- **Email Phishing**: Deceptive emails resembling legitimate correspondence from financial institutions prompt users to click on links and provide their login credentials.

- **Voice Phishing (Vishing)**: Cybercriminals masquerade as customer care agents by telephone to deceive clients into divulging crucial account information.

  Consequences: Financial depletion from bank accounts or associated credit cards, resulting from illicit access.

- **Malicious software and surveillance software**.

Cybercriminals can compromise a victim's smartphone with malware or spyware. These programs are engineered to expropriate sensitive information, such as banking credentials and one-time passwords, or to execute unauthorized transactions on the victim's mobile banking application.

Malware Categories

- **Keyloggers**: Malicious software that captures every keystroke entered by the victim, encompassing usernames, passwords, and PINs.

- **Spyware**: Software that covertly observes the victim's mobile behavior, encompassing their financial transactions and account details.

- **Trojan Horse Applications**: Malicious software masquerading as normal applications that, upon installation, can expropriate data and manipulate phone functionalities.

  Repercussions: Illicit transactions, identity fraud, or compromise of financial information.

- **Account Hijacking**: Account takeover (ATO) transpires when a cybercriminal illicitly accesses a user's mobile banking account. Malefactors frequently exploit compromised login credentials acquired through phishing, data breaches, or inadequate passwords to seize control of the account.

- **Credential Stuffing**: Cybercriminals exploit compilations of compromised usernames and passwords, frequently obtained from data breaches, to infiltrate several accounts.

    Social Engineering: Cybercriminals may coerce victims into revealing login credentials or one-time passwords through telephone or email communication.

- **Counterfeit Mobile Banking Applications**: Cybercriminals frequently develop counterfeit banking applications that closely mimic authentic mobile banking applications. Third-party app shops or phishing websites frequently disseminate these deceptive applications.

- **Counterfeit Payment Links (QR Code Fraud)**: Fraudsters may disseminate counterfeit payment links or QR codes to victims, which, when scanned, redirect the victim to nefarious websites or enable the attacker to obtain payments directly.

- **Loan Fraud and Identity Theft**:

    Fraudsters may utilize stolen identity information to solicit loans or credit via mobile banking applications. This sort of fraud entails the appropriation of personally identifiable information (PII) to fabricate false identities or gain access to victims' bank accounts.

- **Mobile banking fraud using social media and messaging applications**: Social media platforms and messaging applications, such as WhatsApp, are progressively utilized for fraudulent schemes aimed at mobile banking consumers.

Forensic Evidences and Acquisition of Mobile Devices

A subdivision of digital forensics, mobile forensics focuses on the preservation, acquisition, extraction, and analysis of digital evidence found on mobile devices. The primary objective of the scientific procedure is to secure and obtain digital evidence from the mobile device without any modification or alteration of the data. The mobile data-collecting procedure relies on the mobile device's operating system (iOS or Android), OS version, security patch, processor, and hardware (chipset), necessitating highly specialized tools for data extraction.

Methods of Mobile Forensics Acquisition

Physical Acquisition

The physical acquisition is a process of an acquiring data from a mobile phone or tablet. Mobile forensics, digital investigations, or security evaluations frequently execute this procedure. Physical acquisition is a method employed to extract data from a mobile device, particularly when logical collection via normal software or applications is unfeasible or inadequate.

Physical acquisition entails retrieving raw data from a device's memory, encompassing data that may be removed, encrypted, or concealed within system regions.

Physical Acquisitions Process

The technique of data acquisition is contingent upon the mobile device type (e.g., iOS, Android) and the forensic instrument employed. Physical acquisition retrieves all data from the device's storage, including system files, deleted data, application data, and concealed data. The iOS device is incompatible with physical extraction.

Two principal strategies exist for executing physical acquisition:

- **Direct Memory Dump**: This entails acquiring a comprehensive copy (or dump) of the device's physical memory, encompassing all data contained in the device's storage (e.g., NAND flash, ROM, etc.).

This method returns accessible files, deleted data, encryption keys, and system-level data that is generally not apparent during a logical acquisition.

Forensic tools frequently possess the ability to circumvent the device's operating system and access raw memory sectors directly.

Other method is utilized when device is severely locked, broken, or impossible to access normally, the physical memory chips (for example, NAND flash) can be unsoldered and then accessed using special hardware. This is called the chip-off technique.

This methodology requires a sophisticated understanding of hardware and forensic methodologies; yet, it is beneficial when alternative approaches are unsuccessful.

Chip-off acquisition is frequently employed for phones that are extensively damaged or possess robust encryption safeguards. It is a destructive method of data acquisition.

Examine the Collected Data

Upon acquisition of the physical image, forensic analysts might examine the data to determine

- **Deleted files**: Since they may still be in unallocated space after a physical acquisition, they are frequently recoverable.

- **Encrypted data**: Certain forensic technologies can circumvent or compromise fundamental encryption techniques, granting access to data that would often remain unavailable.

- **App data**: Information pertaining to applications that is typically not accessible to the user, including logs, caches, or encrypted communications retained by applications such as WhatsApp, Facebook, etc.

- **Artifacts**: System logs, concealed files, and additional forensic artifacts that may signify device activity or behavior and all other data extracted through logical extraction.

Logical Acquisition

The logical acquisition of data from a mobile device involves extracting available information, such as files, application data, contacts, messages, etc., directly from the device's operating system using conventional techniques like APIs and file system access. In contrast to physical acquisition, which entails accessing the device's raw memory, logical acquisition concentrates on extracting data permitted by the operating system for user or tool access.

Logical acquisition is generally more expedient and less invasive than physical acquisition, making it the preferred method for forensic experts and security professionals during digital investigations. Here is a comprehensive analysis of the logical acquisition process:

Full File System Mobile Extraction

It refers to the process of creating a complete, bit-by-bit copy of a mobile device's entire file system, which includes all accessible files, operating system data, deleted files, application data, and hidden or system-level data that is typically not accessible through logical acquisition techniques. This technique is particularly advantageous for acquiring a comprehensive forensic image of the device, encompassing deleted or concealed data that may be inaccessible by alternative methods, such as logical acquisition. Full file system extraction encompasses the complete contents of the device's storage, rendering it one of the most thorough techniques for mobile device data extraction. This method makes it easier to get back metadata, deleted files that can be recovered and haven't been overwritten, and any fragmented data that might not be easy to find using normal methods.

Timeline Analysis

Mobile Forensics Timeline Analysis pertains to the reconstruction of a chronological sequence of events derived from data retrieved from a mobile device. This analysis helps forensic investigators and security experts figure out the order in which actions were taken on the device, usually within a certain time frame that is relevant to the investigation. Through the analysis of data gathered from mobile devices, one can ascertain the activities of a user (or intruder) and correlate those activities with particular events.

CHAPTER 5   INVESTIGATING MOBILE BANKING AND FINANCIAL APPLICATIONS

In situations like criminal investigations, corporate espionage, or personal data recovery, a timeline analysis is very useful because it shows how devices were used, how communications happened, and what apps were used. This is an outline of the normal process for conducting mobile forensics timeline analysis:

In the first step of a timeline analysis, the necessary data must be extracted from the mobile device using one of three methods: logical acquisition, physical acquisition, or full file system extraction.

Common Data Sources for Analysis for Banking Application Investigation

- **Call records**: Incoming, outgoing, and missed communications
- **Communications**: SMS, MMS, and messaging applications (e.g., WhatsApp, Facebook Messenger)
- **GPS and Location Data**: Geolocation information derived from device sensors, encompassing maps and location history
- **Application Data**: Data pertaining to the application, encompassing usage logs and timestamps for installation and removal
- **Photos/Videos**: Metadata of media files indicating the timestamps of their creation or modification
- **Emails and Contacts**: Metadata derived from email programs and contact lists
- **System Logs**: Logs at the operating system level, including timestamps for device power cycles, application permission and access, and system update events

Application permission: List of permission taken by applications:

- **Browser History**: Visited URLs, search history, cache, and cookies.
- **Device Backups**: Retrieve data from iCloud, Google Drive, or other associated cloud services linked to the device.
- **Erased Data**: A comprehensive file system extraction can restore information that has been erased but not yet rewritten.

A timeline is a visual or chronological depiction of events that transpired on the device, arranged according to their timestamps. This includes application interactions such as launching certain applications, transmitting messages on a social networking platform, and navigating webpages. The process involves the creation, modification, and deletion timestamps of media files.

Location varies according to GPS data or Wi-Fi connections. System events include things like device reboots, application installations, or software updates.

- **Timestamp Organization**: Arrange these occurrences according to their timestamps. This is executed to guarantee that each event is arranged in the proper order.

- **Chronological Arrangement**: After identifying the events, arrange them in chronological order to reconstruct a chronology. This enables investigators to observe the evolution of actions.

- **Data Aggregation**: Timeline analysis generally consolidates data from various sources:

Location Events: Demonstrating the device's trajectory over time, particularly when location data or geotagged images are accessible. Using forensic technologies, this timeline can be turned into a picture, which helps investigators find patterns, oddities, or connections between events.

Examine the Timeline for Trends and Insights

An expert must meticulously examine the constructed timeline to discern noteworthy events, relationships, and patterns pertinent to the study. This may entail

- **Cross-referencing Data**: Investigators can compare the timeline of activities on the mobile device with data from outside sources, like surveillance footage, social media activity, or witness statements, to confirm acts or prove that a person was at a certain place.

- **Finding anomaly**: Timeline analysis can show strange patterns of behavior, like gaps in activity, sudden changes in how an app is used, or geolocation data that doesn't match up. These patterns could be signs of suspicious activity or device tampering.

- **Identifying Deletion**: Analysis can reveal remnants of erased data, especially when using physical or comprehensive file system extraction techniques. These traces may consist of fragmented or residual data that manifest on the timeline.

- **Connecting Events**: Timeline analysis can facilitate the correlation of events. For example, a user might have dispatched an email at a particular time and place, subsequently followed by a phone call or text message immediately thereafter. These interconnected events can offer significant insights about the user's activity.

Examination of Mobile Banking Data or Malicious Applications Using a Case Study

Prior to commencing Forensics analysis, the investigator should verify the following points:

The received mobile device was in pristine condition, akin to being encased in Faraday bags or similar packaging, with airplane mode activated.

Investigation can be carried in two parts like mobile forensics and malicious application forensics. Investigation reveals the following questions:

Table 5-1. Questionaries and Evidence Extraction Strategy

Questions	Evidence extraction strategy
How and when malicious application was downloaded?	Social media analysis, browser analysis to look for suspicious phishing link with time stamp (Mobile Forensics)
What is the behavior of application?	Mobile Malware Analysis
What is the malicious behavior of mobile malware?	Suspicious application permission, SMS forwarder/stealer, hide notification, hidden application, icon change (reveals a legitimate icon)
When and where application has sent a data (C2C) server detail?	Mobile Malware Analysis

A mobile banking application may be a malicious program downloaded by the user via a phishing link or recommended for download during a vishing or social media like WhatsApp, Telegram, or Voip call by an attacker. The following strategies should be utilized to proceed with the investigation:

1. **Identification of the infection vector or pathway**

   This phase encompasses the methods by which an attacker has deployed malware or gained access to a mobile device or banking application to execute a transaction. Occasionally, an attacker obtains access to a mobile device by installing a remote access program such as AnyDesk, fake banking application, etc.

   Cybercriminals utilize bank look-alike domains websites or applications, disseminating their links through adverts and SEO strategies. Consequently, when legitimate users seek banking activity on search engines, these fraudulent sites emerge, often accompanied by contact numbers. Example: cybercriminals use Google ads and SEO technique to disseminate fake domain like www.icicibnk.com, so when any users seek help from google search engine, these ads display first to lure the user.

2. **Detection of harmful files**

   For instance, the attacker transmitted a malicious file via a social networking application on an Android device, and the victim installed it under the attacker's instruction during a phone call after allowing the "install unknown applications" option in the settings.

   During the installation of an Android application, it requests several permissions, like read/write calls, read/write SMS, SD card access, read contacts, and the installation of Android packages, which users typically overlook. The aim of these applications is to retrieve banking credentials, credit/debit card details, and M-PIN or password information. These nefarious applications discreetly deactivate SMS notifications, transmit a message to a website, IP address, or cell numbers upon receipt on the device, and subsequently erase that message from the device. To complete the transaction, the attacker needs only the OTP (one-time password) sent by the program without the user's awareness. Consequently, during the inquiry, when the officer inquires about the sharing of the OTP over a call, the user denies it, since he is unaware of the malware application's behavior.

During forensic examination of these malicious files, a forensics specialist will search for installed applications together with their Android package names in the data collected by a mobile forensics tool. Forensic experts analyze the installation timeline of the application and conduct a timeline analysis to establish the sequence of transactions. Also, the examiner focuses on some key points to identify the suspicious applications like permission, package name, etc. For example, Camera application does not require SMS read/write permission.

3. **Safeguarding harmful files**

   Forensic experts retain the dump file of mobile devices. A dump file is a singular extraction bundle of mobile device storage that encompasses mobile device artifacts. This forensic program will look at the dump file and sort the call logs, application data, SMS, location history, contacts, browser history, event logs, and other data into groups that make sense. This will make the data easier to read for a more in-depth investigation. An alternative option to preserve the malicious file is to extract the Android application from the mobile device using the Android Debug Bridge (ADB) technique, save the .apk file on a local PC, and compute its hash value. Malware files should be password-protected and disarmed by altering their file extension, such as to (.malz).

4. **Analysis of malicious files**

   We now own the Android application package file (APK) for examination. Android files can be analyzed using the three ways enumerated below.

a. **Static Analysis**

It is the procedure of analyzing the Android package file without executing it. A forensic specialist will search for artifacts that indicate the maliciousness of a program, such as odd Android permissions and atypical or obfuscated files containing harmful code.

Typically, a forensic expert employs the following methodology to do static analysis.

1. **Examination of file structure**: Android package .apk files are essentially .zip files; to extract them, one can simply rename the .apk to .zip and then extract it, revealing the underlying file structure of the .apk for examination.

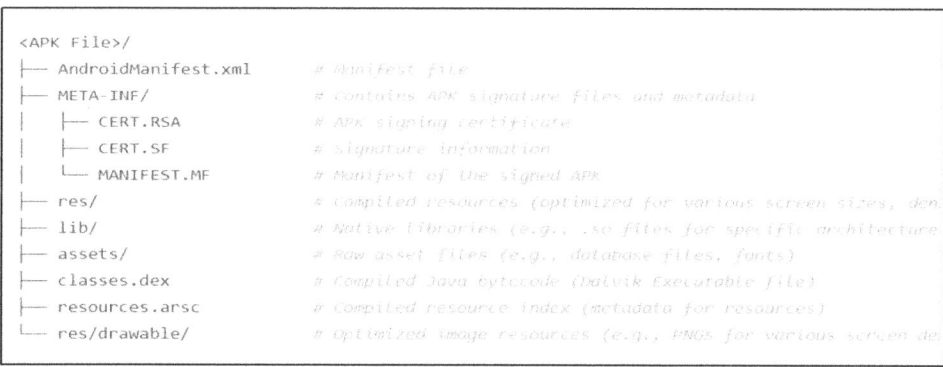

Figure 5-1. *Android file structure after unzipping the apk file*

After unzipping the android apk file, it displays the folder structure which contains android manifest, classes.dex, resources.arsc file, etc.

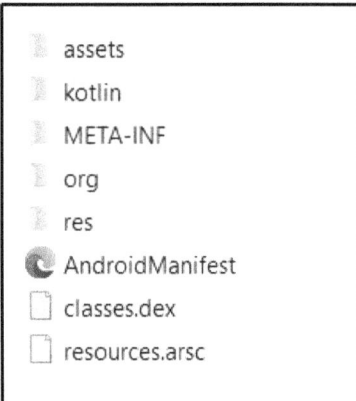

Figure 5-2. *Android file structure after unzipping the apk fileTools like Visual Studio Code, Android Studio, or JADX can examine the AndroidManifest.xml file, which contains all application-level configurations, including Android permissions*

- **META-INF**: This directory contains cryptographic signatures and keys for verifying the authenticity of the Android package on the server.

- **RES/**: Developers store compiled resources such as logos, photos, fonts, layouts, strings, etc. Forensic specialists may discover some artifacts in this location.

- **lib/**: This directory contains native libraries (.so files) tailored for various CPU architectures (e.g., ARM, x86).

- **The assets/**: This directory contains unprocessed files that are included with the APK, such as HTML documents, databases, and fonts.

- **Classes.dex**: This is where developers store the compiled Java code in the Dalvik Executable (DEX) format. Forensic experts want to look at the classes.dex file because it could do a lot of bad things, like spy on users, send data to remote locations, or take advantage of flaws in the operating system.

- **Resources.arsc**: This file contains a binary representation of the resources found in the res/ directory.

Tools for Static analysis and Reverse Engineering: MobSF, Jadx, JdGui, Apktool, VsCode, Android Studio, Byte Code viewer

b. **Dynamic Analysis**

The dynamic analysis of a mobile application entails the real-time monitoring and examination of the app's activity during its operation on a device. Static analysis looks at the code and architecture of an application without running it. Dynamic analysis, on the other hand, looks at how the application works and interacts with its environment while it is running. This includes how it talks to outside servers, uses system resources, and acts in different situations.

Dynamic analysis is an essential procedure in mobile application security evaluations, malware identification, and performance enhancement. It assists in detecting runtime vulnerabilities, security deficiencies, and malicious activities that may not be evident solely from the application's source code.

Objective of Dynamic Analysis

Dynamic analysis is conducted to

- **Analyze application behavior**: Comprehending the app's interaction with the operating system, network, APIs, and other applications.

- **Find weaknesses**: Noticing security issues like insufficient data storage, slow network connectivity, wrong use of rights, and possible pathways for exploitation within the application.

- **Evaluate application performance**: Analyzing the app's functionality under varying situations, encompassing resource use (CPU, memory, battery) and network latency.

- **Identify malware and malicious activities**: Ascertain if the application engages in any unauthorized or dubious acts, including data exfiltration, unencrypted communications, or unauthorized API requests.

- **Emulate real-world usage**: Executing the application under diverse situations to replicate typical user behavior, atypical inputs, and possible attack scenarios.

Dynamic Analysis Process

A variety of technologies are available for conducting dynamic analysis on mobile applications. These tools provide the observation of the application's behavior during execution, examination of system resources, network activities, and additional aspects.

Using Mobile device emulators and simulators

- **Android Emulator**: A tool that replicates the Android device environment on a desktop computer.

Xcode Simulator: A comparable tool for emulating iOS applications on macOS devices.

Network Proxy Tools

- **Burp Suite and HTTP Tool kit**: Commonly employed to intercept and scrutinize HTTP/HTTPS traffic between the mobile application and backend servers.

- **OWASP ZAP**: An open-source tool for assessing the security of online applications, capable of analyzing traffic from mobile applications.

- **Charles Proxy**: This tool intercepts and examines HTTP/HTTPS network data between an application and its backend.

Frida: An advanced instrumentation toolkit for API hooking and runtime modification. It enables analysts to alter or intercept function calls in real time.

- **Xposed Framework**: A utility that enables dynamic runtime alterations to Android applications without requiring source code access.

- **Dynatrace**: A performance monitoring instrument that facilitates the tracking of resource use, application performance, and network activity.

- **Cuckoo and MOBSF Sandbox**: A widely utilized instrument for examining Android and iOS applications within a regulated setting to detect harmful activities and interactions.

- **Android Virtual Device (AVD)**: A utility that enables the creation and execution of virtual Android devices for application testing.

Network Traffic Analysis

- **Wireshark**: A network packet analyzer that captures and examines the traffic produced by mobile applications.

- **tcpdump**: A command-line network packet analyzer frequently employed for capturing network traffic.

- **Traffic Analysis**: Observe the interactions between the mobile application and servers, as well as third-party services. Unencrypted communication refers to the use of HTTP rather than HTTPS. Sensitive data, such as passwords and personal information, is transmitted in plaintext. There may be suspicious or unforeseen network queries (e.g., to unfamiliar IP addresses or domains).

Identifying Behavior of Application

Application Permissions: Ensure that the application solicits only those permissions essential for its operation. Applications soliciting excessive or superfluous permissions (e.g., location, camera, contacts) may be attempting to collect more data than necessary.

CHAPTER 5 INVESTIGATING MOBILE BANKING AND FINANCIAL APPLICATIONS

PERMISSION	STATUS	INFO	DESCRIPTION
android.permission.ACCESS_NETWORK_STATE	normal	view network status	Allows an application to view the status of all networks.
android.permission.ACCESS_WIFI_STATE	normal	view Wi-Fi status	Allows an application to view the information about the status of Wi-Fi.
android.permission.CAMERA	dangerous	take pictures and videos	Allows application to take pictures and videos with the camera. This allows the application to collect images that the camera is seeing at any time.
android.permission.INTERNET	dangerous	full Internet access	Allows an application to create network sockets.
android.permission.MOUNT_UNMOUNT_FILESYSTEMS	dangerous	mount and unmount file systems	Allows the application to mount and unmount file systems for removable storage.
android.permission.READ_EXTERNAL_STORAGE	dangerous	read SD card contents	Allows an application to read from SD Card.
android.permission.READ_PHONE_STATE	dangerous	read phone state and identity	Allows the application to access the phone features of the device. An application with this permission can determine the phone number and serial number of this phone, whether a call is active, the number that call is connected to and so on.
android.permission.REQUEST_INSTALL_PACKAGES	dangerous	Allows an application to request installing packages.	Malicious applications can use this to try and trick users into installing additional malicious packages.
android.permission.WRITE_EXTERNAL_STORAGE	dangerous	read/modify/delete SD card contents	Allows an application to write to the SD card.

Figure 5-3. *List of dangerous permissions*

The above image displays dangerous permissions taken by android application and description contains actual behavior of application if permission is allowed by user.

Some applications create proxy connection to connect C2C server for defense evasion. Figure 5-4 displays architecture of mobile proxy connection made by the android application.

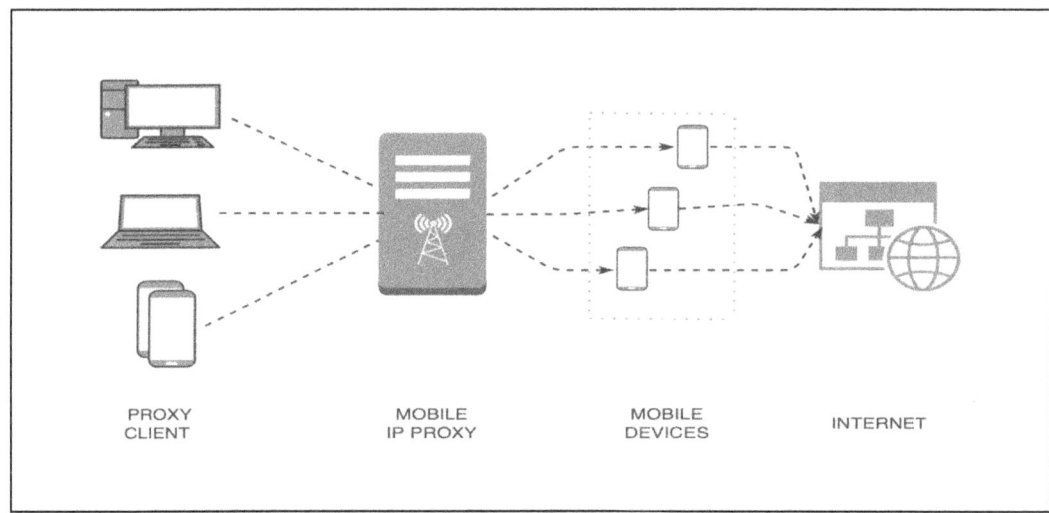

Figure 5-4. *Proxy created by android application*

CHAPTER 5   INVESTIGATING MOBILE BANKING AND FINANCIAL APPLICATIONS

Misusing application permission:

REQUEST_INSTALL_PACKAGES permission used to install malware payload

REQUEST_INSTALL_PACKAGES android permission to download and install the malicious application without user's knowledge by connecting to highlighted link.

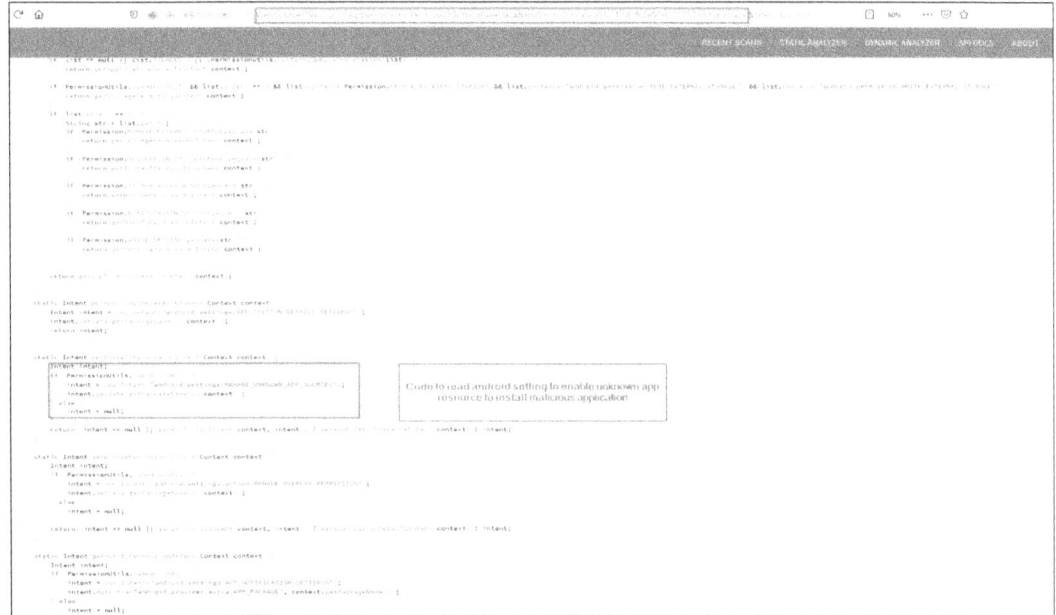

Figure 5-5. *REQUEST_INSTALL_PACKAGES android permission to download and install the malicious application without user's knowledge by connecting to highlighted link*

CHAPTER 5 INVESTIGATING MOBILE BANKING AND FINANCIAL APPLICATIONS

Download and install payload on android device

```
import java.security.cert.CertificateEncodingException;
import java.util.HashMap;
import java.util.List;
import java.util.Set;
import java.util.jar.JarEntry;
import java.util.jar.JarFile;

public class MttLoader {
    public static final String CHANNEL_ID = "ChannelID";
    public static final String ENTRY_ID = "entryId";
    @Deprecated
    public static final String KEY_ACTIVITY_NAME = "KEY_ACT";
    @Deprecated
    public static final String KEY_APP_NAME = "KEY_APPNAME";
    public static final String KEY_EUSESTAT = "KEY_EUSESTAT";
    @Deprecated
    public static final String KEY_PACKAGE = "KEY_PKG";
    public static final String KEY_PID = "KEY_PID";
    public static final String MTT_ACTION = "com.tencent.QQBrowser.action.VIEW";
    public static final String MTT_ACTION_SP = "com.tencent.QQBrowser.action.VIEWSP";
    public static final String PID_ARTICLE_NEWS = "21272";
    public static final String PID_MOBILE_QQ = "50079";
    public static final String PID_QQPIM = "50186";
    public static final String PID_QZONE = "10494";
    public static final String PID_WECHAT = "10318";
    public static final String POS_ID = "PosID";
    public static final String QQBROWSER_DIRECT_DOWNLOAD_URL = "https://mdc.html5.qq.com/d/directdown.jso?channel_id=50079&...";
    public static final String QQBROWSER_DOWNLOAD_URL = "https://mdc.html5.qq.com/mh?channel_id=50079&u=";
    public static final String QQBROWSER_PARAMS_FROME = ",from=";
    public static final String QQBROWSER_PARAMS_PACKAGENAME = ",packagename=";
    public static final String QQBROWSER_PARAMS_PD = ",product=";
    public static final String QQBROWSER_PARAMS_VERSION = ",version=";
    public static final String QQBROWSER_SCHEME = "mttbrowser://url=";
    public static final int RESULT_INVALID_CONTEXT = 3;
    public static final int RESULT_INVALID_URL = 2;
    public static final int RESULT_NOT_INSTALL_QQBROWSER = 4;
    public static final int RESULT_OK = 0;
    public static final int RESULT_QQBROWSER_LOW = 5;
    public static final int RESULT_UNKNOWN = 1;
    public static final String STAT_KEY = "StatKey";

    public static class BrowserInfo {
        public int browserType = 0;
        public String packageName = null;
        public String qushead = "";
        public int ver = 0;
        public String vn = "0";
    }
```

Figure 5-6. *A code with url to download and install the malware application. Here application silently installs malicious application in the background.*

```
public class MttLoader {
    public static final String CHANNEL_ID = "ChannelID";
    public static final String ENTRY_ID = "entryId";
    @Deprecated
    public static final String KEY_ACTIVITY_NAME = "KEY_ACT";
    @Deprecated
    public static final String KEY_APP_NAME = "KEY_APPNAME";
    public static final String KEY_EUSESTAT = "KEY_EUSESTAT";
    @Deprecated
    public static final String KEY_PACKAGE = "KEY_PKG";
    public static final String KEY_PID = "KEY_PID";
    public static final String MTT_ACTION = "com.tencent.QQBrowser.action.VIEW";
    public static final String MTT_ACTION_SP = "com.tencent.QQBrowser.action.VIEWSP";
    public static final String PID_ARTICLE_NEWS = "21272";
    public static final String PID_MOBILE_QQ = "50079";
    public static final String PID_QQPIM = "50186";
    public static final String PID_QZONE = "10494";
    public static final String PID_WECHAT = "10318";
    public static final String POS_ID = "PosID";
    public static final String QQBROWSER_DIRECT_DOWNLOAD_URL = "https://mdc.html5.qq.com/d/directdown.jso?channel_id=50079";
    public static final String QQBROWSER_DOWNLOAD_URL = "https://mdc.html5.qq.com/mh?channel_id=50079&u=";
    public static final String QQBROWSER_PARAMS_FROME = ",from=";
    public static final String QQBROWSER_PARAMS_PACKAGENAME = ",packagename=";
    public static final String QQBROWSER_PARAMS_PD = ",product=";
    public static final String QQBROWSER_PARAMS_VERSION = ",version=";
    public static final String QQBROWSER_SCHEME = "mttbrowser://url=";
    public static final int RESULT_INVALID_CONTEXT = 3;
    public static final int RESULT_INVALID_URL = 2;
    public static final int RESULT_NOT_INSTALL_QQBROWSER = 4;
    public static final int RESULT_OK = 0;
    public static final int RESULT_QQBROWSER_LOW = 5;
    public static final int RESULT_UNKNOWN = 1;
    public static final String STAT_KEY = "StatKey";
```

Figure 5-7. *A code with url to download and install the malware application. Here application silently installs malicious application in the background.*

IPhone Malware

Attackers send a phishing link to download the configuration profile which creates and downloads a malware on the iOS and no legitimate app should require a profile, because any legitimate app can be made available through the app store only. In some cases, these iOS applications can be installed without Appstore also. The only exceptions are enterprise apps created by a legitimate business for its employees, or, rarely, a beta of an app for which you are a registered beta tester. In either case, the profile will be approved by Apple.

Example: attacker has sent a phishing link of mobile banking application on social media which downloads app.mobile.config file. When user opens a config file which adds external profile on the device. While clicking on profile icon on mobile device it downloads a malware on iOS and once installed as it contains a payload and other commands which the attacker or scammer can run on the device.

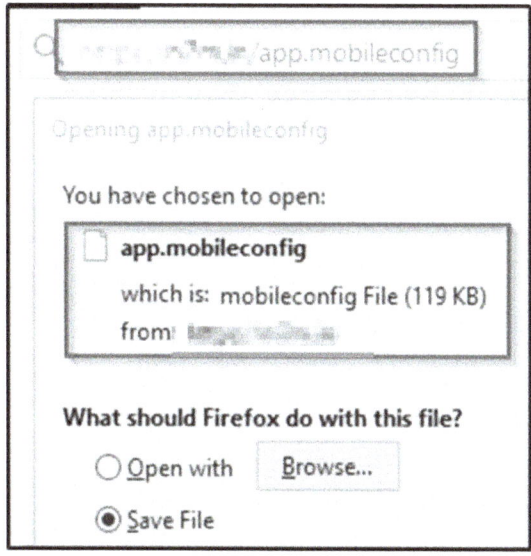

Figure 5-8. After saving and executing the configuration profile, it allows to download the actual malicious application

CHAPTER 5 INVESTIGATING MOBILE BANKING AND FINANCIAL APPLICATIONS

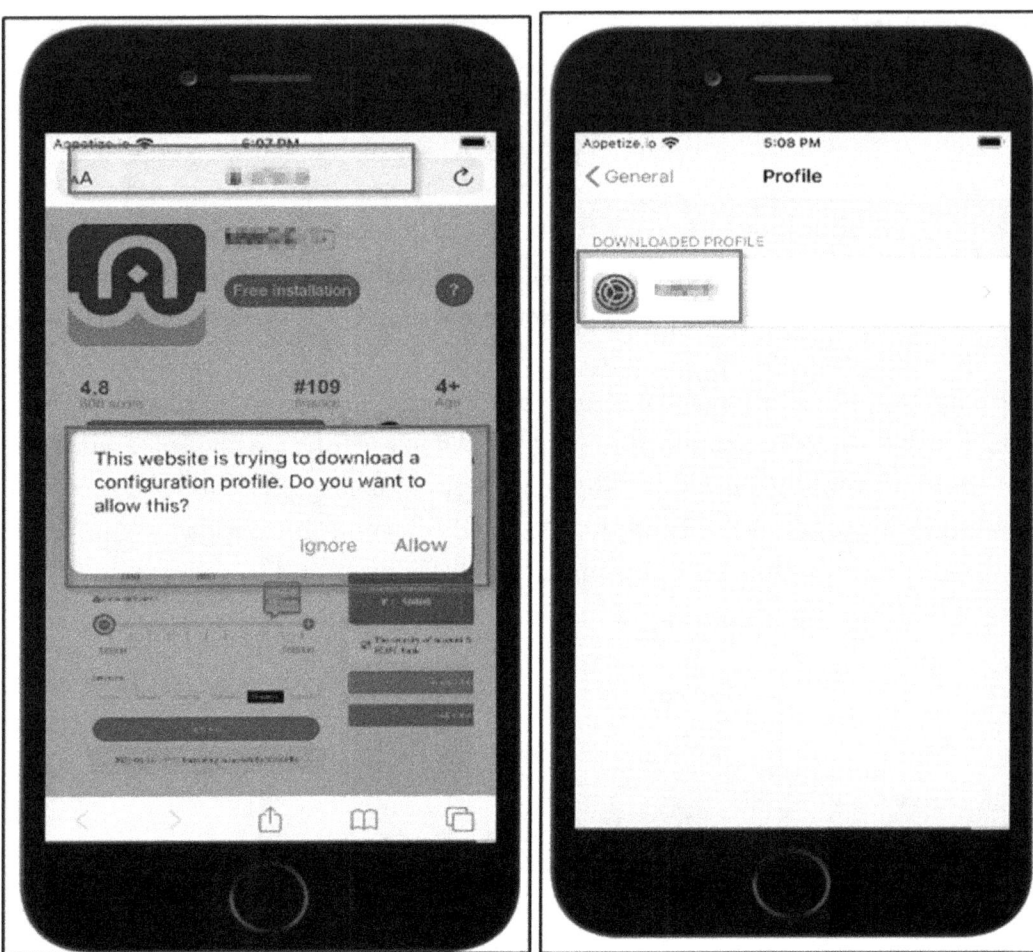

Figure 5-9. *Notification message asking to create a dummy profile to download the malicious application*

Data Exfiltration to Command-and-Control server

CHAPTER 5 INVESTIGATING MOBILE BANKING AND FINANCIAL APPLICATIONS

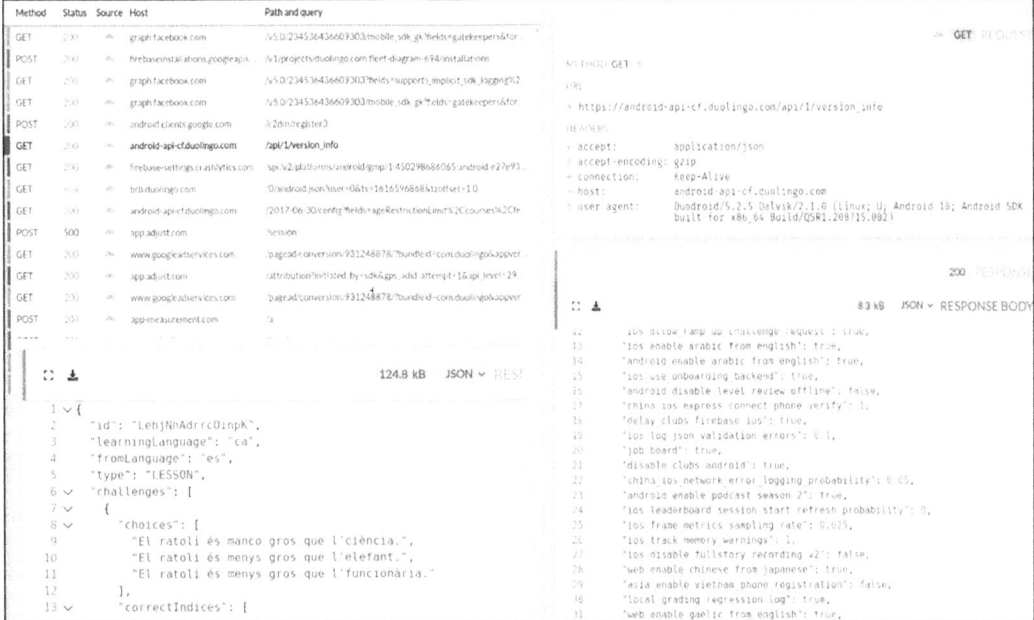

Figure 5-10. *The image has been captured during intercepting of malicious application, resulting sending information to command-and-control server [1]*

Summary

This chapter involves the evolution of mobile-based banking applications their characteristics of mobile and financial applications, their security features, related cybercrimes. It also includes how cybercriminals are misusing financial applications to commit a cyber fraud for extracting money from people. This chapter also explains about artifacts collected during forensic analysis, and mobile forensic acquisition process, current challenges with mobile forensics methodology, advantages and disadvantages of mobile forensics acquisition process, and timeline analysis. Further, it elaborates on forensics analysis of mobile malware or malicious applications, tools, and techniques.

Reference

[1] Toolkit, H., & Perry, T. (2025). Intercept and edit HTTP traffic from (almost) any Android app. *Intercept and edit HTTP traffic from (almost) any Android app.* https://httptoolkit.com/blog/inspect-any-android-apps-http/

CHAPTER 6

Examination of Social Media and Messaging Applications

The growth in smartphone technology has augmented mobile device usage, facilitating not just communication but also a multitude of functionalities for varied activities. Users are storing their personal and professional information on mobile devices. Users are establishing accounts on many social media and messaging applications to interact with their family and friends. Social applications such as Facebook, Instagram, Twitter, Snapchat, and WhatsApp enable users to engage with both familiar and unfamiliar individuals worldwide through the Internet. Cybercriminals are using these social media applications for social engineering and information collection to identify their targets. The utilization of social media platforms has markedly escalated in recent years, as individuals across all age demographics engage in engagement, cooperation, and content sharing. These applications generate vast amounts of data, providing opportunity for cybercriminals to exploit this information for harmful actions. Investigating authorities utilize the social media profiles of criminals to analyze their behavior and ascertain the purpose behind the crime. Mobile forensics and social media analysis are essential in all criminal activities, including rape, murder, robbery, etc. Social media analysis is directly associated with the preferences and aversions of criminals, their interests, psychological condition, affiliations, and more. Numerous terrorist organizations have utilized social media platforms and messaging applications for recruitment and disseminating anti-national operations in the past. This chapter elucidates many categories of data produced by social media and messaging applications, as well as the acquisition and forensic examination of social media platforms. This chapter further elaborates on forensic evidence that can be utilized to identify illegal actions through social media applications. This chapter addresses the technique and challenges of cloud forensics investigation, particularly in relation to data stored by social media platforms.

Types of Data Generated by Social Media and Messaging Applications

Social media and messaging applications can be installed on mobile devices and computer systems, generating various types of data which are crucial for forensic investigations and analyzing criminal behavior to ascertain their motives behind crimes and users' involvement in criminal activities. This data may be saved on users' computer or mobile systems, social media application servers, and third-party application servers with the user's consent.

Data produced by social networking applications can be classified as follows:

1. **User-Generated Content**

   Any type of data or information, including text, images, videos, reviews, or opinions generated by users on specific social media platforms, encompasses the following artifacts:

   - **Text Posts:** Most users post ideas, updates, or opinions on platforms such as Facebook, Instagram, Twitter/X, and LinkedIn.

   - **Comments and Responses:** The means by which people engage with one another and their degree of interaction.

   - **Messages and call:** Direct instant messages and group conversations on social media applications such as WhatsApp, Instagram, Signal, Telegram, Facebook Messenger, etc., encompass private communications among user and groups. VoIP calls using social media applications and its timestamp.

   - **Media Content:** Generated or disseminated.

   - **Photographs:** Shared or uploaded images can disclose additional information such as the time and place of capture, as well as an approximate position.

- **Videos:** Prominent platforms such as YouTube, Facebook, and Instagram permit users to upload and disseminate video material. Which can furnish other information such as the video's creation date and quality.

- **Memes and GIFs:** Various forms of entertaining images and brief videos are frequently disseminated on social media platforms. This can represent prevailing patterns and how users express or react to themselves.

2. **Metadata**

   Metadata is vital information on user-generated content that is automatically produced by platforms or devices, aiding investigative authorities and supplying key details that may be unknown to users.

   - **Device Specifications**: Details on the brand, model, system type, and web browser of the device.

   - **Timestamps**: The specific moments when content was generated, modified, accessed, or disseminated.

   - **File Metadata (EXIF data)**: Concealed information within media files such as photographs and videos, for example. Image resolution and camera model.

   - **Geolocation Information**: Upon the submission of content (text messages, images, or videos)

   - **User Engagement Metrics**: Information regarding user interactions through likes, dislikes, shares, and comments.

3. **User Profile Data—Personal information provided by user to platform**

   - **User Information**: This encompasses the user's name, username, email, phone number, age, gender, and location as specified in account settings.

   - **Profile Image**: User-generated avatars or profile photographs on social media platforms such as Facebook, Instagram, or Twitter.

- **Account Preferences**: The section where users can configure privacy, security, and notification settings.

- **Connections**: The users or groups to which an individual is connected, including friends and followers.

- **Personal Overview (Brief Biography)**: The "about me" section or profile bios where users may disclose personal information and concise descriptions.

4. **Interaction and Engagement Data—User Activity Details**

   - **User Engagement Logs**: These encompass the duration of platform usage, particular interactions with content, both in groups and individually, as well as the frequency of activity.

   - **Followers/Following**: This indicates the relationships between users.

   - **Tagging and Mentions**: Illustrate user interactions and communication behaviors.

   Likes, reactions, and shares indicate the extent to which individuals appreciate, disseminate, or endorse posts on social media platforms such as Facebook or Instagram.

5. **Content Interaction History**

   Numerous cloud services monitor user interactions, such as views, search history, and activity logs, which furnish insights on user interests, intentions, and affiliations.

   Account Activity Logs: Logs of user activities generated within the platform, including logins, signups, logouts, password resets, modifications to account settings, and a list of devices utilized for access.

   User search history is often recorded by most platforms to comprehend user interests, intentions, and affiliations.

Content view histories are contingent upon the platform. For example, YouTube, Facebook, and Instagram retain a record of the videos or content accessed by a user. This may encompass the content viewed, duration of viewing, and any user reactions, such as comments or likes.

6. **User's Network and Relationship details**

   Group Membership: Information regarding user involvement or interest in particular groups.

   Social Graph: It illustrates user connections and relationships, such as friends or followers.

   Event Information: Data regarding events that users have previously attended or expressed interest in.

7. **Access and Authentication Data**

   - Authentication and access data are essential for tracking a user's online presence.

   - **Two-factor authentication logs**: The program records logs of verification actions, for example, verification codes and related data are stored in the application database and on the server.

   - **Login/Logout information:** Including the date and time of each login attempt, utilized IP addresses, and devices employed to access the account, is retained on the application server.

   - **Session Data**: Information pertaining to the presently logged-in user, encompassing session ID, timestamps, and cookies across desktop, laptop, mobile phone, and server.

   The device history records the devices on which the user has previously logged into the account, stored in the cloud and email server.

8. **Geolocation data:**

    Location-based information gathered from applications

    - **Real-Time Updates**: Modifying status to reflect current position across various social sites.

    - **Geotagging**: The automated addition of location information while posting pictures is beneficial for users who tag individuals or provide reviews of posts or photographs.

9. **Additional Data Acquired:** Cross-Platform Sharing.

    - **Social Network Scraping**: Alteration of data across platforms, such as sharing from Instagram to other social media application pages.

    - **External Application Data**: Information derived from non-social media applications that augment the data landscape, such as Spotify sharing music playlists on Facebook.

Social media and messaging involve the generation of several data types, including user-generated content, metadata, and logs. This data will significantly enhance the investigation and behavioral research, hence amplifying the necessity for improved gathering and analytical capabilities. All of these necessitate advanced technologies while upholding privacy and legal obligations.

Forensic Acquisition Methods for Social Media and Messaging Data from Desktop/Laptop or Mobile

The user can access social media applications such as Facebook, WhatsApp, Telegram, and Twitter using a mobile application, web browser, or desktop program. Certain people are utilizing this program in the incognito mode of their web browsers. The forensic examination and e-discovery process of the specific mobile, desktop, or laptop may uncover significant insights regarding the user's social media activity. Forensic acquisition and analysis is a scientifically validated process that preserves data integrity by computing a hash value of digital evidence. The application data may be encrypted during transmission over the network; yet, there are several methods to access the artifacts of this data. This chapter focuses on forensic acquisition, analysis, and significant digital evidence that may assist investigative officers in a case where a user has utilized a desktop or laptop to access social media and messaging applications. All user activities can be kept on the system's hard disk drive. Forensic methodology involves identifying, preserving, acquiring, analyzing, and documenting the entire procedure within the chain of custody. The system's hard disk drive is recognized as evidence. The forensics analyst uses Faraday bags or antistatic bags to safeguard the hard disk drive and meticulously handle the evidence to prevent physical damage. The investigator employs a specialized container and box to transport the evidence to the forensic laboratory.

Forensic Acquisition Procedure

Forensic Imaging/Cloning of a Drive

Forensic imaging is the procedure of generating a bit-by-bit, sector-by-sector replica of physical storage media onto another physical storage medium. This forensic method encompasses all available bits and sectors on the exhibit media to the imaged media drive. The hidden sector, slack space, allocated bits, and unallocated bits are replicated and preserved in a designated format. This differs from standard copy and paste data, which does not encompass hidden space and slack space. Forensic imaging generates a bundle file that can be retrieved by forensic software. File formats include E01, DD, Raw, AD01, among others.

Forensic cloning is a bit-by-bit duplication that allows the system to boot with the identical configuration and settings as the original exhibit. Various imaging programs such as DittoDx, Forensics Falcon, Media Clone, and Tableau offer functionalities for imaging and cloning.

CHAPTER 6　EXAMINATION OF SOCIAL MEDIA AND MESSAGING APPLICATIONS

Some of the tools are as follows:

A. Logicube Forensics Falcon

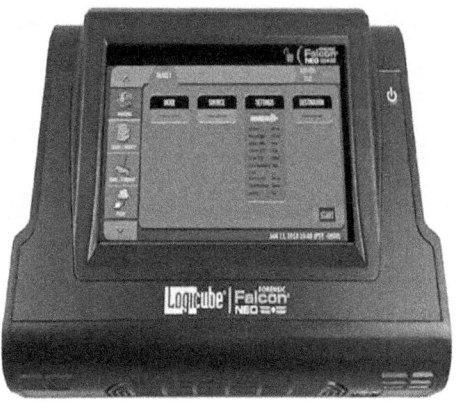

B. Ditto Dx

C. Tableau TX1

D. Super Imager

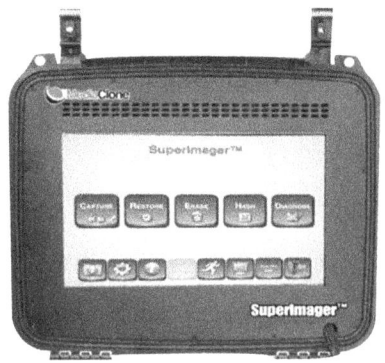

Analysis can be conducted with both open-source and commercial forensic analysis software. Forensic analysis typically uses tools such as Autopsy, Encase, FTK, Magnet Axiom, XRY, and Belkasoft. These tools possess the capability to analyze the forensic image file and offer a segmented view of the data. This data comprises tags, labels, and criteria that categorize and differentiate social media-related images, text, audio, and videos. These forensic tools provide a historical and graphical representation of connected data, allowing analysts to apply various filters to establish connections. These forensic technologies employ numerous filters and parameters to detect images of nudity, threatening statements, objects, narcotics, weapons, and more.

Forensic Artifacts or Evidence
Internet Browsing History

Social networking platforms are frequently accessible via web browsers such as Chrome, Firefox, Edge, Safari, etc. The browser history retains URLs and timestamps of accessed social networking platforms, which can be retrieved.

Cache Files: Browsers store resources (e.g., pictures, scripts, cookies) to enhance page loading speed. Forensic investigators can analyze these cache files to retrieve social media photographs, videos, and other content, regardless of whether the user has deleted their browsing history.

Cookies save user session information, including authentication tokens, which can ascertain a user's login status and timing for a social network account.

The forensic tools also indicate the frequency with which people have accessed the social networking platform.

Social media applications: Numerous social media sites offer desktop programs (e.g., Facebook Messenger, WhatsApp Desktop, Instagram) that retain local data on the system. Investigators can retrieve this data to ascertain social media behavior directly from the program.

WhatsApp Desktop utilizes SQLite database files to store messages and media on Windows computers. These databases encompass conversation logs and related media.

Facebook Messenger Data: Applications such as Facebook Messenger may retain user-specific information locally in designated directories (e.g., cached data, temporary files, logs).

Social Media Digital Artifacts (Images, Videos, and Documents)

File Metadata: Social media networks often permit users to submit photographs, videos, and documents. Forensic investigators can retrieve metadata from these files (e.g., timestamps, file paths, device information) to ascertain the creation or modification time and location of the files.

Deleted Files: Although files (including photographs and videos published to social media) may be removed from the user's device, traces of these files may persist in unallocated space or within the file system (file slack, unused clusters, etc.).

Social Media Artifacts in the Windows Registry: The Windows Registry contains system-wide configurations, encompassing details about installed applications, logon activities, and network connections. A forensic examination of social media artifacts can disclose the Registry:

User Profile Information: The Registry may retain keys associated with online browsers, email clients, and social media applications, providing insight into user behavior.

Web Browser and Application Data: For instance, recent social media interactions may be recorded in the Registry if a user has accessed their accounts via a browser or application.

Prefetch Documents

Windows operating systems utilize Prefetch files to enhance the performance of frequently accessed apps. These files can serve as evidence of social networking programs being accessed, even if the user has deleted their history.

Timestamp Data: Prefetch files frequently include timestamps that denote the last execution time of a particular application, including social networking applications.

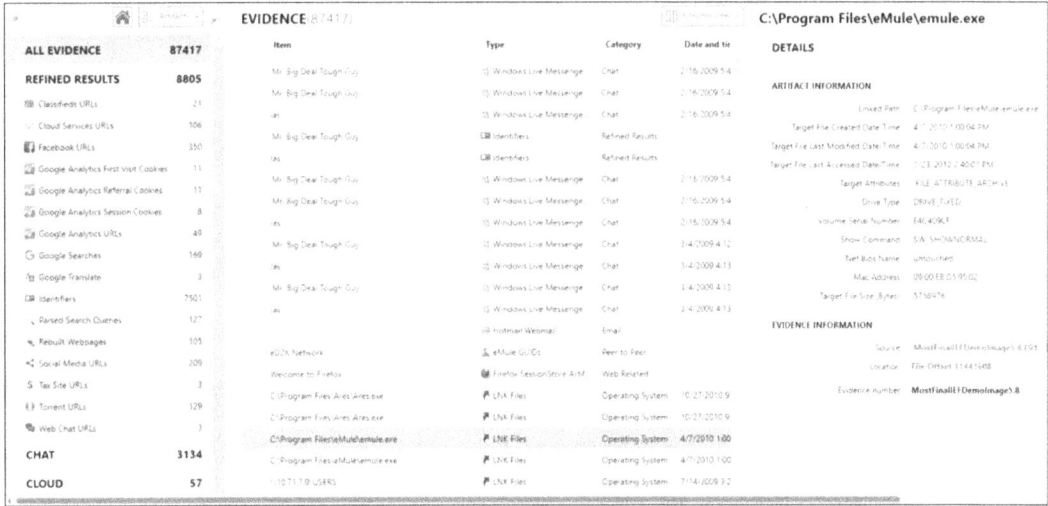

Above screenshot displays artifacts retrieved from image file of user's system which contains social media, browsing history, and cloud-related information.

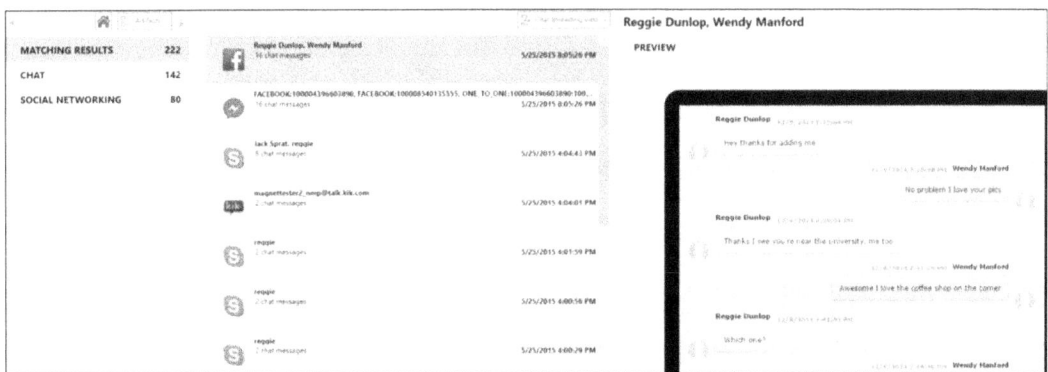

Social media information retrieved from disk image which shows chat data.

Restoration of Erased Social Media Artifacts

All referenced forensic technologies provide the capability to recover lost data or artifacts associated with social media. Forensic analysis occasionally employs open-source software such as Foremost, Bulk Extractor, Recuva, and Scalpel to retrieve artifacts from

CHAPTER 6 EXAMINATION OF SOCIAL MEDIA AND MESSAGING APPLICATIONS

forensic image files. These tools employ data carving and file carving methodologies to extract data from raw data pieces directly from the sector or cluster, bypassing the operating system or file system structure. Occasionally, analysts encounter corrupted artifacts such as incomplete images or text, but partial artifacts could be utilized for investigation.

```
# BANNER FILE NOT PROVIDED (-b option)
# BULK_EXTRACTOR-Version: 1.5.5 ($Rev: 10844 $)
# Feature-Recorder: url
# Filename: C:\Users...
# Histogram-File-Version: 1.1
n=18    100009011284224
n=7     100004810816623
n=5     100009777649491
n=5     100010535416686
n=5     100015282300013
n=5     260212454027987
n=4     100007742445628
n=4     100008168924769
n=4     100009694812741
```

The above screenshots display the output from Bulk Extractor tool list of artifacts such as URL, search data, etc., accessed by users. This text files contains list of Facebook ID and URLs accessed by user.

Social Media Artifacts from Volatile Memory

In forensic analysis, investigators frequently utilize volatile memory (RAM) dumps as a source of evidence. RAM dumps encompass data that was actively utilized during system operation, including information from apps like web browsers, messaging programs, and social networking platforms. Examining a RAM dump might reveal significant forensic data, such as active sessions, messages, login tokens, cached media, and browsing activity associated with social networking networks.

A RAM dump can be obtained from an active system via numerous techniques. Acquiring a RAM dump is crucial, as it retains the ephemeral data that would otherwise be forfeited at system shutdown.

Tools for Acquiring Memory Dumps

1. **FTK Imager/Belkasoft Ram capturer/Magnet Ram Capturer**: A utility capable of obtaining memory dumps from active computers.

2. **DumpIt**: A complimentary, user-friendly utility for generating a comprehensive memory dump from an active Windows system. This tool can acquire RAM in a forensically sound manner.

3. **3 WinPmem**: A memory acquisition utility for Windows that facilitates the capturing of RAM dumps for analytical purposes.

4. **Lime/Avml**: These open-source tools could be utilized to capture the RAM dump of Linux systems.

5. **OSXpmem**: This open-source tools could be utilized to capture the RAM dump of MAC OS systems.

6. **Volatility**: RAM dump analysis tool for Windows/Linux and MAC systems.

Why RAM Dumps Are Important for Social Media Forensics

Volatile Data: RAM is ephemeral, existing solely while the system is energized. This indicates that any social media engagement, including logged-in sessions, messages, and live conversations, may be briefly retained in RAM.

Live Data: In contrast to hard drives, where data is enduring (i.e., persists until erased), data in RAM is ephemeral yet can offer a snapshot of a user's current social media engagement, including browsing or messaging activities.

RAM can store encryption keys, session tokens, and authentication data, which are essential for accessing social media accounts and revealing other pertinent user activity.

CHAPTER 6   EXAMINATION OF SOCIAL MEDIA AND MESSAGING APPLICATIONS

Common Social Media Artifacts Identified in Memory Dumps

RAM dumps may encompass several artifacts associated with social media activities. Below are a few prevalent instances:

- **Web Browser Sessions:**

    Upon logging into social media platforms, such as Facebook, Instagram, Twitter, or LinkedIn, session tokens and cookies may be retained in RAM. These tokens facilitate authentication and may grant access to the user's account without the necessity of re-entering login credentials.

    Active session data, comprising the login and access tokens, can be retrieved from the RAM dump.

- **Chat Records and Correspondence:**

    Numerous social networking networks, including Facebook Messenger, WhatsApp, and Telegram, retain active message data in RAM. This encompasses messages, multimedia information, and user activities within conversations. In certain instances, it may be feasible to retrieve partially loaded or cached messages.

- **Data from Social Media Applications:**

    If the user has accessed social media applications such as Twitter, Facebook, or WhatsApp, data pertaining to such programs, including current interactions, photographs, or videos being downloaded or posted, may be located in the memory.

- **Web Browser Artifacts:**

    Information such as cached information, cookies, and form inputs from the user's activities with social media platforms can be retrieved from RAM. This may encompass stored usernames, passwords, and engagements with social media accounts.

- **Files Pertaining to Social Media:**

    Images, videos, or attachments actively watched or downloaded via social media applications may be briefly kept in RAM prior to being written to disk. This data is available for subsequent analysis.

Examining RAM Dumps for Social Media Artifacts

Upon acquisition of the RAM dump, forensic analysts must scrutinize the memory to retrieve artifacts pertinent to social media. This investigation employs specialized memory analysis algorithms capable of identifying pertinent social media artifacts, reconstructing live sessions, and extracting valuable data from RAM.

Standard Methods for Analyzing RAM Dumps

1. **String Search**: Forensic investigators frequently employ string searching to identify patterns inside memory. Common social networking URLs (e.g., `https://www.facebook.com`, `https://www.instagram.com`) and usernames, as well as application-specific data (e.g., facebook_session_token), may be discernible as readable strings in the RAM dump.

2. **Process Analysis**: By scrutinizing the processes active on the system during the dump, investigators can discern operational social media applications (including Chrome, Facebook, Messenger, WhatsApp, etc.). These operations may include live session data, such as active chats, open web pages, or login tokens.

3. **Network connection**: VoIP social media applications such as Skype and WhatsApp enable users to make calls; thus, the IP address of the recipient may be extracted from the RAM dump.

4. **Memory Carving**: This method entails examining the RAM dump for intact data segments associated with social media material. Memory carving tools facilitate the extraction of files, photos, or chat data that were potentially in use during the RAM dump.

5. **Reconstructing Sessions**: Social media sessions, encompassing user login credentials and chat activity, are frequently retained in RAM during system operation. By extracting these data structures, investigators may frequently reconstitute an ongoing social media session and observe user activities.

Tools for Analyzing RAM Dumps

1. **Volatility Framework**: An open-source utility that is exceptionally beneficial for memory forensics. It can examine RAM dumps and retrieve social media artifacts, encompassing active processes, network connections, and session tokens from browsers or applications.

2. **Rekall**: A memory analysis framework akin to Volatility, Rekall is capable of extracting user activity and artifacts from RAM dumps.

Extraction of Social Media Artifacts from Mobile Devices

Logical, physical, and file system extraction retrieve social media-related artifacts for examination. The following are the artifacts:

- **Messages and Posts:**

  Text messages, chats, or direct messages from applications such as WhatsApp, Facebook Messenger, Instagram, etc. Public posts and status updates from social media platforms such as Facebook, Twitter, or Instagram. Confidential posts, group communications, or engagements on services such as Snapchat or Discord.

- **Metadata:**

  It comprises timestamps denoting the occurrence of a message, post, or interaction. Geolocation data linked to posts, images, or messages, together with device information such as device kind, operating system version, and utilized IP addresses.

- **Deleted Content:**

  Mobile forensics tools can retrieve deleted posts, messages, or media, contingent upon the method of deletion and whether the data has been overwritten.

- **Multimedia Files:**

  It includes photographs, videos, and audio files disseminated through social media applications, screenshots, or screen recordings from these platforms, as well as thumbnails, previews, or cached data retained on the device.

- **Credentials and Account Details:**

  During analysis, cached credentials for social media accounts, encompassing usernames, passwords (if unencrypted), and account recovery information may be discovered.

Data regarding social media account sessions, including IP addresses or access locations, may be employed for subsequent research.

- **Connections and Social Networks:**

  It encompasses contact information, connections from social networking applications, friend lists, follower/following statistics, and interactions such as likes, shares, and comments.

- **Application Utilization Data:**

  Forensic data can disclose the frequency and timing of specific social media app usage, hence offering insights into a user's activity.

- **Notifications and Alerts:**

  Notifications or alerts from social networking applications that manifest in the device's notification history, potentially disclosing conversations or received messages.

- **Web Browser Information:**

  This section includes browsing history or cookies associated with social media websites, accessed using a mobile device's browser application.

CHAPTER 6   EXAMINATION OF SOCIAL MEDIA AND MESSAGING APPLICATIONS

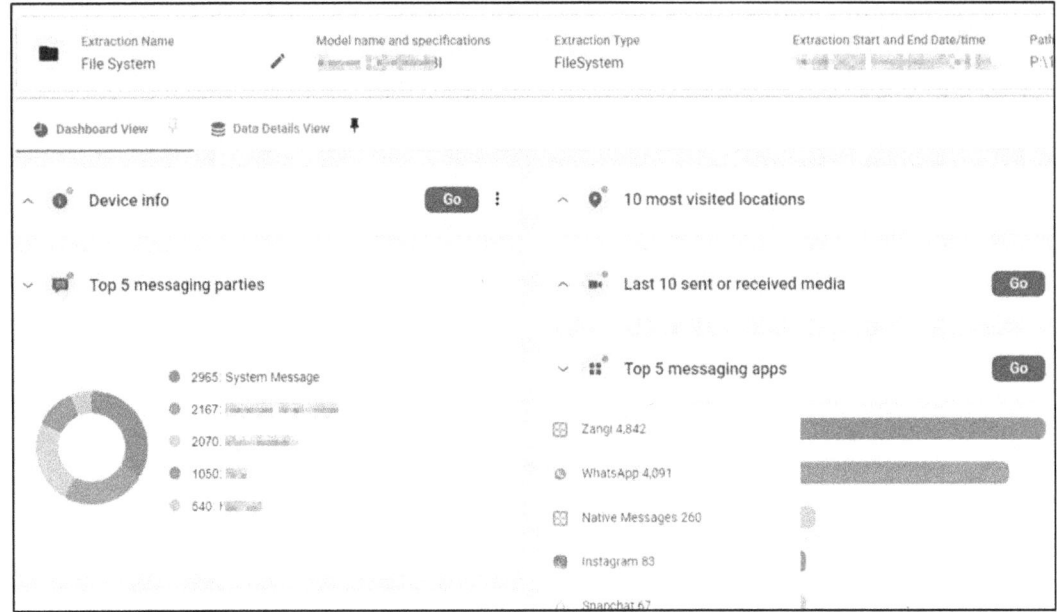

Above screenshot displays social medial application related data extracted from mobile device with file system acquisition method. Forensic tool classifies top used social media application with their access timeline.

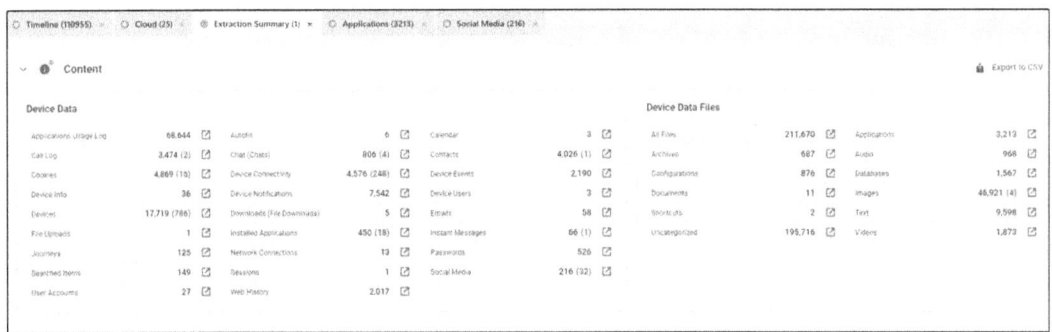

Above screenshot displays overall information retrieved during mobile forensics and numbers in red shows retrieval of deleted information. Forensics tool has extracted a total of 216 social media artifacts from device, in which 32 are deleted but recovered during acquisition process.

Above screenshot displays list and number of artifacts extracted from mobile device.

Social Media Artifacts from the Cloud

The increasing reliance of individuals and companies on cloud storage has heightened the significance of this domain, particularly in the context of investigating cybercrimes, data breaches, and compliance infractions. Social networking programs permit users to keep their data on the cloud; for instance, WhatsApp automatically stores data on Google Drive. Apple's iCloud keeps mobile data in the cloud, which may include data from social media applications.

Cloud storage is an online service enabling users to store data on remote servers, accessible via the Internet using desktop applications, mobile applications, or websites.

Cloud forensics involves the examination of evidence from cloud platforms, including social media sites and instant messaging applications such as Facebook, Instagram, WhatsApp, and Telegram. Primary obstacles in analyzing cloud services encompass data dispersed across multiple locations, encrypted content, continual data changes, and ephemeral information that may not be retained.

Cloud Storage, Synchronization, and Forensics

Cloud storage and file synchronization are integral components of digital services and applications. It enables individuals to access files on any device. Although this is more convenient, it also presents new obstacles for cloud investigators. When examining cloud storage or synchronized files, it is essential to understand the methods of data storage, synchronization, and sharing between devices.

Several renowned services include

- Google Drive
- Dropbox

CHAPTER 6   EXAMINATION OF SOCIAL MEDIA AND MESSAGING APPLICATIONS

- Microsoft OneDrive
- Apple iCloud

Cloud synchronization seamlessly replicates and updates data across various devices linked to a user's account. If a user modifies a document on their smartphone, the alterations will be synchronized on their desktop or tablet.

Common Characteristics

- **Data Accessibility**: Users can retrieve their data from any Internet-enabled device, regardless of location.
- **Real-Time Modifications**: Alterations made to files on one device will be reflected instantaneously across all devices.
- **Automatic backup**: Cloud services often maintain data replicas across many servers.
- **Versioning**: Historical document versions are preserved on cloud servers. Thus, users can revert to previous versions.

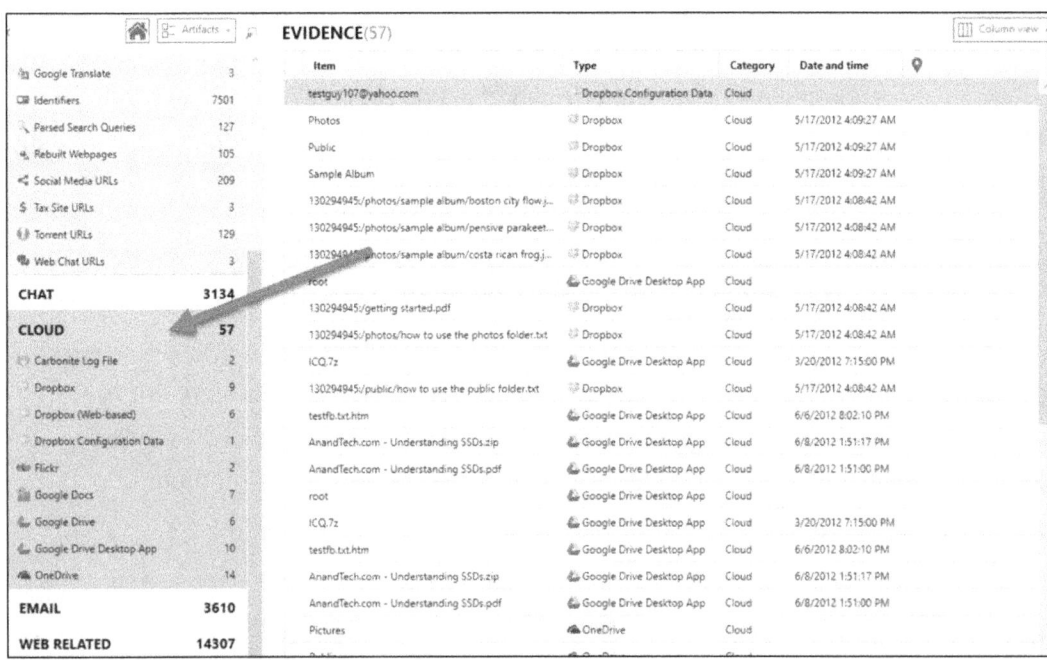

CHAPTER 6 EXAMINATION OF SOCIAL MEDIA AND MESSAGING APPLICATIONS

Above screenshot displays artifacts retrieved during cloud forensics with timestamp.

Above screenshot displays list of cloud application supported by Oxygen forensic tool for data extraction.

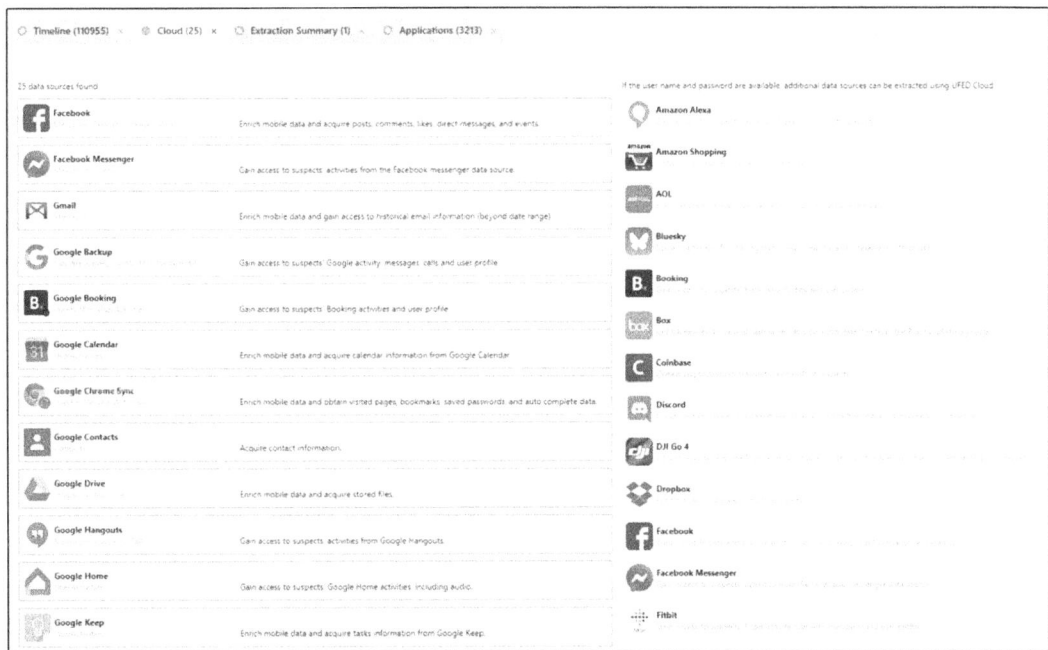

169

List of cloud sync(25) data identified from user's mobile device in left tab extracted using UFED forensics tool.

Challenges in Cloud Storage Investigation

a. **Modifications and Erasure of Data**

- **Modifications to cloud data**: The majority of cloud service providers retain deleted files for around 30 days, after which they are permanently erased.

- **Deletion and synchronization**: If a user removes a file on one device, it will concurrently be removed from all connected devices automatically. It will complicate the retrieval of data for investigators.

b. **Access Regulation and Confidentiality**

- **Account Security**: Cloud services employ robust security measures such as encryption, device authentication, and two-factor authentication. This will complicate access to the user's account during the investigation without the appropriate credentials.

- **User Privacy**: Investigators must have legal authorization to access accounts in order to uphold user privacy. User data may be encrypted and stored in many regions or jurisdictions.

c. **Storage Complications of the Device**

- **Multi-device Storage**: Cloud storage accessible via various devices such as mobile phones and computers. We must examine all connected devices for further data.

- **Data Discrepancy**: Various devices may retain divergent copies of files due to differing usage periods.

d. **Security and Encryption:**

- **Encryption**: The majority of cloud services safeguard data using encryption throughout storage, utilization, and transmission. In the absence of passwords, authenticated devices, or security keys, stored data remains inaccessible.

- **Cloud security**: Regardless of whether customers encrypt their files, cloud service providers employ encryption for protection. This complicates the ability of investigators to analyze data without appropriate access.

Cloud Storage Forensics: Methods for Acquiring Digital Evidence

a. Obtain Data from Cloud Service Providers

To obtain data from cloud services, legal documentation, police complaints such as a first information report (FIR) are required. Cloud services frequently maintain backups, activity logs, and information regarding file access, which might assist in the recovery of erased data in some subscribed services.

They can furnish details such as activity history, emails, photographs, videos, documents, and user information, including login times, duration of usage, and locations.

Legal Challenges: Obtaining data from cloud services might be problematic when it is kept in various countries due to international privacy regulations.

Forensic Tools for Data Retrieval and Examination

Cellebrite UFED, Oxygen Detective, Belkasoft, etc. and similar forensic tools facilitate the collection of significant artifacts from cloud services during forensic investigation. These forensic tools necessitate appropriate credentials/token/account package in case of mobile acquisition to obtain, retrieve, and analyze forensic artifacts. During the actual acquisition of a mobile device, these forensic techniques extract tokens from the social media account package. These tokens may be employed to gain access to cloud storage for the forensic acquisition of email accounts, Facebook, Instagram, etc.

Cloud Data Analysis: Following data acquisition, these forensic tools facilitate analysis.

1. **File System Evaluation**: Assess the organization of files and their most recent modification dates.

2. **Metadata Analysis**: Examine file specifics to ascertain the devices utilized, IP addresses, and user activity records.

3. **Timeline Reconstruction**: Create timelines indicating the timestamps of file uploads, downloads, deletions, or modifications.

4. **File Version Verification**: Examine several file versions to identify modifications.

Forensics Investigation Challenges in Social Media and Instant Messaging Apps

Locating evidence via social media and messaging applications is more difficult due to user communication methods, the multitude of services and platforms, and the manner in which data is kept, synchronized, and secured. The acquisition process necessitates appropriate user credentials and consent. The primary challenges are as follows:

1. Data Distributed Across Various Devices and Locations

    - Typically, users employ multiple devices (phones, PCs, tablets) to access identical data.

    - Synchronization history across different devices complicates change tracking.

    - Data is kept at many server locations.

    - **The data format is diverse**: It encompasses various formats, including text, images, and videos.

- **Data generation occurs in real-time**: Data is continuously produced and can rapidly vanish, be overwritten, or altered.

2. Data Authenticity and Integrity

    - **Chain of Custody**: Investigators must verify that data remains unaltered; otherwise, it may be inadmissible in court.

    - **Metadata Analysis**: Supplementary information regarding posts, uploaded content, and messages is essential.

    - **Preservation for admissibility**: Specialized technologies are necessary to collect, preserve, and analyze evidence securely to ensure data integrity.

3. Challenges of Encryption and Privacy

    - The majority of text messages are encrypted to ensure privacy and security.

    - Social networking platforms inherently safeguard personal information.

    - Privacy settings may limit access to confidential information.

4. Data Modification and Deletion

    - Individuals may modify, revise, or remove their posts and messages at any time.

    - The integrity of the content may be compromised during editing.

    - Certain messages (Recycle Bin or one-time view) self-delete automatically.

    - Deleted data may exist in stored copies or backups.

5. Diverse Platforms

    - Every social media and messaging platform possesses distinct features, security protocols, and data formats, complicating investigative processes.

    - **Distinct Formats**: Numerous platforms save data in proprietary formats that are difficult to interpret without specialized tools or expert proficiency.

6. **APIs and Data Access**: Certain platforms permit individuals or researchers to effortlessly download data through APIs (e.g., Google Takeout, Facebook Data Download), while other services employ convoluted means to obtain user information.

7. **Encryption Is Platform-Specific**: Certain services employ encryption that limits access to stored data. For instance, Apple's iCloud and Google Drive.

8. Disinformation and Extensive Quantity of User-Generated Content

    - The substantial volume of data generated on social media can undermine data authenticity, particularly during significant events such as elections or investigations.

    - **Counterfeit Accounts**: To conceal individuals' identities or disseminate misinformation.

    - **Fabricated Media or Deepfakes**: Users frequently disseminate counterfeit content, such as deepfakes. Determining the authenticity of given content is a significant challenge.

    - **Mass-scale Disinformation Campaigns**: Identifying organized efforts to disseminate misleading information can be challenging due to the rapid dissemination of content across platforms.

    - Examining social media and messaging applications has numerous obstacles, including managing substantial data volumes, absent material, encryption complications, and legal regulations. Investigators must comprehend the characteristics of various platforms and utilize the requisite skills to acquire, manage, and evaluate data proficiently.

Summary

The above chapter thoroughly includes data generated by the social media application on mobile and computer devices. It also includes the forensics acquisition methods and analysis of social media-related application in the mobile and computer system along with some key artifacts and analysis technique. Further, it displays cloud data acquisition methods and tools and services which are supported by the forensics tools. It also highlighted the key challenges faced during social media data analysis and acquisition from cloud service providers.

CHAPTER 7

Location-Based Data Analysis and Geolocation Artifacts

Advancement in mobile technology has provided various opportunities for the end user to use GPS devices. Nowadays, all smartphones, smart watches, cars, tablets, navigation devices, drones, palmtops, etc., have inbuilt GPS device and they store users' movement and travel history on their local database and on cloud application. Cars also comes with inbuilt android systems for navigation. These smart devices could be configured with email ID like Gmail as iPhone required Apple ID(Email) and password after reset the device which could be same ID used first time during first configuration. Global positioning system (GPS) is a satellite-based navigation system which determines locations of GPS device anywhere on Earth. GPS device consists of arrangements of satellite orbiting the planet and their on-earth control tower(station) and GPS receivers. GPS devices receive signals from a group of satellites orbiting Earth, each transmitting its location and time. GPS devices determine location using a technique called trilateration, where they measure distances from multiple satellites to pinpoint an object's position on Earth.

Location-based data analysis and geolocation artifacts are critical components in digital forensics, which includes identification, extraction, analysis of GPS or location-based data from the GPS enabled devices like smartphone, smart watch, vehicle navigation system, etc. These data are crucial, especially when investigating criminal activities, tracking movements, or establishing timelines. Many devices, social media platforms, and messaging applications generate and store geolocation data, providing invaluable insights into the locations of users, devices, and activities. However, analyzing this data comes with several challenges and requires specific forensic methods and tools. Further, this chapter explains about location-based data, forensics acquisition and analysis of location-based data, challenges, and privacy concerns of these data.

What Is Location-Based Data?

Location-based data refers to any information that indicates the physical location of a device or user at a given time. This data can be collected through various sources such as GPS, Wi-Fi, Bluetooth, cell tower triangulation, and geotagging of social media posts, photos, and videos. Some of the terminologies are explained below.

- **GPS Data**: The most precise form of location data, is as data which is collected through satellite signals to pinpoint a device's location in real time.

- **Cell Tower Triangulation**: Used when GPS signals are weak or unavailable, cell tower data estimate's location based on the signal strength between the device and multiple cell towers.

- **Wi-Fi and Bluetooth**: Location can also be derived by scanning Wi-Fi networks or Bluetooth beacons that the device detects, especially in urban areas or indoor settings.

- **Geotagging**: Many social media platforms, messaging apps, and cameras embed location data into photos, videos, or posts. This can include latitude, longitude, altitude, and timestamp.

Geolocation Artifacts

Geolocation artifacts are the data traces left behind by location-based services or features on a device or platform. These can include

- **GPS Logs**: Devices with GPS capabilities (smartphones, fitness trackers, etc.) store logs of geolocation data, which can be used to track movements over time.

- **Wi-Fi Connection Logs**: Many devices automatically log Wi-Fi networks they have connected to. By examining these logs, investigators can infer the device's location when connected to specific networks.

- **Cell Tower Data**: Call logs, text messages, or other forms of communication often include metadata related to which cell tower a device connected to, giving clues to the user's physical location.

- **Geotagged Media**: Photos, videos, and social media posts often contain embedded geotagging information that includes location data (e.g., EXIF data in photos) if location tagging feature is enabled on the device.

- **Location History**: Navigation applications (e.g., Google Maps, Apple Maps) track and store a history of a user's movements. This data can be valuable for tracing a user's whereabouts over time.

Location-based data stored by mobile device applications

Mobile applications store GPS data generally through application permission. Android devices store GPS data using Google play services and location API which allow to fetch the location information and store them on local device and on cloud application. Android devices use two location-based permissions: approximate location (ACCESS_COARSE_LOCATION) or precise location (ACCESS_FINE_LOCATION) to store and collect data. Android-based applications are using Google Play services Location API and **Fused Location Provider** API which intelligently combine data from multiple sources (GPS, Wi-Fi, cellular network) to provide accurate and efficient location data.

These applications can store GPS data in various ways, including

- **Internal Storage:** Android operating system facilitates internal arrangement by dividing data into directories within device storage where an application can organize its files, including persistent and cached files.

- **Database:** Applications are using SQLite databases to store location data and other relevant information.

- **File System:** Applications can store location data in files, such as GPX files, which are commonly used for storing GPS tracks

In iOS, applications uses Core Location API to get GPS data, and then store it using native storage options like UserDefaults, Core Data, Files, etc. On an iPhone, location-based data can be stored in a few different ways—some are handled automatically by iOS, and others are managed explicitly by apps using storage APIs.

Generally navigation apps like Google Maps, Apple maps, fitness apps, car-cab booking services, UPI applications (GPay, Phonepay), Find my device, Google Photos, Snapchat, Browser, weather apps, etc., store location information. Google maps stores travel history and allows users to create a timeline.

Forensic Methods for Collecting and Analyzing Geolocation Data

a. **Acquiring Location Data from Devices**

From Mobile Devices:

A forensics analyst uses the data acquisition methodology like physical, logical, and full file system acquisition. During the parsing phase, an analyst has to focus on the list of applications which store GPS-related data. Forensics software contains specific features to extract location data from the installed applications to create a location timeline. Later the analyst put those GPS coordinates on the navigation application like google maps for details so forensic investigators can use specialized tools to extract location data from mobile devices (smartphones, tablets) such as Cellebrite UFED, Magnet AXIOM, or Oxygen Forensics, etc. These tools can retrieve GPS logs, Wi-Fi connection history, and other location-related artifacts stored in device memory or cloud backups.

Digital Cameras and Social Media: Geotagging data in photos and videos can be extracted using tools like ExifTool, which can read metadata embedded in media files.

Fitness Trackers and Smartwatches: Devices like Fitbit, Apple Watch, or Garmin record location data related to fitness activities, such as steps taken and routes traveled. These devices often sync data to cloud services, allowing investigators to request or extract location history.

Data Collection and Preservation

- **Live Acquisition**: In some cases, data should be acquired while the device is still powered on, especially if it is actively receiving location updates or transmitting data (e.g., cloud syncing or GPS logging).

- **Data Copying**: A bit-for-bit copy of the data should be created to preserve the original device state, minimizing the risk of contamination or alteration.

- **Cloud Synchronization**: Data should be retrieved as soon as possible before it is overwritten or deleted from the cloud.

- **Acquisition Techniques**

- **Physical Extraction**: A method where the entire device storage is copied, ensuring that all available data, including deleted items, can be accessed. This method is commonly used for mobile phones and storage devices.

- **Logical Extraction**: Involves extracting data at the file system level (e.g., GPS logs, photos, or text messages) without accessing the entire device storage. Logical extraction can be less invasive but may not capture all data.

- **File System Extraction**: This extraction focuses on accessing specific file systems (e.g., media or application data) to obtain location-based data related to specific applications.

- **Cloud Data Extraction:** Acquiring data directly from cloud servers using tools like Cellebrite, Magnet AXIOM, or through legal requests to service providers. This method allows investigators to access location data stored remotely.

Data Integrity and Verification

Data Integrity: It's important to ensure that location data is not altered during the acquisition process. This can be verified by generating hash values (e.g., MD5, SHA-1) for the acquired data and comparing them with the hash of the original source.

Documentation: Every step of the acquisition process must be well-documented to ensure transparency and accountability in legal proceedings. This includes details of the methods used, tools employed, and personnel involved in the acquisition process. Challenges in Forensic Acquisition of Location-Based Data

- **Encryption**: Many devices and cloud services encrypt data, making it difficult to access without the correct decryption keys. This may require additional efforts like legal orders to unlock or decrypt devices.

- **Data Volatility**: Location data, particularly GPS logs or app-based location history, may be overwritten or erased after a certain period, reducing the time frame in which data can be recovered.

- **Cloud Synchronization Issues**: Cloud services continuously synchronize data, which can result in the loss of valuable location history if not collected promptly.

- **Device Fragmentation**: Different devices and applications store and format location data in various ways, requiring specialized forensic tools for each platform.

- **Location Spoofing**: Users may use software or hardware to fake their location, making it necessary to validate the authenticity of the acquired data.

b. **Acquiring Location Data from Cloud Services**

Cloud storage services such as Google Drive, iCloud, or Dropbox may store backups of geolocation data or allow access to geotagged photos or videos. Investigators may request a Cloud service provider with the help of nodal officer appointed by Law Enforcement to collect this data directly for further analysis.

Google Location History:

Google Maps offers location history tracking for Android and iOS devices where data is synced to associated Google account. Investigators can access detailed timelines of users' locations by retrieving location data directly from their Google account with the help of law enforcement agency via the legal channel.

Apple iCloud Backup: If an iPhone user has iCloud Backup enabled, the backup may include location data stored within apps or system settings (e.g., Maps, Photos, etc.).

Mobile and cloud forensics tools like UFED, Oxygen detective, Belkasoft, etc., supports cloud data extraction when credentials are available of cloud account or may be provided by accused with consent.

c. **Analysis of Geolocation Data**

Timeline Reconstruction: Geolocation data can be used to build a timeline of events based on the time and location of data points. This can help in verifying alibis or establishing movements during a specific time frame.

Geospatial Mapping: Investigators can use Geographic Information Systems (GIS) tools to map geolocation data. By plotting coordinates from GPS logs, Wi-Fi, or cell tower data, they can visualize the subject's movements across various locations over time. GIS tools such as **ArcGIS**, **Google Earth**, or forensic software like **Magnet AXIOM** and **X1 Social Discovery** can help investigators map out the locations of interest.

d. **Cross-Referencing Geolocation Data with Other Evidence**
Location-based data can be cross-referenced with other types of digital evidence, such as

- **Phone Logs**: Call data or messaging logs that include timestamps and associated cell tower data can help establish whether a device was at a specific location.

- **Surveillance Footage**: CCTV or private security camera footage, when combined with geolocation data, can verify movements or interactions at specific locations.

- **Social Media Posts**: Investigators can correlate timestamps and geotagged content posted on social media platforms with location data, confirming the subject's presence at certain locations.

e. **Challenges in Analysis**

- **Time Zones**: Different devices may store timestamps in different time zones. Investigators need to account for this discrepancy when correlating data from multiple devices.

- **Data Gaps**: Geolocation data is not always available continuously, and gaps in data collection can make it difficult to establish an accurate timeline. Investigators may need to account for possible missing or incomplete data.

Location Data Storing Path from Android and IOS Device

	Type of location	Path
IOS	All Location Requests on iOS	/private/var/mobile/Library/Caches/com.apple.routined/Cache.sqlite
IOS	Location(Good)/Speed(Ok)	/private/var/mobile/Library/Caches/com.apple.routined/Cache.sqlite
IOS	Visits	/private/var/mobile/Library/Caches/com.apple.routined/Cache.sqlite
IOS	Parked Vehicle Events	/private/var/mobile/Library/Caches/com.apple.routined/Local.sqlite
IOS	Network Locations, NOT Device Location	/private/var/root/Library/Caches/locationd/cache_encryptedB.db [WiFi]
IOS	Approximate Cell Site Location, NOT Device Location	/private/var/root/Library/Caches/locationd/cache_encryptedB.db [Cell]
IOS	Interior Details of Buildings	/private/var/root/Library/Caches/locationd/indoor_tiles/availability.db
IOS	Reminder Locations, Store Locations, Frequent Location Monitoring (BundleId and Name)	/private/var/root/Library/Caches/location/consolidated.db

(continued)

CHAPTER 7 LOCATION-BASED DATA ANALYSIS AND GEOLOCATION ARTIFACTS

	Type of location	Path
IOS	GPS Location of Captured Media Items	/private/var/mobile/Media/PhotoData/Photos.sqlite
IOS	GPS Location of Captured Media Items	/private/var/mobile/Media/DCIM/*
IOS	GPS Location of Captured Media Items (iOS 15)	/private/var/mobile/Media/PhotoData/Photos.sqlite
IOS	GPS Location of Captured Media Items	/private/var/mobile/Media/DCIM/*
IOS	Photo Location Analytics (Core Location Services (CLS))	/private/var/mobile/Media/PhotoData/Caches/GraphService/*
IOS	List of Placemark details referred by all other databases	/private/var/mobile/Media/PhotoData/Caches/GraphService/CLSLocationCache.sqlite
IOS	Area of Interest	/private/var/mobile/Media/PhotoData/Caches/GraphService/CLSBusinessCategoryCache.AOI.sqlite
IOS	Region of Interest	/private/var/mobile/Media/PhotoData/Caches/GraphService/CLSBusinessCategoryCache.ROI.sqlite
IOS	Place of Interest	/private/var/mobile/Media/PhotoData/Caches/GraphService/CLSBusinessCategoryCache.POI.sqlite
IOS	Nature Area	/private/var/mobile/Media/PhotoData/Caches/GraphService/CLSBusinessCategoryCache.Nature.sqlite
IOS	Personalization Portrait, gathers locations from email signatures, map location searches, photo locations and more	/var/mobile/Library/PersonalizationPortrait/PPSQLDatabase.db
IOS	MapSync	/private/var/mobile/Containers/Shared/AppGroup/.*/Maps/MapsSync_0.0.1
IOS	Find My Devices	/private/var/mobile/Library/Caches/com.apple.findmy.fmipcore/Devices.data
IOS	Find My Airtags	/private/var/mobile/Library/Caches/com.apple.findmy.fmipcore/Items.data

Android

OS	Type of location	Path
Android	Google Maps turn-by-turn directions	/data/com.google.android.apps.maps/app_tts-temp/tts-%UNIXEPOCH%
Android	Photos	/data/media/0/DCIM/Camera/*
Android	Home address/Work address/Labeled addresses	/data/com.google.android.apps.maps/databases/gmm_myplaces.db
Android	Search history Maps	/data/com.google.android.apps.maps/databases/gmm_storage.db
Android	Timezone	/data/com.google.android.apps.turbo/databases/turbo.db
Android	Favorites places	/data/com.google.android.apps.maps/databases/gmm_sync.db
Android	Google Maps, Way To Go, Favorites, Starred Places, Labeled	/data/data/com.google.android.apps.maps/new_recent_history_cache_search.cs
Android	Google Maps Search History	/data/data/com.google.android.apps.maps/gmm_storage.db
Android	Tracks Photos and Locations of apps that use device camera	/data/data/com.sec.android.app.camera/databases/core2.db
Android	Creates a geofence around each known/connected network	/data/system/wifigeofence.db
Android	Creates a geofence around each known/connected network	Dumpsys_wifi
Android	Tracks wireless networks, frequency of connection and lost connection status	/data/log/wifi/iwc/iwc_dump.txt
Android	Tracks wireless networks, frequency of connection and lost connection status	/data/misc/apexdata/com.android.wifi/WifiConfigStore.xml

Location Data from Social Media Applications

Social media platforms and messaging apps (e.g., Facebook, Instagram, Twitter, WhatsApp) often allow users to share their location through geotagging in posts, photos, or status updates.

Posts and shared media can provide insights into the locations where events occurred, where an individual was at a certain time, or when they were at a location. Investigators can extract geolocation data embedded in posts or photos from social media platforms. This is particularly useful in cases where an individual's public online activities need to be correlated with real-world locations.

IP Address Geolocation

An IP address, when a device connects to the Internet, can be used to approximate the device's geographic location. This is usually based on the location of the Internet service provider (ISP) or the server the device is connecting to. IP-based geolocation provides a rough estimate of location, typically down to the city or region level. It is often used to identify the general area a user was accessing the Internet from. While IP geolocation is not as precise as GPS or cell tower data, it can still provide useful information, especially in investigations involving online activities, cybercrimes, or tracking the origin of digital communications.

Forensic Challenges in Analyzing Geolocation Data

While geolocation data is incredibly valuable in forensic investigations, several challenges arise during collection, preservation, and analysis which are listed below:

a. **Privacy Concerns**

   Geolocation data can be highly sensitive and may reveal a significant amount of personal information, such as an individual's home address, places they frequent, and even their daily routines. As a result, accessing location data without proper legal authorization (warrants, subpoenas, or consent) can lead to privacy violations.

b. **Data Volatility and Loss**

Geolocation data can be volatile and sometimes inaccessible due to following reasons:

Device Deletion: Location-based data, particularly logs from apps or device settings, may be deleted either by the user or by automatic processes, making it hard to recover.

Data Overwriting: Some devices or applications may overwrite geolocation data after a certain period, reducing the time frame available for investigation.

Limited Data Retention: Many services (such as Google Maps or Apple iCloud) only retain location history for a specific period (e.g., 30 days or a few months), and older data may be permanently deleted.

c. **Accuracy and Precision of Data**

GPS Limitations: While GPS is accurate, it is not infallible. Poor weather, tall buildings, or other environmental factors can degrade the signal, leading to inaccurate or missing data.

Cell Tower and Wi-Fi: Cell tower triangulation or Wi-Fi-based location services may offer less precision than GPS, and in rural or less-developed areas, data can be less reliable.

Device Settings: Devices may not always have location services enabled, or the device might be in airplane mode, which can affect the collection of accurate geolocation data.

d. **Multi-Device Synchronization**

Many individuals use multiple devices (e.g., smartphones, tablets, laptops, wearables) that synchronize location data across different platforms. Tracking a suspect's location may require extracting data from all of these devices, which can be time-consuming and complicated. Different devices may also store geolocation data in different formats or locations, requiring the use of different forensic tools to collect and analyze it.

e. **Geolocation Spoofing**

   Geolocation spoofing involves falsifying a device's location, either by modifying system settings or using third-party apps. This can mislead investigators and may require careful analysis of the device's data to distinguish between real and fake geolocation information.

f. **Geolocation Data from Third-Party Services**

   Many social media and messaging applications store and share geolocation data in the form of posts or photos with embedded GPS information. Investigators must not only retrieve data from the device itself but may also need to request data from third-party services, which may complicate the process due to legal and jurisdictional issues.

Analysis Techniques for Geolocation Artifacts

The analysis of geolocation artifacts involves examining location-based data recovered from digital devices, cloud services, social media, and other sources to identify patterns, verify timelines, and corroborate evidence in forensic investigations. These artifacts typically come from GPS logs, Wi-Fi data, Bluetooth, geotagged media, and other sources that provide spatial and temporal information. To draw meaningful conclusions from this data, forensic investigators employ various **analysis techniques** that help reconstruct events, track movements, and understand the context behind location-based evidence.

Here's a breakdown of key analysis techniques for geolocation artifacts:

1. **Timeline Analysis**

   This technique involves arranging geolocation data in chronological order to reconstruct an individual's movements over time. Reconstructing the route a suspect took during a crime. Verifying a reason by matching known events with location data (e.g., photos taken at specific places or messages sent from particular locations).

CHAPTER 7   LOCATION-BASED DATA ANALYSIS AND GEOLOCATION ARTIFACTS

Correlating timestamps from different devices and sources (e.g., cell towers, GPS logs, and social media posts).

2. **Mapping and Geospatial Analysis**

   Geospatial analysis involves plotting geolocation data on maps to visually track movements, identify hotspots, or analyze spatial patterns of behavior. Identifying locations of interest (e.g., crime scenes, meeting points, or areas frequently visited by a suspect). Analyzing the path traveled by a suspect during specific events. Detecting clusters of activity (e.g., areas where an individual spent significant time).

   Investigators use Heatmap and Route Mapping to visualize areas with frequent or high-density activity and Plotting the exact routes taken based on GPS coordinates and timestamps

   Tools like **ArcGIS** (Advanced geographic information system (GIS) tools to visualize and analyze spatial data) and **Google Earth** allow investigators to overlay GPS coordinates or traces on real-world maps.

3. **Correlation with External Data Sources**

   This technique involves correlating location data with other available information, such as event logs, surveillance footage, or witness statements, to confirm or challenge a hypothesis. Validating location data from a device with surveillance camera footage to see if a suspect was at the scene at the same time. Correlating timestamps of geotagged media (photos/videos) with external sources like security camera timestamps or call logs. Comparing social media activity and its associated geolocation with nearby Wi-Fi networks or cell towers.

4. **Pattern and Behavior Analysis**

   This technique involves analyzing movement patterns and behaviors inferred from geolocation artifacts. It seeks to uncover habitual patterns or suspicious behaviors based on location data.

Detecting routine behaviors (e.g., travel routes, daily commutes, or frequent visits to certain locations) that might indicate a suspect's typical movements or activities. Identifying unusual or suspicious behavior, such as visits to remote or unexpected locations at unusual times. Identifying frequently visited locations (e.g., a suspect's home, office, or crime scenes). Highlighting deviations from a person's usual location patterns, which could indicate criminal activity or an attempt to avoid detection.

5. **Reverse Geocoding and Address Matching**

    Reverse geocoding is the process of converting geographic coordinates (latitude and longitude) into human-readable addresses or location names. Identifying specific locations (e.g., a suspect's home, a business address, or a crime scene) from GPS coordinates. Matching the coordinates from geolocation data to known addresses or places of interest.

6. **Location Validation Using Multiple Data Sources**

    Location data from different sources, such as GPS logs, Wi-Fi signals, and Bluetooth, can be cross-validated to improve the accuracy of location analysis. Ensuring that location data from one source (e.g., GPS) matches or corroborates data from another source (e.g., Wi-Fi or Bluetooth signals). Verifying the accuracy of location points using different geolocation technologies to eliminate potential errors or false readings is important.

Privacy and Security Implications

Privacy and security implications in the context of geolocation artifacts and digital forensics are critical considerations in any investigation involving location data. The sensitive nature of geolocation data, which often includes real-time information about an individual's whereabouts, poses both privacy risks and potential security threats. In digital forensics, investigators must balance the need for location-based evidence with the ethical and legal boundaries that govern the collection, analysis, and dissemination of such data.

CHAPTER 7   LOCATION-BASED DATA ANALYSIS AND GEOLOCATION ARTIFACTS

Below are some of the key privacy and security implications related to the acquisition and analysis of geolocation artifacts:

1. **Privacy Concerns**

    Geolocation data provides detailed insights into a person's movements, activities, and behaviors, which can be incredibly invasive if misused or accessed without proper consent. These concerns are heightened in digital forensics, where investigators need to ensure that only the relevant data is collected and that privacy rights are respected.

    a. **Unwarranted Surveillance**

    Continuous tracking of a person's location through their mobile device, social media, or GPS devices can lead to unwarranted surveillance. Investigators must ensure they are not overreaching or accessing more data than necessary. Legal protocols such as search warrants should be followed to ensure proper authorization before acquiring geolocation data. Additionally, the data should be limited to only what is necessary for the investigation.

    b. **Access to Private Locations**

    Geolocation data can reveal private locations such as home addresses, workplaces, or personal destinations. If this data is accessed without authorization, it could lead to invasions of privacy. Investigators should avoid accessing geolocation data that is irrelevant to the investigation and should follow legal restrictions that prevent the collection of unnecessary location data.

    c. **Geotagging of Media**

    Photos and videos often include geotags that indicate where and when the media was created. These tags may reveal sensitive locations such as a person's home, vacation spots, or other private events. When analyzing geotagged media, investigators should be cautious about accessing and disclosing personal data embedded in images, especially if it is not relevant to the case. Additionally, individuals may disable geotagging in their devices to protect privacy.

d. **Legal Boundaries and Consent**

Collecting geolocation data without proper legal authorization (e.g., a court order, FIR, or complaint) can lead to violations of privacy laws and infringe on individuals' rights. Investigators must adhere to data privacy laws such as the **DPDP Act and** other data protection regulations. Consent must be obtained before accessing location data from individuals or service providers when required by law.

Data Security Risks

Geolocation data, when improperly handled, poses significant security risks both during collection and storage. This data can be targeted by malicious actors or exposed unintentionally during forensic investigations.

a. **Unauthorized Access**

Geolocation data is often sensitive, and unauthorized access could lead to security breaches. If this data falls into the wrong hands, it could be exploited for malicious purposes, such as stalking, theft, or exploitation. All geolocation data should be encrypted during storage and transmission. Strict access control protocols should be enforced to ensure only authorized personnel handle the data. Digital forensics labs should implement secure systems and procedures to protect data integrity.

b. **Data Corruption or Tampering**

Data corruption or tampering can occur if location data is not properly preserved during acquisition or analysis. If location data is altered or modified, it can compromise the validity of the evidence and affect the outcome of the investigation. Ensuring the integrity of geolocation data is critical. Investigators should employ cryptographic hash functions (e.g., SHA-256) to generate hash values for all acquired data to verify its integrity throughout the forensic process. Additionally, investigators should use write blockers and other tools to prevent the alteration of original data.

Ethical Issues in Geolocation Data Analysis

The use of geolocation data for forensic analysis raises several ethical concerns, particularly around the balance between gathering evidence and protecting individuals' rights.

   a. **Intrusiveness of Location Tracking**

   Tracking an individual's movements over extended periods can be perceived as highly invasive, especially when done without consent or proper authorization. The analysis of detailed location data can also reveal sensitive information about personal habits and relationships. The principle of proportionality should guide the use of location data in forensic investigations. Investigators should ensure that the extent of data acquisition and analysis is proportionate to the severity of the case and the specific legal issues being addressed.

   b. **Bias and Discrimination**

   The use of geolocation data in forensic investigations can potentially lead to biased or discriminatory outcomes. For example, individuals from certain geographic areas or communities may be disproportionately targeted based on location data. Investigators must remain objective and avoid over-relying on location data as the sole basis for drawing conclusions about an individual's behavior or involvement in a crime. Geolocation evidence should be combined with other types of evidence to avoid bias.

Summary

Location-based data and geolocation artifacts play an essential role in modern forensic investigations. They provide critical information about a subject's movements, interactions, and activities over time. However, analyzing geolocation data involves overcoming challenges such as privacy concerns, data loss, device fragmentation, and accuracy issues. Digital forensic investigators must use specialized tools and techniques to collect, preserve, and analyze this data while navigating legal and technical complexities. By combining location data with other forms of digital evidence, investigators can piece together compelling narratives and establish key facts in criminal or civil investigations. It involves retrieving data from multiple sources like mobile devices, cloud services, social media, and GPS devices, each with unique challenges. Ensuring data integrity, maintaining an unbroken chain of custody, and adhering to privacy laws are essential to ensuring that location-based evidence can be successfully used in legal contexts. The forensic acquisition process helps build accurate, reliable timelines and geospatial reconstructions, making location data an invaluable tool in digital investigations.

CHAPTER 8

Mobile Device Network and Forensics

This chapter explains how mobile networks like 3G, 4G, Wi-Fi, and Bluetooth are built and work. It looks at how we can study network traffic to find useful patterns. The chapter also talks about digital clues, called forensic artifacts, that are left in network data. Different types of attacks and threats on mobile networks are described. Finally, it explains how to respond to such incidents and investigate them properly.

Overview of Mobile Networks and the Importance of Forensic Analysis in Mobile Network Environments

This section explains various mobile networks like 3G, 4G, Wi-Fi, and Bluetooth, which connect people and devices everywhere. Since these networks can be attacked or misused, forensic analysis helps in finding digital evidence and understanding what really happened. This makes it an important part of keeping mobile communication safe.

Introduction to Mobile Networks

Mobile networks have transformed quite profoundly over the past few decades. It's now an integral part of modern communication and connectivity. The technology of mobile has changed the ways in which people communicate with each other, from 1G analog systems to the current fifth-generation (5G) network. Mobile networks form the backbone of a host of services ranging from voice calls to text messages, high-speed Internet, streaming, and IoT. These networks consist of several communication standards, such as 3G, 4G, LTE, Wi-Fi, and Bluetooth, suitable for different use cases with different speeds, coverage, and reliability. A mobile network basically consists of the following two parts:

1. **The Core Network**, which deals with routing as well as control of communications.

2. **Radio Access Network (RAN)**, which connects various devices, such as handsets, to the core network through base stations or towers.

This advancement in technology has, in turn, led to the creation of several devices and services that rely on mobile networks. Currently, applications supported by mobile networks include banking, e-commerce, remote work, and social networking, among others. These networks carry huge chunks of data, where billions of devices worldwide are connected. The information and data held in such networks can therefore become highly substantial sources of evidence that can be used in forensic investigations.

The Growing Complexity and Security of Mobile Networks

With an increased propagation of mobile networks, complexity has also increased. Today's networks have to accommodate exponentially growing numbers of devices that get connected due to an ever-increasing rate of IoT adoption. Moreover, such networks face increasing demands for "big data" privacy and security, making mobile networks inherently complex systems. Examples include encryption protocols, multifaceted communication technologies, and dynamic switching between heterogeneous types of networks, such as transitions between 4G and Wi-Fi.

Key obstacles confronting mobile networks include cyber threats and unauthorized access. Adversaries employ such network vulnerabilities to attack sensitive information, intercept communications, or carry out debilitating attacks on users and organizations. Threats have evolved as network technologies have evolved. Now, they include more sophisticated techniques, such as Man-in-the-Middle (MITM) attacks, malware diffusion, SIM swapping, and rogue base station capture, including IMSI catchers or Stingrays.

Importance of Mobile Device Network and Forensics

Mobile networks have become a significant part of day-to-day life, and security threats are also increasing with growing mobile network utilization. Therefore, it necessitates forensic analysis on mobile network environments. Mobile network forensics is a subset of digital forensics that involves the collection, preservation, analysis, and reporting of data related to mobile communication network traffic through various forms. These forms include call and SMS logs, network traffic, metadata, location data, and much more.

The importance of mobile network forensics can be understood by considering several key factors:

a) **Ubiquity of mobile Devices**: Mobile devices are ubiquitous, and for most people, smartphones are at the center of their digital lives. Forensic investigations mostly require including mobile devices because they contain rich and valuable data content that can play a determining role in criminal, civil, and corporate cases. Mobiles contain information about communications (calls, texts, and social media interactions), location (GPS data and cell tower triangulation), browsing history, and application use. All this information can be used to reconstruct events or establish intent.

b) **Mobile devices as evidence hubs:** Mobile devices serve as hubs for evidence in criminal and corporate investigations. In criminal offenses, suspects or victims may employ their cell phones for communications, planning, or record-keeping related to the crime. In corporate investigations, a suspect's cell phone can be searched for evidence of insider threats, information leakage, or theft of intellectual property. Forensic analysis by professionals can extract and review such data, frequently providing evidence that investigators would otherwise not have had access to.

c) **Location-based forensic evidence**: Cell networks produce vast amounts of location-based evidence, including call detail records (CDRs) that log when a device was at a particular location when it made or received calls or texts. Wi-Fi and Bluetooth networks can provide location-based information on proximity to access points and devices. Information gained from these streams can be used in forensic analysis to place individuals at specific locations at specific times, which is important in establishing alibis or placing suspects at crime scenes.

d) **Discovering cybercrime**: Mobile phones have become a preferred platform for cybercrime. Forensic analysis can uncover any cybercrime that may be in the form of phishing, ransomware attacks, identity theft, or data exfiltration. Network flow capture and analysis across network traffic allow forensic experts to determine anomalies in data transfer, malware communications, or unauthorized access to sensitive information. The captured data can aid in tracing the criminal's actions and the extent of the breach.

e) **Incident response and investigation**: Incident response through mobile network forensics is a part of organizations' protocols. In cases where a network breach has occurred, forensic experts must determine how the intrusion occurred, which systems were affected, and the extent of the damage done. This may involve analyzing network logs, observing traffic patterns, and identifying signs of an active attack. Telecommunication network forensics also prevents future threats by identifying network infrastructure vulnerabilities and undertaking improved security measures.

f) **Legal and regulatory compliance**: Many industries, especially those handling confidential or regulated information (such as health and finance sectors and government ministries), possess distinct legal and regulatory mandates regarding data protection and security. Mobile Network Forensics ensures adherence to legal and regulatory compliance by detecting violations of laws, such as the General Data Protection Regulation (GDPR) or the California Consumer Privacy Act (CCPA), ensuring correct handling of users' information, and providing forensic proof when violations occur.

Challenges of Mobile Network Forensics

Mobile network forensics is the process of gathering, collecting, preserving, and analyzing data from mobile devices and related network infrastructure to support investigations into crimes, security incidents, or data breaches. However, forensic investigators' concerns regarding mobile networks are indeed much more challenging than traditional digital forensics. This mainly relates to the inherent characteristics of mobile environments, such as mobility, encryption, and the sheer volume of data generated. These factors complicate the technical and legal issues surrounding the processes of gathering and analyzing forensic evidence from mobile devices and their respective networks.

a) **Mobility and dynamic network environment**: Being inherently proliferative, mobile devices' movement from one location to another, and inevitably from one network provider to another, creates a problem that is somewhat difficult to tackle for forensic purposes. It is even more complicated in terms of data tracking and correlation related to a particular device or user.

b) **Encryption and data protection**: The growth of individual users' privacy and security for data has resulted in its introduction into mobile networks. While encryption boosts data protection for users, it makes it extremely difficult for forensic investigators to access and evaluate network traffic or data cached on mobile phones.

c) **Data volume and complexity**: Mobile networks generate humongous amounts of data daily. As smartphones, IoT devices, and other connected appliances grow in usage, network traffic and data volumes have grown multi-fold, making it challenging for forensic analysts to differentiate valuable information from such a massive data set.

d) **Legal and ethical limitations**: Mobile network forensics, like most other types of investigations, is susceptible to various legal and ethical restrictions that may deter the investigative process. Mobile data access usually involves strict legal measures, such as court orders or warrants, especially when handling encrypted data or private communications. International investigations face the issue of multiple jurisdictions, as data is typically stored or transmitted across several countries. Forensic investigators must strike a balance between the various requirements of evidence collection and the requirement of not violating user privacy while responding to all applicable legal frameworks, such as the General Data Protection Regulation in Europe and the CCPA in the United States. Techniques used also raise ethical concerns, including communications intercepts or decrypts that may be considered invasive and a violation of civil liberties.

Mobile Network Architecture (3G/4G/Wi-Fi/Bluetooth)

This section describes the architecture of mobile networks such as 3G, 4G, Wi-Fi, and Bluetooth. It explains how these networks are designed, how they work, and how they enable communication between people and devices. Understanding this architecture is important for analyzing security and forensic aspects of mobile networks.

Mobile Networks Overview: Cellular Network Architecture

Cellular networks form the heart of existing mobile communications, connecting billions of devices worldwide for voice, text, and data. The underlying principles of these networks include a few closely related elements that are crucial to the proper operation of mobile communication. The key constituents of cellular network architecture are main base stations, Mobile Switching Centres (MSCs), and databases, such as Home Location Registers (HLRs). Collectively, these features allow for the smooth provision of services to mobile users, as well as control of user mobility, data traffic, and network resources.

Base stations: Base stations are central in cellular network architecture since they form a communication link between mobile devices and the rest of the network. A base station includes the following: Base Transceiver Station (BTS) and Base Station Controller (BSC) that handle radio communications with mobile devices. Modern LTE

(4G) and 5G networks integrate the functionality of BTS and BSC together in a single component, known as eNodeB within LTE or gNodeB in 5G. A single unit can provide both radio access and control functionalities.

Mobile switching center: A mobile switching center or MSC is an important component in the structure of a cellular network, playing an essential role for smooth communication. The MSC is therefore able to efficiently route voice calls, SMS, and all other forms of communication services between mobile devices and external networks, which include the PSTN and the Internet. The MSC controls network traffic, thus allowing smooth interconnectivity. Consequently, with reliability and without any break in communication services, the gap between mobile networks and other external communication systems is closed.

Home Location Registers (HLRs) and other databases: Databases are one of the major subcomponents of cellular networks, and essentially, they act as centralized repositories of critical information regarding mobile subscribers and their real-time status within a network. They are storage components that allow the efficient running of the network, smooth management of subscribers, and communication services. There are two broad databases in cellular networks: the Home Location Register (HLR) and the Visitor Location Register (VLR). The HLR is a master database hosting permanent information about subscribers, such as profiles, phone numbers, and service plans. In contrast, the VLR acts as a temporary database where information regarding roaming subscribers is maintained, providing uninterrupted communication services to subscribers when they are away from their home network.

The cellular network architecture is composed of several interacting parts that work together to provide seamless communication services. Base stations are responsible for the wireless link between a mobile device and the network, while MSCs ensure the switching and routing of calls and data. Another set of databases, such as HLR and VLR, ensure that the system keeps track of where each device is located and the type of services it can access. Such complex architecture makes it possible for cellular networks to successfully integrate the dynamic and mobile nature of communication found in present times.

3G and 4G Networks

Third-generation (3G) and fourth-generation (4G) mobile networks have significantly enhanced mobile communications. Unlike traditional networks, these mobile networks offer higher data speeds and better services than their predecessors. Now, although

CHAPTER 8   MOBILE DEVICE NETWORK AND FORENSICS

3G was the catalyst in ensuring Internet access, 4G was significant in changing the face of mobile networks due to its high-speed data and advanced services, which include HD video streaming and IoT applications. As such, it is important to understand the architectures and key components of these networks, particularly in forensic challenges, that professionals in mobile network forensics need to be aware of.

The main Components of 3G and 4G Networks. The 3G and 4G networks are essentially composed of two main components: the Core Network (CN) and the Radio Access Network (RAN). The architecture adopted in every generation of mobile network has focused its attention on different aspects of functionality and efficiency enhancements.

The Core Network (CN) is the main component that handles fundamental services in a mobile network, such as user authentication, routing, billing, and communication between mobile users and access networks like the Internet and PSTN. In 3G networks, the core uses two distinct domains: Circuit-Switched (CS) for voice call handling and Packet-Switched (PS) for data transmission. These include major components such as the Mobile Switching Centre (MSC) for circuit-switched voice services, the Serving GPRS Support Node (SGSN) for packet-switched data services, and the Gateway GPRS Support Node (GGSN) for Internet connectivity. In contrast, 4G networks, also known as Long-Term Evolution (LTE), utilize an all-IP core network that supports both voice and data on the same platform. The central component is the Evolved Packet Core (EPC), which manages data routing, authentication, mobility, and access to external networks. The EPC comprises four main elements: the Mobility Management Entity (MME) handles mobility and authentication, the Serving Gateway (SGW) handles data routing, the Packet Data Network Gateway (PGW) links users to the Internet, and the Home Subscriber Server (HSS) stores user subscription data and authenticates users.

Radio Access Network (RAN) connects mobiles to the core network wirelessly through various base stations. In 3G networks, this is done through the Universal Terrestrial Radio Access Network, consisting of NodeBs and Radio Network Controllers. NodeBs manage the radio communication with devices, and RNCs control handovers between cells. In 4G, E-UTRAN fully integrates the architecture, eliminating RNCs. The advanced eNodeB base station conducts both radio communication and control functions, resulting in relatively faster data transfer rates and lower latency compared to 3G. Mobile network forensics in 3G and 4G environments are highly challenging due to the intricacy of the networks and strong security implementations.

Forensic Challenges in 3G and 4G Networks

Mobile network forensics in 3G and 4G environments pose a lot of challenges mainly because they are complex and have many security features. Some of the main forensic challenges in such networks include encrypted communications, handovers, and data retention laws.

- **Encrypted communication** is a major forensic challenge in both 3G and 4G networks because it protects not only the user's data but also their privacy. In 4G, encryption applies to both data and signaling traffic to prevent eavesdropping and unauthorized access; IPsec (Internet Protocol Security) and AES (Advanced Encryption Standard) are examples of such technologies. This widespread encryption presents significant challenges in forensic analysis. Network traffic is generally captured and decoded in minimal proportions due to the nature of the encrypted data. Due to the inability to obtain decryption keys, investigators can only access metadata, such as IP addresses and timestamps, but not the content of communications. Even upon legal procedures, service providers may not readily comply and hand over decryption keys to access the encrypted content.

- **Handovers and mobility management**: With the 3G and 4G networks, mobile devices frequently hand over to other base stations as users move around. This complicates forensic examinations while allowing for a seamless service leading to uninterrupted connectivity. Hence, investigators have the challenge of correlating information from different sources to come up with the timeline of activities of a suspect. The complexity of handovers can introduce some gaps in data, making it much harder to reconstruct the events. This challenge underscores the need for more sophisticated forensic tools and techniques to analyze and connect the fragmented data to finally put together the comprehensive picture of the suspect's movements and activities.

- **Data retention law:** Various countries and regions have stringent data retention laws, which determine the amount of time a service provider is supposed to preserve sensitive information. Forensic investigators find this problematic because, generally, critical evidence that might be obtained may be deleted according to local law in those places. In 4G networks, data generated during handovers or VoLTE calls is preserved for short times, so action to retrieve evidence must be prompt. In any case, investigators must act quickly to retrieve critical information, which, if not done in time, would be erased, emphasizing the need to collaborate with service providers and understand jurisdiction-specific data retention policies.

Modern mobile communication is actually supported by third-generation and fourth-generation networks, built with backbones from architectures that contain key elements such as the core network and radio access network. However, beyond these networks, several forensic challenges arise, including difficulties in data attribution, cross-jurisdictional legal issues, data encryption, anti-forensic techniques used by attackers, and the volatility of digital evidence, all of which complicate the collection, analysis, and preservation of forensic data. Beyond encryption, handovers, and data retention laws, these laws continue to complicate the gathering and analysis processes of forensic evidence within mobile networks. Forensic investigators must continue to adapt their methods and tools based on the constantly changing tides of mobile technology and then use this emerging methodology in technology to overcome the challenges involved with such technologies to conduct investigations effectively

Mobile Wi-Fi Networks

Mobile Wi-Fi networks have developed into the very core of modern wireless mobile communications, whereby mobile devices connect, whether to the Internet or to local networks, through wireless access points (APs). By nature, these networks are dynamic and operate in environments where devices constantly move from one Wi-Fi zone to another. Such networks make this a very valuable and challenging area for forensic investigations. The coverage of the structure of mobile Wi-Fi networks includes structural components, such as access points (APs), SSIDs, and channels. Data collection is done through packet capture and interception techniques.

Mobile Wi-Fi Network Structure

A mobile Wi-Fi network is similar to any other type of traditional Wi-Fi network, but it is tailored to meet the requirements of mobile devices. These types of mobile devices include smartphones and tablets, as well as other wearables. The infrastructure of the network comprises a number of integral parts that manage communication between devices and the Internet or other network entities. In a Wi-Fi network, several key components work together to enable wireless communication between devices and the Internet. Understanding these components is essential for analyzing network performance and identifying potential security vulnerabilities. The primary components of a Wi-Fi network include

- **Access Points (APs)**

    Without a doubt, the Access Point (AP) is an indispensable part of any mobile network, as well as in any Wi-Fi network. The access point acts as a bridge between wireless devices and the wired backbone of the network. In mobile environments, it is common to have multiple APs placed strategically to ensure continued coverage when users move from one location to another.

    a) **Infrastructure Mode vs. Ad hoc Mode**: Mobile Wi-Fi networks often operate in infrastructure mode by allowing APs to connect mobile devices to a larger wired network. However, in ad hoc mode, devices can communicate with one another in the absence of an AP. Infrastructure mode is the usual type of implementation in mobile Wi-Fi networks deployed in public places, such as cafes, airport hotspots, and mobile hotspots.

    b) **Roaming and Handoffs**: Mobile Wi-Fi networks support roaming, which is the ability of a mobile device to roam between different APs without losing its connection. The term handoff usually refers to this process. It is an important design requirement for mobile Wi-Fi networks that allows for transparent connectivity. Handoff is an area important for forensic analysis because investigators often need to investigate the movement of a device and the flow of data between APs.

- **Service Set Identifier (SSID)**

    The SSID is the name of a Wi-Fi network and is used for distinguishing one network from another. Each AP broadcasts its SSID; mobile devices recognize and connect to it in its presence.

In fact, a typical mobile Wi-Fi network would include multiple APs broadcasting the same SSID to allow seamless roaming.

a) **SSID Broadcast**: APs send beacon frames, which contain a period for sending these frames and contain the SSID, periodically. The beacon frames thus enable mobile devices to discover available networks. Such beacon frames also present information on the network's capabilities and details on support with speed, supported encryption protocols, and radio channel usage.

b) **SSID Conflicts and Rogue APs**: There may exist several networks with the same SSID in dense mobile Wi-Fi. For example, public Wi-Fi providers usually tend to use generic SSIDs across various locations (e.g., "Public_WiFi"). This can lead to confusion or even worse, security risks, as rogue APs tend to mimic legitimate networks and start sniffing sensitive data. Therefore, such rogue APs have always been very important to investigate in Wi-Fi forensics, as an attacker might perform Evil Twin attacks to capture sensitive data.

- **Channels**

Wi-Fi networks operate over specific channels in allocated frequency bands. Mobile Wi-Fi networks typically use the 2.4 GHz and 5 GHz frequency bands, with newer devices also utilizing the 6 GHz band introduced in Wi-Fi 6E for enhanced performance, each divided into several different channels.

a) **Channel Selection and Interference**: In the 2.4 GHz band, there are typically 11 to 13 overlapping channels, which can lead to interference in densely populated areas. The 5 GHz band offers a greater number of non-overlapping channels, reducing interference and improving performance. More recently, the introduction of the 6 GHz band—used in Wi-Fi 6E and supported by newer mobile devices—has significantly expanded the available spectrum, offering even more non-overlapping channels and enabling faster speeds, lower latency, and improved overall efficiency in high-density environments, hence even less interference and better performance. Mobile Wi-Fi networks have to intelligently select the right channels to avoid congestion and optimize network performance.

b) **Channel Hopping and Forensic Implications**: Some Wi-Fi networks use the channel hopping feature, where devices change channels dynamically to avoid interference or lose the connection. In forensic investigation, this poses a major challenge because packet capture tools required to track those changing channels make capturing relevant data from the system very hard.

Mobile Wi-Fi Traffic and Forensic Collection

This may be the starting point for volumes of traffic that may turn out valuable for forensic investigations. The gathering and analysis of this traffic play a crucial role in determining whether security breaches, unauthorized access, or malicious activities are present. This chapter discusses techniques and tools for capturing and intercepting mobile Wi-Fi traffic for use as forensic evidence.

Packet Capture (PCAP)

Packet capture is the most basic evidence-gathering technique in Wi-Fi network forensics. The process involves intercepting the data packets communicated through the network and then analyzing them to obtain useful information regarding devices and data. Packet Capture Working Steps:

a) **Monitor Mode**: Monitor mode is the ability of the wireless adapter in mobile Wi-Fi networks to capture traffic by putting it into monitor mode, so it listens on all packets at a specified channel, regardless of source or destination.

b) **Tools For Packet Capture**: Some common tools utilized for packet capture in mobile Wi-Fi forensics include Wireshark and Airodump-ng. Wireshark uses a deep graph for an interface by which analysts can view and filter packets in real-time. On the other hand, Airodump-ng is a command-line tool, convenient for gathering Wi-Fi traffic, such as beacon frames, management frames, and data packets.

c) **Captured Data**: It provides all types of data, like management frames, control frames, and data frames. This management frame provides network structure to investigators, containing SSID, AP MAC addresses, encryption protocols. Data frames contain user payloads, such as web traffic and messages, provided the network is not encrypted.

Capture of Wi-Fi Traffic

Intercepting Wi-Fi traffic captures communication between mobile devices and APs is usually done without the target device's knowledge. While useful for forensic purposes, it incurs legal and ethical implications, as interception without authorization may violate privacy laws in many jurisdictions. Techniques for Interception:

a) **Passive Sniffing**: With passive sniffing, the investigator or analyst monitors network traffic by silently listening to all communications on a specific channel without actively interacting with the network. This method allows the capture of unencrypted traffic and observation of network behavior without alerting users or impacting network performance. Common tools used for passive sniffing include **Wireshark**, **tcpdump**, and **Kismet**, which help in collecting and analyzing wireless network data.

b) **De-authentication Attacks**: The most common method to cause a device to reconnect to a given Wi-Fi network is by sending de-authentication frames to the targeted device, allowing investigators to capture the handshake process and decrypt future communications.

c) **Rogue AP and Evil Twin:** Forensic investigators should locate these access points, as rogue APs refer to those bearing the same name as a legitimate network without authorization. An evil twin attack uses a rogue AP with an identical SSID to dupe users into accessing a different network. Investigators can then capture and potentially steal credentials.

Mobile Wi-Fi networks are mobile, convenient, and comfortable, but they impose significant burdens on forensic investigations. It is important to know the configuration of how the network is structured, especially APs, SSIDs, and channels, in order to make efficient traffic monitoring, along with forensic analysis. Investigators must utilize packet capture techniques to gather useful information, but encryption and dynamic changes in mobile devices make these issues even more complicated. With the help of sophisticated forensic equipment and approaches, it is possible to intercept and analyze Wi-Fi traffic, thus discovering critical evidence that might help trace security breaches, unauthorized access, or malicious activity.

Mobile Bluetooth Networks and Forensic Collection

One of the most important enabling technologies for mobile communication is Bluetooth technology, short for short-range wireless connectivity, for a wide array of devices, such as smartphones, tablets, laptops, and wearables. For several years, new Bluetooth standards, like BLE, have started to gain widespread adoption in mobiles due to low power consumption. This section critically analyzes the architecture of mobile Bluetooth networks—particularly Bluetooth Low Energy (BLE)—along with the associated forensic artifacts of Bluetooth communications.

Short Range Wireless Architecture

Bluetooth operates on short distances, between 10 to 100 meters, depending on the class of device, where power output is also very important. It uses the 2.4 GHz ISM band and frequency-hopping spread spectrum (FHSS) to reduce interference from other wireless technologies.

a) **Bluetooth versions and range:** Classic Bluetooth (BR/EDR) supports data-intensive applications like file transfers, audio streaming, and hands-free calling. Traditional Bluetooth is what most people normally call whenever they mention Bluetooth, but it mainly refers to the earlier versions until 4.0. Bluetooth Low Energy (BLE), which was introduced in Bluetooth 4.0, is low power and thus suitable for mobile devices, wearables, and Internet of Things (IoT) uses. Later editions, including Bluetooth 5.0 to 5.3, have carried BLE further still, boosting data transfers, range, and reliability, and energy efficiency—functionalities that are more and more being exploited in contemporary smart appliances. .

b) **Bluetooth Architecture:** Several groups of Bluetooth networks are known as a piconet. A piconet consists of a master and one or more slave devices. The master initiates communications, setting a timing for the request to be transmitted, while slave devices respond to the request from the master. These piconets can aggregate into a larger network, called a scatternet, as some devices act as intermediaries for different piconets.

c) **Pairing and Bonding:** Devices in a Bluetooth network use pairing to create a secure connection, usually by using PIN codes exchanged between devices or acceptance of a pairing request. Once devices are paired, they form a bond enabling direct reconnection without repeating the pairing process.

d) **Bluetooth Low Energy (BLE):** BLE technology dominates mobile Bluetooth connections due to its efficiency and wide support in mobile operating systems, including Android and iOS. It is frequently used in fitness trackers, smartwatches, or other applications where device proximity plays a role in services like contact tracing or location beacons.

e) **GATT Protocol:** BLE devices use GATT to communicate, arranging data into services and characteristics. Each service is like a function, such as heart rate monitoring, and characteristics hold the actual values of that data.

f) **Low Power Consumption:** BLE significantly extends battery life because devices do not constantly use energy unless actively transmitting data. Thus, this technology is best suited for mobile devices and IoT.

Identifying Mobile Bluetooth Forensic Artifacts

Bluetooth communications leave various forensic artifacts on devices and networks, which are crucial as evidence in investigations involving device connections or proximity-based interactions and data transfers.

a) **Paired Device Logs:** Smartphones keep logs of Bluetooth-enabled devices paired with the phone, containing names and MAC addresses associated with interactions, including timestamps.

b) **Connection History**: Log files consist of active and inactive connection information, timestamps, and data transferred during communication. This helps determine devices near the target device at any point.

c) **Bluetooth MAC Addresses**: Each device has a MAC address that can be sniffed from wireless traffic or extracted from logs. This helps investigators correlate devices with suspicious activity.

d) **MAC Address Spoofing**: Attackers can spoof MAC addresses to conceal devices. Forensic analysis seeks patterns of spoofed addresses or connection behaviors.

e) **GATT Data and BLE Characteristics**: With BLE devices, analysts can fetch GATT data, relating services and characteristics of connected devices. This may include sensitive information like fitness data, location history, or health metrics.

f) **Proximity-Based Evidence:** Proximity to a Bluetooth device can be critical evidence. Contact tracing apps use BLE to detect proximity between devices. Logs provide investigators with understanding of movement and proximity.

With the advent of BLE, mobile Bluetooth networks have become a ubiquitous part of day-to-day mobile communications. Forensic analysis of Bluetooth artifacts, such as paired device logs, MAC addresses, and GATT data, can provide incisive evidence in mobile investigations. Understanding how Bluetooth works, how it is used within its short-range wireless architecture, and the types of artifacts it leaves behind enables forensic analysts to track connections, data transfers, and sometimes even the proximity of suspects or devices in criminal cases.

Capturing Network Traffic from Mobile Devices and Infrastructure

Mobile device network traffic can be captured via a variety of technical methods and tools, each suited for different environments and access levels—Wi-Fi and cellular networks. What follows is an overview of the most commonly employed techniques:

CHAPTER 8   MOBILE DEVICE NETWORK AND FORENSICS

Packet Sniffer

Packet sniffing is a process that intercepts and analyzes packets transmitted via a network. It provides visibility into raw data exchange between a device and the network through flow information understanding. The process can be done with the help of software-based or hardware tools and may vary depending on the type of network being observed, whether it is a Wi-Fi or cellular network.

On Wi-Fi Networks: Capturing traffic is pretty easy when a mobile device connects to a Wi-Fi network, as the open nature of wireless communication lends itself to interception. The task is done by setting up a network interface card (NIC) on a laptop or another device in one of the modes listed below:

a) **Promiscuous Mode:** In this mode, the NIC captures all packets of a network segment, not only those addressed to the monitoring device. This is mostly used in wired networks' sniffing but can also be used in some setups over wireless networks. Tools such as Wireshark, Tcpdump, and Airodump-ng are frequently used.

b) **Monitor Mode:** In wireless, the NIC can be set to monitor traffic on a channel without source or destination. This mode is extremely useful for catching communications between other devices on the network, and it is the common method used for Wi-Fi sniffing in applications such as Airodump-ng.

On Cellular Networks (3G/4G): Capturing cellular network traffic is much harder since it forms a centralized and encrypted environment, compared to local, decentralized settings of Wi-Fi networks. It is routed through the core network infrastructure of the mobile carrier, incorporating various layers of encryption, as well as protocol-specific communication.

a) **Encryption Protocols:** Cellular networks use complex encryption algorithms, such as A5/3 and Snow 3G, to shield voice, messaging, and data. Because encryption is fully maintained end-to-end by the carrier, it becomes extremely hard to capture or decrypt cellular traffic without special hardware.

b) **IMSI Catchers and Stingrays:** Counterinsurgency agencies obtain specialized IMSI catchers, or stingrays, to tap into cellular traffic. These tools act like a cell tower and cause nearby mobile devices to start connecting to them. After connecting, the IMSI catcher can intercept and analyze streams of data coming from a particular device. However, these devices are strictly regulated by very strict regulatory laws, and an individual needs proper legal authorization before using them. Moreover, modern cellular standards, such as 4G and 5G, enforce higher security standards, making it hard to tap in using such methods.

c) **SIM Cards and Tracking**: A SIM card (Subscriber Identity Module) contains essential identification, such as IMSI, which can trace and intercept communications if proper hardware is available. It also makes traffic interception through the secure encryption of the SIM card difficult, because the decryption keys required are within the SIM itself and cannot be easily accessed without the carrier's cooperation.

Direct Capture on Mobile Devices: Another method is capturing directly from the mobile device itself. To bypass the in-built security mechanisms on the mobile operating system, elevated privileges are necessary.

a) **Rooting or Jailbreaking:** In order to have low-level access to the mobile network stack, both Android and iOS devices should be rooted and jailbroken, respectively. These two processes involve evading manufacturer-set security to allow third-party software installation, which can capture and analyze network traffic. Packet capture can be done on rooted Android devices by applying tPacketCapture or Shark for Root. Although it's a more complex process, Burp Suite can intercept traffic with proper configurations on iOS devices.

b) **VPN-based Capture:** This is a more complex procedure for acquiring mobile device traffic without rooting or jailbreaking the device. All traffic will be routed through the VPN server. By routing all traffic through the VPN server, the server can capture and store all packets for analysis. This is helpful because it won't require changes to the mobile device's operating system; encryption can be managed at the VPN server level, making traffic analysis much easier.

c) **TLS/SSL Interception:** When using VPN-based capture or any proxy tool, such as Burp Suite, TLS/SSL traffic may be decrypted if a custom CA is installed within the capture environment. In other words, encrypted traffic, such as HTTPS, could be intercepted without either the user or server knowing, making this a potent technique for granular traffic analysis.

Network Tap Devices

A Test Access Point, or network tap, is a passive type of hardware that makes a direct copy of the data passed between two points in a network. These devices are critically important for monitoring and troubleshooting communications on a network in real time without interfering with the normal flow of data. Network taps can be installed in a number of different wired and wireless environments; their application reaches every point of both traditional and mobile network infrastructures. A few network taps can be installed at strategic points for analyzing traffic throughput within the infrastructure. The architectures of modern mobile networks are quite complex and multilayered, both LTE and 5G. A network tap can also be used in these systems for tapping traffic at all vital junctures to gain complete insight into interactions involving mobiles and other network components.

a) **Evolved Packet Core**: EPC is an advanced LTE network architecture. The Evolved Packet Core controls data sessions, user authentication, and mobility. Three primary components in EPC—including the Serving Gateway (SGW), Packet Data Network Gateway (PGW), and Mobility Management Entity (MME)—are involved in other processes in data flow from mobile devices and the external network.

b) **Serving Gateway (SGW):** The SGW carries out the routing of packets to be forwarded from the mobile device to the external packet data network (PDN). The SGW manages user mobility as well as acts as an anchor point of data connections locally once the device moves between various eNodeBs (base stations). One can install a network tap between the SGW and the eNodeB to intercept all user traffic for data analysis related to handover processes, session management, and data forwarding.

c) **Packet Data Network Gateway (PGW):** The PGW provides connectivity to the mobile network with external IP-based networks, such as the Internet or private enterprise networks. It handles QoS enforcement, packet filtering, and allocation of IP addresses for devices. Installing a network tap in the PGW allows outbound and inbound traffic to be captured, necessary to monitor data usage, track traffic patterns, and catch potential security breaches or performance bottlenecks.

d) **S1 and S5 Interfaces**: Interfaces S1 and S5 of LTE networks link multiple EPC elements, such as between MME and SGW through the S1 interface or SGW and PGW through the S5 interface. Both user plane and control plane traffic are carried over these interfaces. Network taps may be installed at these interfaces, where both signaling and user data traffic can be captured, including session management and authentication messages and Internet browsing or application data.

Tools and Software for Traffic Capture

There are numerous software and hardware open-source tools for capturing mobile network traffic in various repositories or their official website. The most useful tools have been shortlisted and described below:

a) **Wireshark**: One of the most popular packet sniffing tools, used to capture and analyze traffic in any wired and wireless network. It can be used in conjunction with Aircrack-ng and Airodump-ng to capture Wi-Fi traffic and then decrypt it if the PSK is known.

b) **Tcpdump**: A command-line packet analyzer that captures and analyzes TCP/IP traffic. Proving quite handy, Tcpdump can capture raw traffic in both Wi-Fi and cellular networks with the right network interfaces.

c) **tPacketCapture**: An Android app that can record network traffic without requiring root access via a local VPN. tPacketCapture proves useful in an investigator's arsenal for capturing traffic from individual devices in user space without compromising device security.

d) **MITM Proxy/Burp Suite:** These interceptors become middlemen in capturing traffic in HTTPS sessions, creating a man-in-the-middle (MITM) environment. A proxy certificate can be issued to decrypt and inspect HTTPS traffic from mobile applications.

Capturing Infrastructure-Level Traffic

Traffic capture in mobile infrastructure refers to the interception and monitoring of data in transit between mobile devices and a further network, followed by its analysis. This usually occurs in enterprise, ISP, or mobile carrier environments, which process and route large amounts of users' data. Monitoring mobile infrastructure's ability to capture traffic necessitates knowledge of the network architecture, the high-level components involved, and the protocols used in managing its communication. Below, we delve into the technical mechanics and challenges of traffic capture in mobile infrastructure environments.

a) **Access Points and Routers:** Capture tools can be installed on enterprise routers or access points to monitor all traffic into and out of the mobile device. A number of routers have built-in traffic monitoring tools that easily capture network data, for example, Cisco NetFlow, Juniper JFlow, etc.

b) **Network Forensic Appliances:** NetWitness, Solera Networks, or other network forensic appliances can capture and store all network traffic for later analysis, mainly in large-scale network environments. These appliances are installed at strategic points in the network (for instance, gateways, firewalls) and automatically capture and index traffic for examination.

c) **Cellular Infrastructure Monitoring:** Service providers can monitor large-scale cellular traffic on 3G/4G networks through services such as NetFlow or sFlow. The EPC infrastructure in LTE networks provides many points where traffic monitoring is possible. For example, capturing at the level of PGW allows monitoring of traffic either entering or leaving the mobile operator's network toward the Internet.

Types of Network Forensic Artifacts in Mobile Devices

Digital forensic artifacts involve pieces of digital evidence that may be collected and analyzed to help investigate criminal activities, breaches, or incidents relating to mobile networks. Sources of artifacts in mobile network forensics include log files from network infrastructure, mobile devices, and subscriber information related to the identity and activities of users. The art of collecting and analyzing these artifacts helps determine the level of an incident and attribute actions to a specific individual or device. Here is a more elaborate breakdown of key types of forensic artifacts and their practical importance in mobile network investigations:

Log Files

Log files are probably one of the most important pieces of forensic evidence in any investigation. Network activity, device-to-device interactions, and a myriad of user behaviors—all may be traced through log files. In the mobile network ecosystem, a mix of devices, including routers, access points, and mobile devices themselves, could generate these logs.

Router and Network Logs: Routers, switches, and network infrastructure devices maintain logs about network activity and communications between devices. This often includes details such as Source and destination IP addresses; Timestamps of events at which communication took place; Number of connection attempts, successes, or failures; Network protocols being utilized (TCP, UDP, and others); Bandwidth usage; any anomalies (such as sudden surges in traffic reflecting attacks). In terms of forensic contribution, these logs can be used to determine when a particular device accessed the network, what type of communication is happening, and what types of abnormal activities were possibly initiated, including possibly unauthorized accesses or DDoS attacks. Investigators can correlate abnormal activity with a particular router through a specific device or user.

Access Point Logs: Access Points (WAPs) create logs of connected devices as well, and they contain information about MAC addresses and times when they are connected or disconnected. Such logs may be useful during forensic investigation work in making inferences about the presence of a given device within a given geographic area through the use of timing information and the prior location of the access point.

Nevertheless, MAC address randomization, the privacy option to change the MAC address of the device at regular intervals, is commonly used in many current mobile devices, particularly in probing and joining Wi-Fi networks. This adds considerable complications to the consistency of identifying a MAC address with a particular device or user, and thus makes tracking of movement or guaranteeing device presence complicated further in forensics.

Mobile Device Logs: By themselves, mobile devices create internal logs that capture a wide array of activities. Examples include, but are not limited to: Call and SMS history, Data usage, App activities, System events and much more. In reality, extracting and analyzing these logs is critical in building a thorough account of a user's actions, communications, and suspicious activities.

IP Address

An IP address is an identifier for each device on a network. Thus, IP addresses play a very significant role in the conception of network forensics. However, most mobile networks use dynamic IP address assignments and will frequently change the assigned IP address for devices as they move between different parts of the network or reconnect.

IP Address Assignment Logs: There will be IP address assignment logs maintained by mobile network operators, which will map the IP addresses used by mobile devices back to their corresponding IMSI or IMEI at particular points in time. It is crucial because dynamic IP addresses may change continuously. Investigators must have a correlation of specific acts to a particular device.

NAT and CGNAT Correlation: Most mobile networks employ NAT and CGNAT just for IP address conservation in a technique adopted by the carrier's network for mapping a public IP to more than one device or user behind a NAT. Thus, more than one device may be allowed to share a given IP address such that it cannot be easily determined which one actually performed those activities. Investigators might have to rely on the mobile network operator's logs to trace back through the NAT to the originating device. These logs have port numbers used during the NAT process, and investigators can correlate with IP address assignment logs, establishing specific traffic to a specific device.

Forensic Use: IP address correlation forms an important forensic step in identifying the actual devices involved in malicious activities, such as illegal website access, malware distribution, or cyberattacks. Without accurate IP-to-device mapping, one cannot point fingers at any malicious activity.

Subscriber Identity Module (SIM) and International Mobile Subscriber Identity (IMSI) Tracing

A Subscriber Identity Module and its IMSI constitute unique parts which uniquely identify the subscriber of mobile networks. The SIM contains vital information about the subscriber, where one specific mobile apparatus is connected with a user account at the operator of a mobile network. The IMSI constitutes an exclusive number identifying the subscriber in a mobile network.

SIM and IMSI Fundamentals: The IMSI is stored in each SIM card that links a cell device to the network. Upon connection of a device to the network, the IMSI is sent from the device to the network for authentication of the subscriber to access services. IMSI enables tracking by the network of the subscriber's activity—not only calls and messages but also data usage, among many others.

IMSI Tracing: IMSI tracing tracks the location and activities of a specific mobile device in a forensic investigation. Various mobile network operators keep logs of IMSIs; that is, the cell towers or eNodeBs in LTE technology that a device has contacted over time. If one could extract these logs, one could build a timeline of where the device was. This would assist in tracing a suspect, building alibis, or tracing an accomplice. IMSI catchers, commonly known as "stingrays," are devices that can disguise themselves as a cell tower to trick nearby mobile phones into connecting to them. Law enforcement agencies have been known to use these devices to intercept IMSIs and gain information about those around in real time.

SIM Cloning and Fraud: This is the creation of a duplicate SIM, which holds the same IMSI as that of the original. It is mostly utilized by thieves to intercept communications or impersonate the original users. Forensic investigators must be aware of SIM cloning in order to visualize how it may obscure the identity of the user connected with an investigation. Indeed, the service provider will be in an excellent position to compare network logs found at the mobile network operator with the physical SIM card in hand as a way of detecting such fraud.

SIM Card Analysis Through Forensic Device Exams: Data stored, including details of phone numbers and SMS messages, can be retrieved by conducting a physical examination on SIM cards. Data extraction tools, such as SIM card readers, can extract some of this data for further analysis to identify activities and connections in the network. By correlating the IMSI found on the SIM card, a conclusion can be drawn through comparisons with network logs indicating the existence of the user at a particular time and date.

Forensic artifacts in mobile network investigations are important pieces of evidence to uncover criminal or suspicious activities. Network devices, like routers and access points, maintain logs regarding their own activities. Logs from these devices contain the most information about network traffic and network connections—IP resolution helps associate activity with certain devices and tying actions with individual users, using IMSI tracing and SIM card analysis. Through such analysis, systematic analysts can construct comprehensive timelines, associate actions to devices and users, and construct concrete cases of legal precedents. Such processes, however, sometimes have to engage mobile network operators and should align with technical protocols to ensure evidence collection follows the right legal channels in court.

Evidence Preservation

Mobile network forensics should preserve evidence so that volatile data does not lose its integrity and can be used for analysis and jurisprudence purposes. Recommended practices to preserve volatile network data are as follows:

Post-incident Data Collection: Upon learning of the incident, immediately use network traffic capture tools like Wireshark or Tcpdump to collect the respective volatile network data before it is lost due to system reboots or network changes.

Preservation in Read-Only Format: Document information gathered in a read-only format, which cannot be altered, such as the .pcap file format. Remove external storage devices to prevent contamination of live systems.

Documentation of Actions: All data-gathering activities should have detailed records displaying dates and tools used.

Chain of custody is an essential aspect of evidence preservation. It refers to

Documentation: Record all individuals handling the evidence, including the date, time, and purpose of transfer. This ensures accountability and transparency.

Secure Storage: Evidence must be stored in a secure, inaccessible location to prevent interference or unauthorized access.

Verification: Ensure data integrity by routinely computing cryptographic hashes, such as SHA-256. This ensures effective preservation of volatile network data and maintains its reliability for forensic analysis and legal use.

Mobile Network Attacks and Threats

Mobile networks are always under a barrage of various attacks and threats on user devices as well as the core infrastructure. Some of these include man-in-the-middle attacks used in intercepting communication and SIM swapping attacks. Such attacks enable hackers to hijack phone numbers. Vulnerabilities of a signaling system enable attackers to intercept calls or track the user. Attackers can employ a Denial-of-Service attack to overwhelm network resources, making it impossible for a service to be accessed. Rogue base stations, also referred to as IMSI catchers, identify as real towers while intercepting your data. Other types of threats include malware and ransomware targeted at mobile devices, weak encryption, and outdated security practices on the network.

Common Mobile Network Threats

Mobile networks are increasingly targeted by various security threats that can compromise user privacy, data integrity, and system availability. Below are some of the most common threats encountered in modern mobile network environments:

> **Man-in-the-Middle (MITM) Attacks:** A Man-in-the-Middle attack occurs when someone intercepts a communication between two parties without their knowledge. In a mobile network, MITM attacks can occur either due to insecure Wi-Fi connections or because mobile infrastructures were breached. The attacker may then observe, change, or inject malicious data into the communication. For example, in a financial transaction, the attacker could change payment details, forward funds, or steal passwords and credit card numbers. Figure 8-1 [1] gives basic understanding about the MITM attack.

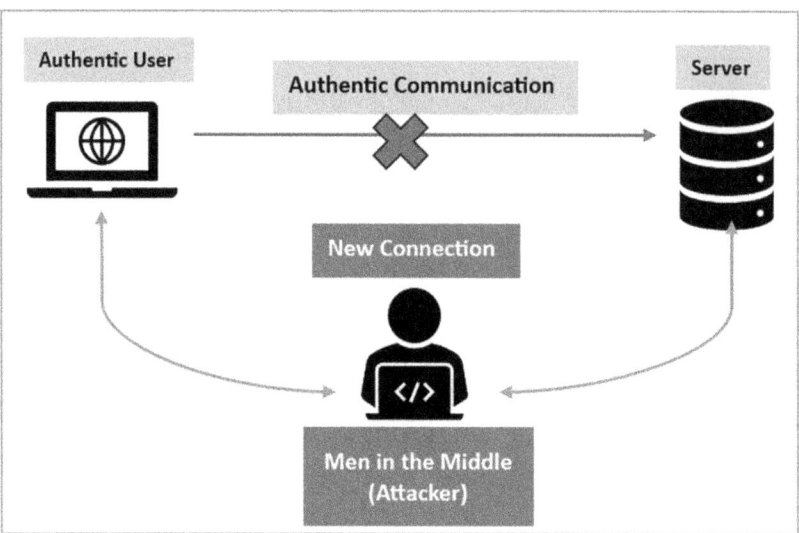

Figure 8-1. *Men-in-the-middle attack*

- **SIM Jacking:** SIM jacking, also known as SIM swapping, is a form of identity theft wherein an attacker persuades a mobile carrier to change a victim's telephone number over to the attacker's SIM card. This enables the attacker to access calls, messages, and authentication codes meant for the victim, thereby enabling the attacker to gain unauthorized access to the victim's accounts, including bank accounts and social media.

- **IMSI Catchers (Stingrays):** IMSI catchers, also branded as "Stingray," are devices that masquerade as cell towers to intercept mobile communications. An IMSI catcher can seize IMSI numbers, along with all phone calls, SMS messages, and data traffic of mobile phones in proximity. Both governments and threat actors use IMSI catchers for surveillance purposes.

Wi-Fi-linked Threats: Wireless devices mainly connect to the Internet using Wi-Fi. This means there are special attacks, especially rogue access points and Evil Twin.

   a) **Rogue Access Point**: Rogue access points are unauthorized or deployed access points by hackers simulating legitimate networks. Most times, victims connect unintentionally to these rogue access points. Hackers can sniff data and attack victims.

b) **Evil Twin Attacks**: This is an advanced rogue access point style in which the attacker creates a rogue access point that looks like any other authenticated one. Once the connection is established, the attacker can capture everything, including login credentials and more sensitive data.

Advanced Persistence Threats (APTs)

Advanced persistent threats (APTs) are complex and long-term cyberattacks that intend to steal data or monitor activity over time. Mobile networks are at high risk from APT threats, mainly considering how important the data is and what it may hold for pervasive surveillance. APT threats are usually well-funded and deployed by orchestrated attacks that can either be nation-state or advanced cybercrime organizations. Below, we discuss how APTs get into mobile networks:

a) **Social Engineering**: APT cyber-attacks initially use social engineering to gain access. Spear-phishing campaigns include various tactics wherein tailored messages trick users into downloading malware or making sensitive information known. Such individuals impersonate trusted entities, such as colleagues or vendors, to increase the likelihood of success.

b) **Exploitation of Known and Unknown Vulnerabilities**: Among the most common types of APTs is the exploitation of known vulnerabilities in mobile operating systems and applications. Attackers can utilize zero-day vulnerabilities, or previously unknown flaws, to bypass whatever security measures are in place. As an example, some applications may enable the user to access the data on the device even without the correct username and password by making use of the vulnerabilities in the mobile web browsers or applications. An example is the Pegasus spyware, which had access to zero-day exploits in Apple iOS, including the WebKit vulnerability, which was used in the "FORCEDENTRY" exploit. This meant that iPhones could be hacked remotely and without any involvement of the user by a malicious iMessage, with complete access to the messages, calls, emails, and even the microphone and the camera of the device. Supply Chain

Attacks: Attackers also obtain illegal access to mobile devices by sabotaging the software supply chain or by making malicious applications available on trusted platforms in other cases. As an example, the 2018 ASUS Live Update Utility was the target of a complex supply chain attack during which fraudsters incorporated malware into genuine updates to software. This allowed attackers to have remote access to thousands of systems, which automatically downloaded a backdoored update without users ever noticing. Malicious apps that are freely distributed through the Google Play Store have also exhibited similar risks, as even with vetting on security, they have caused millions of devices to get infected by being undetected by security mechanisms. Malicious Apps: This approach also entails disguising malware as authentic mobile applications. Attackers will create fake apps similar to popular apps and have users download them. Once installed on a victim's mobile, these apps can send attackers the person's information, track their behavior, and send data to a remote server.

c) **Network Intrusion**: APTs can penetrate mobile networks directly through network weaknesses. That can come in the form of vulnerabilities in infrastructure elements, such as routers, base stations, or other network devices. Once on the network, attackers can then use lateral movement to reach sensitive data.

Case Study: High-Profile Attacks

T-Mobile Data Breach (2021): T-Mobile, one of the best mobile operators in the United States, suffered a major data breach in August 2021, leaving more than 40 million affected customers. A hacker gained unauthorized access to T-Mobile's server by targeting vulnerabilities within the API weaknesses of the network. Names, dates of birth, social security numbers, driver's licenses, and account PINs were accessed. This was a serious breach that primarily implicated the security of personal information, with all its derivable risks, including identity theft. This breach gave T-Mobile massive regulatory scrutiny and reputational damage. In addition to strengthening API security, the firm should ensure that affected customers are notified promptly to mitigate the risks.

Pegasus Spyware (2020): One of the most notorious attacks in mobile networks is the case of Pegasus spyware, which occurred in 2020. In itself, the attack already speaks to the risks of advanced surveillance technology. Conceived by the Israeli company NSO Group, Pegasus is a highly influential tool in the form of software intended to hack and infiltrate mobile devices and extract sensitive information from these devices without the owner's knowledge. It may be used through zero-click exploits where, at times, messages or the device's software may provide openings for the attacker to break into the web application of a messaging app like WhatsApp and the device's software, without any action on the part of the owner. Journalists, activists, and political dissenters were apparently surveyed by governments using Pegasus. The case has raised huge issues concerning privacy and the ethical use of surveillance technology. Among the lessons drawn are ongoing software updates and increased user awareness of what their apps are permitted to do.

Simjacker Attack (2019): The Simjacker attack brought about some basic flaws in SIM cards, and attackers used the attack to track and control mobile devices remotely over some mobile networks. The attack was conducted by sending a specially created SMS which triggers commands on the victim's SIM card, hence giving attackers control over the removal of location data from it. This attack reached users in different countries and brought attention to concerns over the security of mobile infrastructure. The incident generated a new call for stronger SIM card security, compelling carriers to implement stronger measures of protection and continue to monitor and detect suspicious SMS traffic to better deter similar threats.

WhatsApp Spyware Incident (2020): In May 2020, hackers exploited a flaw in WhatsApp to install Pegasus spyware on the devices of targeted individuals. The attack was said to have been made possible because the attackers exploited a VoIP flaw in WhatsApp, allowing them to deploy the malware through missed calls, even when a victim does not answer. This attack targeted human rights activists, journalists, and political dissidents. There were also other issues of government surveillance. The attack raised wide concerns regarding the vulnerability of popular messaging applications. It emphasized the importance of constant software updates, as well as updates to the said application. The need for actual awareness among users concerning app permissions and security updates was also shown.

CHAPTER 8   MOBILE DEVICE NETWORK AND FORENSICS

Incident Response and Investigation

Effective response and investigation to incidents in mobile network security evolving over time mitigate damage, protect sensitive data, and then restore trust in such dangerous scenarios. This section is all about the phases of incident response, techniques for investigating mobile network incidents, tools for forensic investigations, and information about reporting and legal compliance. The phases of incident response and investigation should be followed as mentioned below:

Phase 1—Detection: Detection is the first critical step in the incident response process, with a focus on identifying potential security incidents. In any case, timely detection relates to the effective management of the impact of an attack. One employs various techniques regarding effective detection. Monitoring tools, such as intrusion detection systems and security information and event management systems, analyze network traffic for anomalous activities that may indicate a breach is in progress. Reporting is highly valued, as users can flag suspicious activities, such as unusual account access or unexpected behavior from their device. Also, threat feeds assist organizations in keeping abreast of and preparing to deal with emerging threats or vulnerabilities that may impact their mobile networks.

Phase- 2—Containment: Once an incident is discovered, the crucial response activity involves containment, which is the restriction of the impact of the incident as well as the halt of further damage. Containment in the short term includes, for example, immediate actions in the form of isolation of affected devices or network segments so that the attack spreads no further, and includes the disconnection of compromised devices so as not to allow exfiltration of data. This progresses to long-term containment in the event that one can have a minute study of the situation. During this phase, patches or fixes will be applied to vulnerable systems available with plans set for permanent remediation to finally work on the root cause and securely bring back normalcy.

Phase 3—Eradication and Mitigation: Eradication forms a crucial part of the incident response process, aiming to completely eliminate the root cause of the incident from the affected environment. It is one of the most critical steps in preventing future occurrences of similar breaches. The eradication phase begins by identifying the threat through detailed analysis of logs and forensic data to understand how the compromise occurred and which vulnerabilities were exploited. After identifying the threat, the next step is to try cleaning malware out using the assistance of antivirus and anti-malware tools to eradicate any malicious software running on compromised devices. Finally, vulnerability patching has to be carried out by giving systems security updates and patches, closing any exploited gaps during the attack, and also improving the overall security posture.

Phase 4—Recovery: The Recovery phase aims at rebuilding and restoring affected systems to a normal state such that there is maximum security even as normal operations resume. Rebuilding/Recovery: Rebuilding is often one of the very first steps, with restoration from clean backups and complete eradication of remnants of the attack. This phase will require constant monitoring for anomalies after the restoration process, close observation of systems that may show symptoms of reinfection or lingering threats. Communication will also be effective; stakeholders and users must be involved throughout this recovery process concerning the changes made to improve the security mechanisms used. This is important in promoting trust and promoting transparency as operations return to normal.

Tools for Mobile Network Forensic Investigation

Mobile network forensic investigation applies various tools for analyzing and extracting data within mobile devices and networks. Some of the common tools among others include

a) **Cellebrite UFED**: This is an extraction tool that conducts data extraction and decoding analysis from mobile devices. It supports many devices, allowing the recovery of data on both iOS and Android devices, including deleted files and application data.

b) **Oxygen Forensics Suite**: You receive a comprehensive set of functionalities provided by Oxygen Forensics for mobile data extraction, such as call logs, messages, and other applications. It also has cloud data extraction and analysis functionality.

c) **Paraben's Device Seizure**: Paraben's Device Seizure provides the highest productivity in extracting and analyzing smartphone, tablet, and all other mobile device data. A tool to recover deleted data and reporting tools are also available.

d) **XRY**: XRY is an application from MSAB that extracts data from mobile devices and analyzes it. It is very applicable for both physical and logical extractions and has tools to analyze the extracted data.

e) **FTK Imager**: FTK Imager is primarily a tool for creating images for general forensic purposes; however, it also has utility in mobile forensics to image storage in mobile devices and extract data.

f) **Wireshark**: This network protocol analyzer is crucial for packet capture and analysis of network traffic, which might be used to probe mobile network communications for suspicious activities.

g) **Axiom**: Developed by Magnet Forensics, Axiom is a digital forensic tool that allows investigators to analyze data from various sources, including mobile devices. It comes with powerful data analysis capabilities and rich visualization features.

h) **Sleuth Kit and Autopsy**: These are open-source forensic tools for file system analysis. They can easily be adopted for mobile forensic purposes. Autopsy presents the most user-friendly interface for analyzing data that could be extracted.

i) **NetWitness**: This tool provides high-level network visibility and analysis, capturing network traffic coming from a mobile device and allowing forensic investigators to analyze it for any security incidents.

j) **MSAB XRY Cloud**: Its target is extracting data from cloud services connected to a mobile device. Hence, it allows an investigator to extract data from the cloud.

k) **Magnet Cyber Tools**: Magnet is a company specializing in cyber-related investigations and has a variety of tools to help in cyber research on cloud, network, and mobile devices. Magnet AXIOM, Magnet GRAYKEY (in partnership with Grayshift), and Magnet AUTOMATE allow forensic investigators to gather, examine, and match digital evidence in various sources on one platform. AXIOM is devoted to evidence retrieval and matching in immediate traffic and cloud-based information; GRAYKEY is specialized in accessing and unlocking advanced mobile devices; and AUTOMATE is devoted to workflow orchestration and optimization in laboratory settings. Which tool is used will vary due to the nature of the investigation and its needs and due to the needs of the forensic crime specialist and the requirements of that particular investigation; investigators can be effectively equipped to fit in a variety of forensic situations with a single vendor ecosystem. Burp Suite can also be used to analyze mobile application traffic. It allows investigators to intercept, inspect, and modify the exchanged data between mobile apps and servers.

Summary

This chapter gives a general overview of the field, with considerable emphasis on the particular challenges and complexities that need to be addressed with respect to research on mobile devices and networks. It describes the growing use of mobile technology in everyday life and the spread of crimes and security incidents related to mobiles. Some of the core topics are as follows: types of data that exist on mobile devices, legal and ethical concerns within the process of forensic investigation, and methodologies concerned with collecting and analyzing evidence. This section also highlights the importance of using specialized tools and techniques tailored to the complexities of mobile network forensics. These tools and methods are essential for extracting reliable, consistent, and legally admissible evidence that can withstand scrutiny in a court of law.

Reference

[1] Samad, Abdul. "MITM Man-In-The-Middle Attacks (ARP-Spoofing, Http/Https Sniffing)." *Medium*, 3 Nov. 2021, abdul-samad.medium.com/mitm-man-in-the-middle-attacks-arp-spoofing-http-https-sniffing-dd093050c63. Accessed 27 July 2025.

CHAPTER 9

Mobile Browser Forensics

Introduction to Mobile Browser Forensics

In the age of mobile phones and increased mobility and connectivity, mobile browsers have become the very access to the digital world we treasure. More than ever, people are using mobile web browsers to read emails, shop, stay up-to-date on various social media platforms, and now access business apps, websites, and documents. As result, mobile browser forensics has gained attention as a vital field of study under digital forensics, aimed at detection, extraction, preservation, and examination of web browsers' data that were installed on mobile devices [1].

Overview of Mobile Browser Forensics and Digital Investigation Relevance

Mobile browser forensics is the application of science to the process of recovering digital evidence present in the browser of mobile devices such as smartphones and tablets. These artifacts range from browsing history, cache pages, cookies, downloads, bookmarks, saved session remnants [2], saved credentials and auto-fill data, among others. Analyzing such data systematically, provides useful clues regarding the user behavior, user intent, timeline reconstruction, and the interaction patterns online.

The significance of mobile browser forensics in digital forensics has been rising sharply with the fast adoption of smart phones in our lives. Mobile phones [3], that now contain more data than some of our older PC's, are therefore again at the heart of digital forensic investigations in criminal, civil, and corporate investigations. In addition, with the increasing popularity of web-based cloud applications such as Google Docs, Facebook, and online banking service, mobile web browsers act as the main front-door to the benign and malicious world [4].

In cases including cybercrime and financial fraud as well as insider threats and terrorism, a suspect's browser history can be a kind of virtual map of intent and a guide to a suspect's activities. For example, web search histories may tell what crimes were researched and cached, login pages and cookies may inform which services were accessed [5]. Furthermore, metadata, including timestamps, IPs, and geospatial metadata, can prove very beneficial in materializing timelines, supporting other digital evidence, etc.

Common Mobile Browsers and Platforms

Understanding mobile browser forensics requires familiarity with the dominant mobile platforms and the variety of browser applications available on them. Android and iOS are the two leading mobile operating systems globally, each supporting several browser options with differing architectures, storage systems, and security models.

- **Android Browsers:**
  - **Google Chrome:** The most widely used browser, integrated tightly with Android and Google services. Chrome stores data in SQLite databases and uses sandboxed directories, which are often located under /data/data/com.android.chrome/.
  - **Samsung Internet Browser:** Pre-installed on Samsung devices; based on Chromium but stores data differently.
  - **Firefox for Android:** Uses different storage formats (like places.sqlite) and is less tightly integrated with the OS.
  - **Opera Mini and UC Browser:** Popular in emerging markets, often use cloud acceleration or proxy-based systems, which have unique implications for forensic data availability.
- **iOS Browsers:**
  - **Safari:** The default browser on iPhones and iPads, built on the WebKit engine. Stores data in plist files and uses iOS's native data protection and encryption.

- **Chrome and Firefox for iOS:** Though available, these browsers are required by Apple to use WebKit rather than their native engines, which may influence the type and location of stored artifacts.

Each browser's architecture significantly affects how and where it stores data, which in turn impacts the forensic process [6]. Differences in cache handling, encryption, user session management, and privacy features (such as incognito or private mode) require investigators to tailor their techniques based on the specific browser in use [7].

Importance in Criminal, Civil, and Corporate Investigations

Mobile browser forensics holds immense importance across various investigative domains:

- **Criminal Investigations:**

  Mobile browsers can reveal incriminating searches, access to illegal marketplaces, communication with co-conspirators through web-based messaging apps, or even interactions with dark web resources. In cases of online grooming, child exploitation, or cyberstalking, browser data can corroborate or contradict witness and suspect statements.

- **Civil Litigation:**

  In divorce cases, custody battles, or harassment claims, browsing activity can serve as evidence of misconduct. For example, visits to dating websites, gambling portals, or explicit content sites may influence case outcomes. Browser logs may also reveal breach of contracts or terms of service in employment-related lawsuits.

- **Corporate Investigations:**

  Companies use mobile forensics to detect data exfiltration, intellectual property theft, and non-compliance with corporate Internet usage policies. In Bring Your Own Device (BYOD) environments, analyzing browser usage can reveal access to competitor sites, unsanctioned file transfers, or unauthorized cloud storage access.

Given the breadth of possible scenarios, forensic professionals must treat browser data as one of the most informative and versatile sources of digital evidence. Properly contextualized, browser artifacts can provide narrative threads that are difficult to dispute [8].

Forensic Acquisition Challenges: Sandboxing, Encryption, and Permissions

Despite its critical role, mobile browser forensics is fraught with technical and procedural challenges. Unlike traditional desktop environments, mobile operating systems impose several security layers designed to protect user data from unauthorized access. These same layers often complicate forensic acquisition [9].

1. **Sandboxing:**

    Apps are sandboxed, meaning they function in an enclosed environment which prohibits them from accessing data from any other apps. This is obviously more secure; however, behind the scenes, forensic tools must overcome tough OS controls in order to scrape data from each browser. For example, on Android's data/data directory, without root or specific acquisition tool, is usually inaccessible.

2. **Encryption:**

    Current mobile operating systems impose full-disk encryption or file-based encryption. On iOS, Secure Enclave and the data protection classes together certainly make it hard to brute force data without the right credentials or access tokens. Even Android, especially on newer versions with File-Based Encryption (FBE), gates access to data based on device form (e.g., locked vs. unlocked).

3. **Permissions and App Restrictions:**

    Forensic acquisition access requires super user permissions or device vulnerability to avoid system lockout. Legal and ethical concerns arise when using an exploit for investigative purposes. In addition, app updates often modify the storage layout or integrate new privacy features (e.g., encrypted local storage or self-clearing history) which requires regularly updating the tool and knowledge.

4. **Incognito and Private Modes:**

   While these modes claim not to save history or cookies, forensic evidence often shows residual traces—particularly in temporary files, system logs, or swap areas. However, retrieving such data is complex and often time-sensitive, requiring in-depth knowledge of the browser's inner workings.

5. **DataVolatility:**

   Mobile browsers frequently update their data caches and clean up temporary storage to optimize performance. This makes some artifacts highly volatile and susceptible to being overwritten if not promptly acquired.

To overcome these challenges, forensic professionals rely on a variety of acquisition strategies, including

- Logical acquisition (app data backup)
- File system acquisition (rooted/jailbroken access)
- Physical acquisition (bit-by-bit clone of storage)
- Cloud artifact retrieval (browser sync data from Google or Apple accounts)

Selecting the right strategy requires balancing evidentiary goals, device type, OS version, and legal constraints.

Analysis of Browser Artifacts

The foundation of the forensic investigation left upon a mobile browser is to conduct a thorough analysis of browser artifacts—digital artifacts that were created as a result of users' online activities. These artifacts can be crucial in building browsing histories, times of access, possible use of accounts, and web-based threats [10]. Mobile web browsers, similar to their desktop counterparts, produce an abundance of information that can be stored up locally, and reveal information about the user's intent, action, and digital presence. Knowing types, deposits, volatility, and tools used to extract the artifact is very necessary for investigation.

Types of Browser Artifacts

Mobile browsers produce various categories of artifacts, many of which are structurally similar across browsers but differ in format, storage mechanism, and volatility. The key artifact types include

- **Browsing History:**

  Records of visited URLs, timestamps, visit counts, and page titles. History logs enable chronological reconstruction of user activity and can highlight attempts to access illicit or suspicious content. In many cases, timestamps can help align browser usage with other mobile events such as messages, GPS data, or app usage.

- **Cookies:**

  Small data files stored by websites to retain user preferences, session states, and login details. Cookies can indicate the presence of user sessions on particular services, revealing credentials, user behavior, or tracking elements.

- **Cache Files:**

  Cached pages, images, scripts, and media that can be used to help load faster during the next visit. These files can be extracted by forensics tools to show the web page exactly as it was when visited rather than exactly as it appears at the moment of analysis. This is particularly valuable when it comes to proving web-based content displayed to or opened by a suspect.

- **Saved Credentials and Autofill Data:**

  Consists of usernames, passwords, email addresses, payment information, and other form submissions saved by the browser. Such artifacts are useful in associating a suspect with particular accounts or services, and in following authentication patterns.

- **Downloads and Bookmarks:**

  Downloads can contain case important documents, pictures, and executables. The bookmarks represent the interests or the long-term intent of a user and can indicate a list of visited or trusted domains.

- **Session and Tab Data:**

  Open tabs. Active sessions and recently closed windows usually are not cleared after the browser is closed. They can be employed by investigators to harvest the browsing environment surrounding the event (during or soon after).

- **Indexed Databases and Web Storage:**

  A plethora of web apps do store substantial amounts of data client side (IndexedDB, LocalStorage, etc.) nowadays. Such storage containers may house structured data of any of an app like the social network, banking application, e-commerce application.

Artifact Storage Paths on Android and iOS

Understanding where mobile browsers store these artifacts is crucial for effective acquisition. The locations vary by operating system and browser, and they are subject to access restrictions and sandboxing policies [11].

Android Storage Paths:

Android browsers store data in app-specific directories within the /data/data/ path. These directories are sandboxed and typically require root access for direct extraction. Common storage paths include

- **Chrome:**
  - /data/data/com.android.chrome/
  - History: app_chrome/Default/History (SQLite DB)
  - Cookies: Cookies file
  - Cache: Cache and Code Cache/directories
  - Login data: Login Data (encrypted)

- **Firefox:**
  - /data/data/org.mozilla.firefox/
  - History and bookmarks: places.sqlite
  - Session info: sessionstore.jsonlz4
  - Cookies: cookies.sqlite
- **Opera/UC Browser/Samsung Internet:**

  Typically follow a similar pattern under their respective package directories, with Chrome-based browsers storing artifacts in the Chromium model.

iOS Storage Paths:

iOS apps store data under containerized directories, with unique identifiers per app. Access typically requires jailbreaking or logical extraction via trusted tools.

- **Safari:**
  - History: /private/var/mobile/Library/Safari/History.db
  - Bookmarks: /Bookmarks.db
  - Cookies: /Library/Cookies/Cookies.binarycookies
  - Cache: /Library/Caches/com.apple.WebKit/
- **Chrome for iOS:**
  - Stores data in a path like /private/var/mobile/Containers/Data/Application/[UUID]/Library/Application Support/Google/Chrome/.

It's important to note that iOS utilizes file-based encryption and Data Protection APIs, complicating artifact extraction if the device is locked or inaccessible without credentials.

Volatile vs. Persistent Artifacts

Browser artifacts can be broadly classified based on their volatility—how long they remain accessible and under what conditions they are deleted or overwritten.

Volatile Artifacts:

These are transient in nature and often reside in RAM or temporary storage. Examples include

- Open tabs and session files
- Recently closed tab logs
- Temporary files during active browsing
- Incognito/private browsing data (partially recoverable)

Volatile data may be lost if the browser or device is closed, restarted, or left idle. Forensic tools that support live memory capture can sometimes retrieve this data, although this is more challenging on mobile devices.

Persistent Artifacts:

Creating .ktx files, which are snapshot images made by the system when a user visits a webpage in the Safari browser, is an important but often overlooked forensic artifact on iOS devices. These files are basically cached screenshots of the web content, which means they keep a visual record of what the user saw. This mechanism is not only for static browsing; .ktx files are also made when the user switches between apps or swipes through different Safari tabs, which records the activity's transitional states. Forensic investigators can get these .ktx files and change them into common image formats like PNG. This lets them figure out how the user browsed, switched apps, and even what specific web pages were showing at a certain time. This makes them a useful piece of evidence in digital forensic investigations because they give contextually rich information that adds to browser history or cache analysis, giving a more complete picture of what users are doing and why on iOS devices. Persistent include data written to disk that remains available even after the device reboots or the browser is closed. Examples are

- Browsing history
- Cached media
- Downloaded files
- Cookies and saved login credentials
- Bookmarks and web storage

Persistent artifacts are the primary targets in forensic investigations and can be preserved using disk imaging or logical acquisition methods.

CHAPTER 9   MOBILE BROWSER FORENSICS

Tools and Methods for Extraction

Forensic tools used in mobile browser artifact analysis must support a wide range of devices, OS versions, and browser types. Some key tools and techniques include

- **Cellebrite UFED and Physical Analyzer:**

  Industry-standard tool for mobile forensics, capable of logical, file system, and physical extractions. Offers browser artifact parsing across various apps.

- **Magnet AXIOM:**

  Comprehensive digital investigation suite that extracts and categorizes browser data, visualizes timelines, and correlates artifacts across multiple sources.

- **Oxygen Forensic Detective:**

  Particularly useful for accessing and decoding encrypted or app-specific browser data from Android and iOS.

- **ADB and Rooted Access (Android):**

  For devices where rooting is permissible, Android Debug Bridge can be used to extract entire app directories for manual analysis.

- **iTunes Backup and Jailbreaking (iOS):**

  Logical backups can reveal partial browser artifacts. Jailbreaking allows deeper access, although it must be handled carefully for evidence integrity. A major problem with iOS forensic analysis is that jailbreaking is possible, which lets you get to system-level artifacts more easily. Jailbreaking can give investigators access to files that would otherwise be off-limits, but this option is mostly only available for older iOS devices and older firmware versions. Newer iOS devices have better security systems and get updates more often, which have closed off many of the holes that jailbreaking used to take advantage of. Because of this, jailbreak-based methods aren't always useful in modern investigations. Instead, analysts need to focus more on other ways to get information that work with the latest iOS versions.

CHAPTER 9   MOBILE BROWSER FORENSICS

- **SQLite and plist Viewers:**

    Since many artifacts are stored in .sqlite and .plist formats, specialized viewers are essential for manual data analysis.

In conclusion, the examination of browser evidence constitutes the primary focus of mobile browser forensics. With clear information on what to pull out and where to find and understand meaning, it is quite possible for investigators to recreate digital behavior with accuracy. Subsequent sections will demonstrate that this analysis provides the grounds for threat detection, data recovery, and cross-browser comparisons that serve to bolster investigative findings.

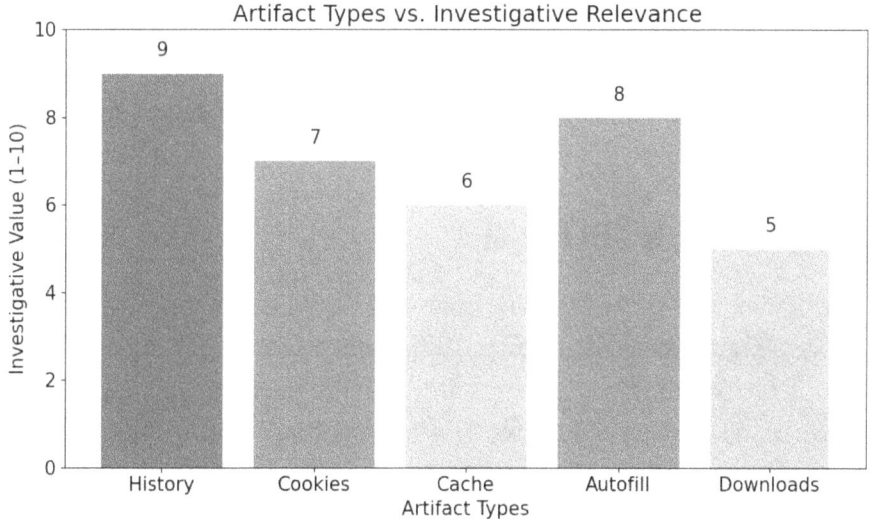

Figure 9-1. *Artifact types vs. investigative relevance*

This bar chart (as shown in Figure 9-1) compares different mobile browser artifact types based on their average investigative value, scored from 1 to 10. History and autofill data rank the highest due to their ability to directly reveal user behavior and intent, while downloads and cache offer moderate forensic insight.

Artifact Location Comparison: Android vs. iOS

Artifact Type	Android Path	iOS Path	Accessibility	
History	/data/data/com.brows	/var/mobile/Container	High	
Cookies	/data/data/com.brows	/var/mobile/Library/Co	Medium	
Cache	/data/data/com.brows	/var/mobile/Library/Ca	Low	
Autofill	/data/data/com.brows	/var/mobile/Library/Pr	Medium	
Downloads	/sdcard/Download	/var/mobile/Media/Do	High	

Figure 9-2. Artifact location comparison: Android vs. iOS

This comparative table (as shown in Figure 9-2) presents typical file paths and accessibility levels for key mobile browser artifacts across Android and iOS platforms. It enables forensic practitioners to quickly reference and locate critical data, accounting for OS-specific storage conventions and permission constraints in investigations.

Recovery of Deleted Data

The retrieval of deleted browser data from mobile devices is an important aspect of digital forensics. Although active artifacts can provide quite a lot of information about a user's browsing history, deleted remnants are often the most revealing—particularly for a user that has tried to cover their tracks. This chapter discusses technical methods for retrieving these data, the challenges for incognito browsing or history clearing, and the role of the underlying file system on data recoverability.

Techniques for Browser Data Recovery

Recovering deleted data from mobile browsers involves understanding how different browsers manage, store, and delete user information. Three primary techniques have become standard in modern forensic practices: SQLite recovery, file carving, and analysis of write-ahead logs (WAL) and journal files.

SQLite Recovery

Mobile browsers—especially on Android—also frequently use SQLite databases to retain structured data (e.g., history, cookies, and bookmarks). In SQLite, when you delete a row, that isn't necessarily immediately getting deleted from the disk. Instead, the entry is marked for removal and the space is recycled for reuse. This makes it possible to undelete rows using forensic SQLite analyzers or hex editors [12].

The erased records might still be present on the unallocated space of the SQLite database file. The bones can be scanned by specialized devices that can recreate partially or totally erased records. Some forensic tools (Magnet AXIOM and Oxygen Forensic Detective [8]) come with modules that are built to parse and recover deleted SQLite records (records that are no longer shown through the browser interface can be viewed by the investigators).

File Carving

The low level file carving technique recovers lost or deleted data by looking for known file signatures in raw disk images and memory dumps. Cached fragments or complete copies may still be on disk, even if the browser deleted the data and the filesystem lost track of the file, in unallocated and slack space [13].

The investigators can use tools like Scalpel or Autopsy to form specific custom carving rules for browser-related file types, such as cached images, HTML documents, SQLite databases, and cookie files. The method works well to storages that store binary objects or media, but additional benefits are its ability to find a deleted entire database or session file that was once related to the storage file structure in the browser.

WAL and Journal Analysis

SQLite databases in recent versions are used with one of two mechanisms in order to ensure that the database contents are consistent: Write-ahead log (WAL) or rollback journaling (RB). These files are used to remember uncommitted or backup snapshots but are not necessary to apply changes to the main database file [14].

There are likely commands in WAL files that never made it to the primary database, but were saved as a result of crashes, forced-shutdowns, or partial-deletions. Likewise, journal files may have old versions for database rows before they were modified or deleted. These files can be extracted and processed to show missing rows that are deleted from the current database files, providing a good source of recovered data.

Challenges in Recovering Incognito or Cleared History

Browsing in incognito mode, or private mode, is geared to circumvent digital document trail. For these modes, mobile browsers will inhibit artifacts such as history, cookies, form data, and cache from being written to disk. This creates important forensic challenges [15].

And while some browsers promise you won't be able to trace their movements in a private mode, there is often leftover evidence in specific circumstances. Temporary file or cache could be still created during the session, when the browser communicates with third party libraries, downloads and caches contents, or creates logs without falling back to being fully controlled. In addition, it has been observed that mobile operating systems and browser engines do not necessarily clear memory or cache spaces completely between sessions, and may have glitches leading to fragment contamination [16][17].

For the retrieval of incognito session artifacts, forensic tools focus on the analysis on the volatile memory, such that they scan the RAM snapshot of the device to identify all the in-use files or the open database handles. On mobile, acquiring memory is challenging because of security considerations and due to the volatility of RAM contents.

Cleared history poses a separate issue. Although deletions are generally restricted to the visible records in the SQLite database table, physical data can still exist on disk until it's replaced. The knock against the development has been one of timing: If the deletion has occurred recently, forensic tools are able to retrieve the entries. If the time lag is too large, the same disk sectors may have meanwhile been reallocated for different data so that the artifacts no longer exist.

File System Behavior and Data Remanence

The behavior of mobile file system is an important factor in the success of recovery of deleted browser data. The storage layout on Android and iOS differ significantly, and even the corresponding deletion method, which will have effects on forensic significance of evidence.

On Android, which uses the ext4 file system almost exclusively, a delete marks a file's inode (a structure in which its metadata is stored) as available for reuse, but does not overwrite and erase the actual data blocks themselves. This provides file carving and unallocated space examination. Though new Android iterations and versions have added more protection in the form of file-based encryption and scoped storage, these obsessive measures could restrict direct access to browser folders or hide deleted data from direct view without root.

On iOS, the Apple File System (APFS) focuses on file system performance, such as cloning files and freeing up space on an APFS-formatted disk. When a file is deleted, in many situations, the underlying blocks are immediately reused, either overwriting or encrypting them, making recovery much harder. Also, encrypted files maybe inaccessible by a device that is not unlocked and not awake, for example, protected by a sleep mode or similar, due to data guard classes. These prevent the recovery of deleted Safari or Chrome for iOS data unless a complete file system image is known, which is not available through jailbreaking.

Another important consideration is the data remanence, which refers to the residual memory of the storage media that remains after the entry was deleted. Solid state storage, which is now standard in nearly all mobile devices, makes this worse with wear-leveling algorithms. These algorithms try to spread the writes and erasures over the storage in order to increase lifetime; however, they also cause non-linear overwriting of data. Accordingly, deleted data may be retained longer or irreversibly destroyed by block erasure.

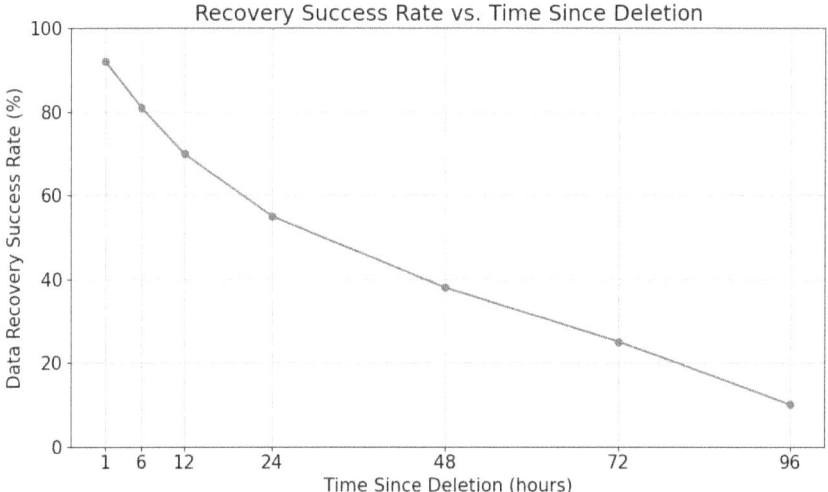

Figure 9-3. Recovery success rate vs. time since deletion

Figure 9-3 shows how the time that has passed since data deletion affects the chance of successful recovery. It shows how important it is to act quickly in forensic cases. The curve shows that the chances of recovery are very high at first—almost 90% in the first hour after deletion—but they drop quickly over time, falling below 40% after 48 hours and getting closer to 10% by the 96-hour mark. These numbers usually come from controlled tests in which files of different types (like text documents, images, and logs) were deleted and recovery tools were used at different times to see how well they worked. The relevance in real-world forensic situations comes from the fact that new system activity gradually overwrites deleted data, making it harder to recover over time. The exact percentages may change based on the device, file system, workload, and recovery tool used, but the trend always shows that getting evidence as soon as possible is the best way to keep it safe. For investigators, this shows how important it is to act quickly after suspected tampering, deletion, or cyber incidents to make sure that important artifacts are kept safe before they are lost forever.

To make things clear and not too simple, it's best to show how to recover deleted data in a structured way instead of using abstract visualizations. The three most common methods—File Carving, SQLite WAL, and Logical Backup—were chosen because they are all different types of forensic techniques: raw storage-level recovery, application-level recovery, and system-level recovery. Each method has its own pros and cons, and how well they work depends a lot on the version of Android that is being used. The table below shows a detailed comparison of important areas like process, strengths, weaknesses, OS dependency, and real-world relevance.

CHAPTER 9　MOBILE BROWSER FORENSICS

Table 9-1. *Comparison of Deleted Data Recovery Methods*

Recovery method	Process	Strengths	Limitations	OS dependency	Real-world relevance
File carving	Scans raw storage for file signatures and reconstructs files	High completeness, works even if metadata is lost	Slow, fragmented files may be incomplete, not efficient on encrypted storage	Impacted by modern encryption in Android 15/16; less effective	Useful in older devices or unencrypted systems, limited today
SQLite WAL	Extracts residual data from SQLite Write-Ahead Logs	High success rate for app-related data (e.g., chats, browser history)	Limited to SQLite-based apps, availability depends on OS version	Affected by database changes in newer Android versions	Still relevant for app-level forensic analysis, but diminishing with updates
Logical backup	Uses device's built-in backup/export features	Fast, reliable for accessible files, minimal technical expertise needed	Restricted by OS encryption and permissions, low completeness	Heavily constrained by scoped storage and encryption in Android 15/16	Best for quick extractions in live cases, less useful for deep recovery

Parameter settings: Time intervals: 1–96 hours; simulated recovery method: SQLite file carving and journal parsing; environment: Android 11 filesystem simulation; data type: browsing history.

249

CHAPTER 9 MOBILE BROWSER FORENSICS

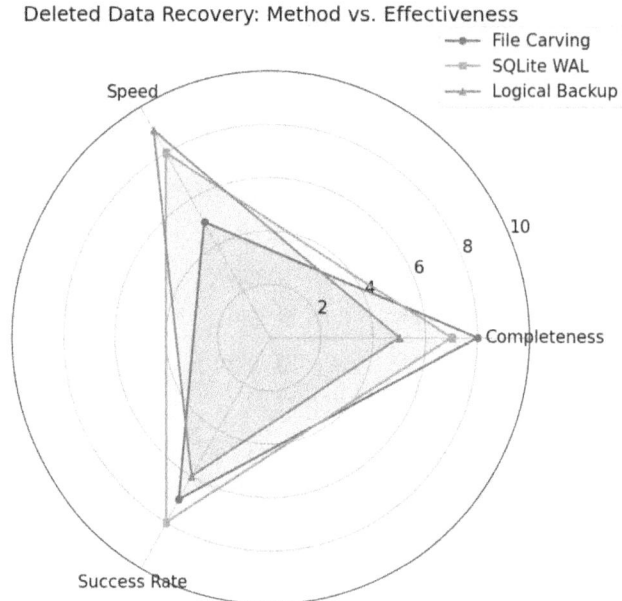

Figure 9-4. Deleted data recovery: method vs. effectiveness

This radar chart (as shown in Figure 9-4) compares three browser data recovery techniques across three performance metrics. SQLite WAL recovery shows the best balance of speed and success, while file carving yields the most complete results but is slower. Logical backup excels in speed but has lower success and completeness.

Parameter settings: Rating scale: 1 to 10; methods compared: 3; metrics: Completeness, Speed, Success Rate; environment: mobile filesystem emulation on Android 10/11; recovery simulations performed using standard forensic tools.

Identification of Web-Based Threats

It does more than enable viewing—it serves as a conduit through which multifaceted cyber threats spread into the personal and business environments. In mobile browser forensics, it is essential to determine and reconstruct web-based threats to understand how an attack was initiated, what vectors were used, and whether data was exfiltrated. This chapter investigates type of attacks from web-based threats; how high- and low-level artifacts are used as indicators of compromise (IoCs) in order to detect and correlate the attacks in a forensic timeline.

Mobile Browser Threat Vectors

Mobile browser attack vectors from phishing and fraud to man-in-the-middle and malware distribution, modern cyberattacks have found proliferating abuse vectors in the mobile browser. The most frequent web-based threats include malicious redirects, phishing sites, and drive-by downloads.

Malicious redirects are misleading redirects and forwards to spam sites in compromised or malicious sites. If a user browses to an infected website, they will be redirected to another site that contains malware. These kinds of redirects can leverage the vulnerabilities of the mobile browser's rendering engine or can work based on social engineering—they can fool, or force, the victim into clicking a deceptive link.

Mobile is a perfect target for phishing due to small screen real estate and the absence of full URLs. For example, a fake banking or login site could be presented, and a user, not being able to scrutinize the address bar properly, could fill in their details. This danger is heightened by mobile browsers that save your personal information with autofill.

Drive-by attacks happen when a good and trusty mobile web browser is deceived into downloading and running a number of malicious scripts without the knowledge of the owner. Such attacks typically use exploit kits that are built to exploit browser flaws or outdated Adobe Flash versions. For Android, drive-by payloads could be used to prompt the downloading of APKs; for iOS, users could be tricked into clicking links that exploit WebKit vulnerabilities and run code inside the browser sandbox.

Those same threat vectors also leave subtle (but key) evidence behind in browser artifacts—but only if they are examined in a timely and informed manner.

Artifact-Based Threat Detection

The forensic analysis under mobile browser attacks is fixed on discovering certain patterns and remnants of traces. URLs, cache files, JavaScript logs, redirect chains, download histories, etc., are all carefully examined to reverse engineer the timeline of events which led to compromise.

The URL patterns are among the simplest and yet most powerful signals. Regular visits to domains blacklisted for phishing kit hosting, or malware uploading, would send immediate red flags up the pole. Dodgy redirect chains are another forensic sign. By analyzing browser history or network logs, investigators can follow how a user was redirected from a legitimate starting domain to a nefarious ending domain. Redirects

CHAPTER 9   MOBILE BROWSER FORENSICS

with ad (or tracking) domain and/or multiple redirects with unrelated domains are a bad sign, as they tend to mask referrer chain and lead to a suspicious final URL. By matching these with timestamps, user actions, and page load events, the nature of the threat can be inferred.

Encrypted JS and HTML files present in a cached state might contain encoded/obfuscated scripts were also used in the attack. These scripts can be reverse-engineered or executed in sandboxed environment to determine their purpose (credential theft, redirection, download).

Indicators of Compromise (IoCs)

IoCs are the physical tracks that are the result of nefarious actions. For mobile browser forensics, these signs are the correspondences and variations of URLs, IP addresses, cookies exceptions, changes in user-agent, downloads, and DNS query.

One such IoC is having a blacklisted/already reported malicious URL on the browsing history. Entries like these can be automatically chased down by things like VirusTotal results or domain reputation services. When the visited domain is one of known phishing infrastructure or malware spreading sites, this is a proof of the attack.

Another important IoC (Indicator of Compromise) is to look for strange user-agent strings or browser actions. When the mobile browser starts impersonating other devices or versions, possibly as part of some cloaking, this might be indicative of exploitation or script injection.

Also, the illicit downloads, especially if it's APKs or encrypted payloads, lend credence to drive-by or social engineering. Download directories, permission logs, and file metadata can be examined by investigators to ascertain the validity of such files.

Corrupted cookies or cookies falsification (especially cookies containing encrypted or Base64 payloads) can also be used as tracking beacons or session hijacking means. Such artifacts can frequently link a browser session to a broader command-and-control (C2) effort.

Timeline Correlation of Malicious Behavior

Four of the five payloads have task manager-related classes and a temporal correlation is important to fully comprehend the browser-world threats. A single anomalous event is insufficient to indicate attack, but when viewed as context for the overall system, there is a story to be told.

CHAPTER 9 MOBILE BROWSER FORENSICS

Compiling the browser history in parallel with the timestamps of downloads, logins, and system logs essentially provides a full story. For example, one can visit some suspicious website at 10:03 am, and in the next minute get redirected, and the minute later has downloaded some unknown APK... At 10:06 am the system logs say an app was installed and permissions granted. This schedule demonstrates the success of the drive-by download attack.

To improve visualization, IoCs can be graphed on a timeline. This can tie observations—such as redirects, downloads, and operating system modifications—visually to show a cause–effect relationship between a browser action and the production of a threat.

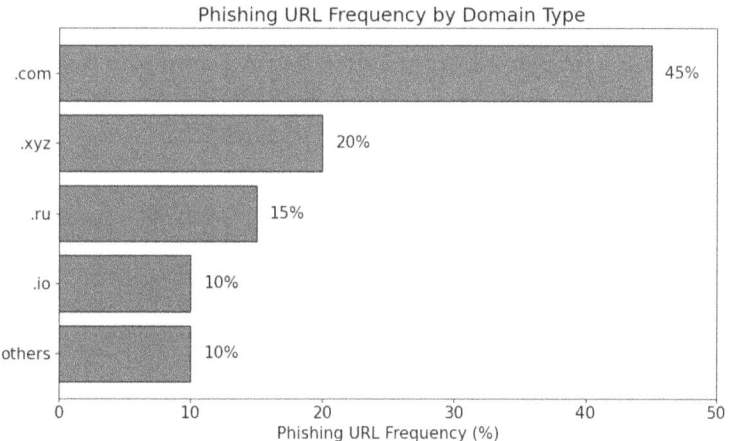

Figure 9-5. *Phishing URL frequency by domain type*

This bar chart (as shown in Figure 9-5) shows the relative frequency of phishing URLs encountered in mobile browser traffic, grouped by top-level domain. Domains like .com and .xyz are commonly abused due to ease of registration and low cost, highlighting patterns useful for forensic alerting and threat intelligence.

We used a domain sample size of 1,000 phishing URLs gathered over 30 days to set the parameters for the experiment. These URLs came from simulated browser traffic logs, which are records of how users interact with potentially harmful websites in a controlled setting. Using a combination of blacklist correlation and redirect analysis, the detection was able to find suspicious domains that tried to get around static blocking methods. To make the results clearer and trustworthy, it's important to say when the logs were collected and where the dataset came from. Letting readers know if the collection period matches recent phishing campaigns or well-known repositories (like PhishTank,

253

OpenPhish, or in-house simulated traffic) will help them understand the data better. This clarification not only makes it easier to reproduce the results, but it also shows how the dataset is useful in the real world when it comes to changing phishing tactics.

Cross-Browser Forensic Analysis

In the fluid environment of mobile browsing, users frequently run multiple browsers on their mobile devices out of preference, functionality, or privacy considerations. This diversity adds a great deal of complexity for forensic examiners, who need to process and interpret data from many different browsers, with different architectures, artifact storage practices, and security implementations. Cross-browser forensic analysis is therefore a crucial part of the forensic investigation of mobile browsers, which can be used to reconstruct a full profile of user behavior as well as to reveal coordinated threats and suspicious activities in a multi-platform setting.

The Need for Multi-browser Correlation

A single browser doesn't often reflect the totality of a user's online footprint. For example, someone might log into personal accounts in Chrome, do work in Firefox Focus and look at vulnerable websites using DuckDuckGo or Safari's "private" mode. And in cases involving criminal activity, cyberstalking, insider threats, and digital fraud, the evidence from a single browser may tell an incomplete or deceptive story.

Cross-browser correlation allows forensic analysts to associate common activity, discrepancies, or even attempts to hide behavior. So for example, someone might visit a dark web site in Firefox and download content, then open that file in Chrome or share it on another site. Through reviewing the histories, downloads, and cached data within various browsers, investigators can follow these interwoven activities and put together a timeline.

Structural and Encryption Differences

Each mobile browser implements its own data storage structure, often with proprietary formats, encryption mechanisms, and update cycles that affect forensic visibility.

- **Google Chrome**, prevalent on Android, stores its history and cache data in SQLite databases located in sandboxed directories (/data/data/com.android.chrome/). While most data is stored in plain SQLite format, recent versions have increasingly incorporated encryption, especially for cookies and saved credentials, complicating direct access without root privileges.

- **Safari**, the default browser on iOS, leverages Apple's Core Data and WebKit-based architecture. Artifacts are typically stored in binary plist files, often compressed or encrypted. Accessing these requires specialized tools and, in many cases, physical acquisition or backup analysis, especially with iOS's strong sandboxing and encryption protections.

- **Mozilla Firefox** uses its own profile structure with multiple SQLite databases (like places.sqlite and webappsstore.sqlite) for storing history and local storage. Unlike Chrome, Firefox profiles may retain more granular site metadata and permission settings, offering richer forensic value when accessible.

The divergence in these formats and protections necessitates customized parsing strategies for each browser. General-purpose forensic tools may support limited browser types, and investigators must often rely on browser-specific scripts or forensic suites with modular plugin support.

Cross-Browser Behavioral Profiling

While architectures can vary, common behavior patterns can appear in user activity across these different browsers, such as regular access to a site, or repeatedly requesting a search term, or overlap between a user's bookmarks. Through this standardizing browser artifacts (timestamp sorting, URL decoding, session metadata, etc.), analysts are able to create behavioral patterns that indicate intent, habits, or unusual activities.

For example if one browser does not report access to a phishing domain but another browser has saved credentials or a cache of relevant files, the merged view can reveal intentional bypass attempts. Additionally, activity synchronization in time among browsers also can bring out browsering session activities, aimed to avoid identification, such as cutting electronic illicit exposure across browsers.

Challenges in Dataset Merging

There are data normalization issues when it comes to cross-browser analysis. Timezones, timestamp formats (Unix timestamps vs. Mac Absolute), varying degrees of metadata granularity, and various policies of retention of FSEvents points all conspire to impede us from easily merging the sets of FSEvents. The block to omniscience is made worse if you've used incognito modes, regularly nuked caches for apps, or used log-free secure browsers as these become data carrots sitting in the data landscape.

Investigators must also face the differences in encryption and protections, which become relevant in permission restrictions (specifically in iOS) and the risk of running into some missing artifacts should the ingested browser be uninstalled or wiped.

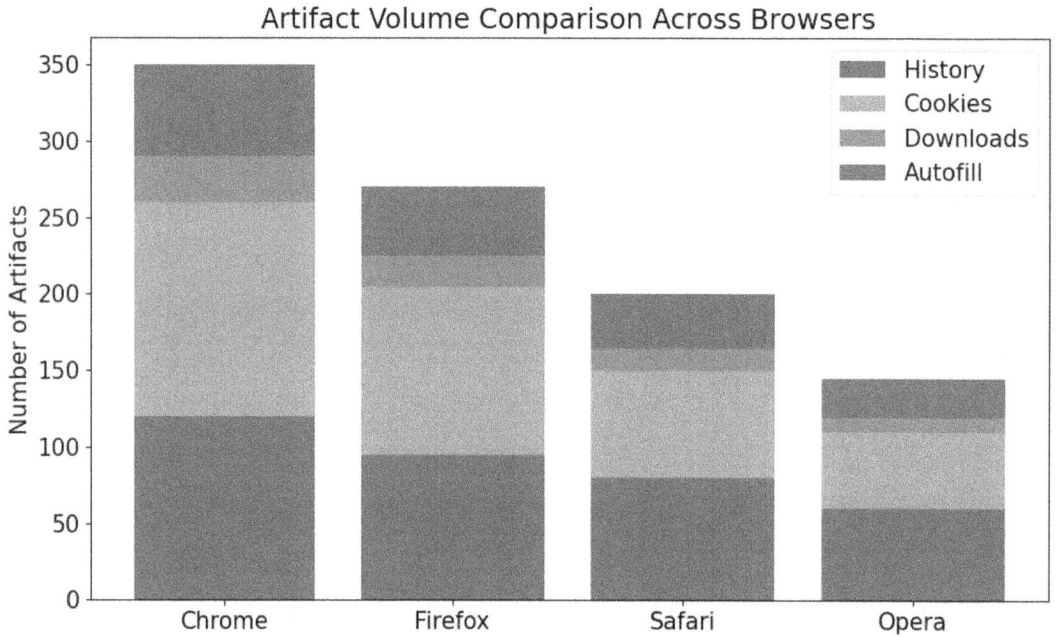

Figure 9-6. *Artifact volume comparison across browsers*

This stacked bar chart visualizes (as shown in Figure 9-6) the distribution of browser artifacts (history, cookies, downloads, and autofill entries) across Chrome, Firefox, Safari, and Opera. Chrome retains the highest volume of forensic data, followed by Firefox, illustrating significant differences in data retention across mobile browser platforms.

Parameter settings: Data source: emulated browser usage sessions; session length: 3 hours per browser; number of visited sites: 50; artifact counts extracted via manual forensic tools and SQLite parsing; environment: Android 11 and iOS 14 testbeds.

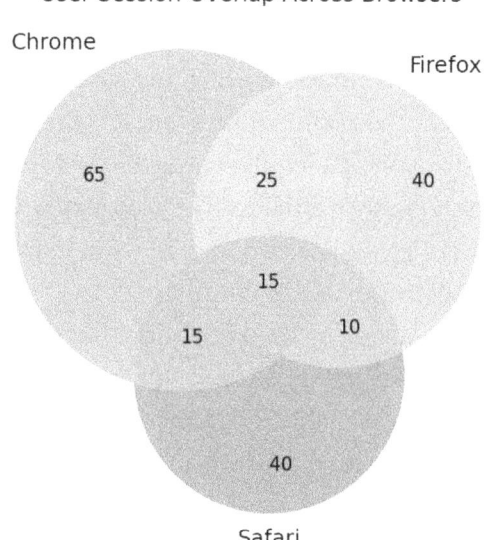

Figure 9-7. User session overlap across browsers

This Venn diagram (as shown in Figure 9-7) depicts the intersection of user browsing sessions across Chrome, Firefox, and Safari on mobile devices. It highlights unique and shared sessions, helping forensic analysts piece together a more complete picture of user behavior across multiple browsers for comprehensive investigations.

Parameter settings: Sample size: 200 total unique sessions across browsers; overlap estimated from synchronized timestamps; data source: synthetic multi-browser session logs collected over 48 hours on Android and iOS devices.

Conclusion

Due to the increasing reliance on mobile devices for both personal and professional transactions, mobile browser forensics has emerged as a significant topic of study within the discipline of digital forensics. On the basis of browser artifacts, including history logs, cookies, cached files, and saved credentials, investigators are able to build a profile of a user's online activities with a substantial degree of precision. This evidence, frequently spread across several browsers, and further confounded by session (or private) browsing, is crucial in civil, criminal, and corporate investigations.

CHAPTER 9   MOBILE BROWSER FORENSICS

Throughout this chapter, we have seen how browser-based evidence is created, maintained, removed and, in some cases, revocable despite deliberate user attempts to delete it. In conjunction with methods like SQLite record recovery, file carving, and write-ahead-to-log (WAL)/journal examination methods, these techniques make it easy for forensic professionals to recover deleted records relevant to a case. Moreover, being able to pinpoint threats from the web artifacts like suspicious redirection or patterns related to drive-by downloading lets browser forensics organically serve user's behavior analysis as well as cybersecurity threat detection.

Yet, these investigational advances are not without issues. Platform-level sandboxes, strong file encryption, privacy-preserving browser architectures, and legal constraints all limit how much depth data is available. Then there is the very process of the forensic acquisition itself, which has to be weighed up against legal compliance, again legislation isn't always on the side of legal professionals, privacy laws and rights, as well as GDPR, CCPA, and other local laws to consider. Whether evidence can be admitted in court depends very much upon the impeccable legality and transparency of the investigative and analytical chain.

Toolkits such as Cellebrite, Magnet AXIOM, Oxygen Forensic Detective, and even open-source browser artifact parsers continue to be critical in mobile browser investigations. However, no single tool provides complete coverage for all browsers and platforms; there seems be a necessity for the investigator to acquire experience and keep track of the browser progress.

New developments are changing the face of forensics. Private browsers, encrypted DNS (DoH/DoT) protocols, hardware-based data protection are complicating visibility of artifacts. Furthermore, the growth of incognito or private browsing modes is stimulating the research on residual evidence patterns, memory-based artifact recovery, and behavioral inference models.

Mobile browser forensics is a fast-growing area of high stakes that requires a mix of technical, legal, as well as adaptive approaches. As privacy-oriented design becomes increasingly common in mobile ecosystems, the fate of browser forensics will rely on advances in inconspicuous investigation, behavioral linking, and responsible investigations.

References

[1] Chand, R. R., Sharma, N. A., & Kabir, M. A. (2025). Advancing Web Browser Forensics: Critical Evaluation of Emerging Tools and Techniques. *SN Computer Science*, 6(4), 355.

[2] AlOwaimer, B. H., & Mishra, S. (2021). Analysis of web browser for digital forensics investigation. *International Journal of Computer Applications in Technology*, 65(2), 160–172.

[3] Joshi, N. N., & Bajeja, S. L. (2024, October). Enhanced Web Browser Forensics: Innovative Methodologies for Evidence Collection and Analysis. In *International Conference on Advancements in Smart Computing and Information Security* (pp. 139–164). Cham: Springer Nature Switzerland.

[4] Boucher, J., Choo, K. K. R., & Le-Khac, N. A. (2022). Web browser forensics—a case study with chrome browser. In *A Practical Hands-on Approach to Database Forensics* (pp. 251–291). Cham: Springer International Publishing.

[5] Younis, L. B., Sweda, S., & Alzu'bi, A. (2021, May). Forensics analysis of private web browsing using android memory acquisition. In *2021 12th International Conference on Information and Communication Systems (ICICS)* (pp. 273–278). IEEE.

[6] Aji, M. P., & Hakim, D. K. (2023, November). A digital forensic analysis on mozilla firefox browser in Android operating system. In *AIP Conference Proceedings* (Vol. 2702, No. 1). AIP Publishing.

[7] K. Kaushik, "Investigation on Mobile Forensics Tools to Decode Cyber Crime," Secur. Anal., pp. 45–56, May 2022, doi: 10.1201/9781003206088-4.

[8] Arshad, M. R., Hussain, M., Tahir, H., Qadir, S., Memon, F. I. A., & Javed, Y. (2021). Forensic analysis of tor browser on windows 10 and android 10 operating systems. *IEEE Access*, 9, 141273–141294.

[9] Hariharan, M., Thakar, A., & Sharma, P. (2022, June). Forensic analysis of private mode browsing artifacts in portable web browsers using memory forensics. In *2022 International Conference on Computing, Communication, Security and Intelligent Systems (IC3SIS)* (pp. 1–5). IEEE.

[10] M. Garg, S. Dahiya, and K. Kaushik, "International Cyberspace Laws: A Review," Unleashing Art Digit. Forensics, pp. 15–28, Jun. 2022, doi: 10.1201/9781003204862-2.

[11] Sanghvi, H., Rathod, D., Shukla, P., Shah, R., & Zala, Y. (2024). Web browser forensics: Mozilla Firefox. *International Journal of Electronic Security and Digital Forensics*, 16(4), 397–423.

[12] Chrome, F. (2023, August). Check for updates Web Browser Forensics: A Comparative Integrated Approach on Artefacts Acquisition, Evidence Collections and Analysis of Google. In *Computing Science, Communication and Security: 4th International Conference, COMS2 2023, Mehsana, Gujarat, India, February 6–7, 2023, Revised Selected Papers* (p. 204). Springer Nature.

[13] A. Garg and A. K. Singh, "Internet of Things (IoT): Security, Cybercrimes, and Digital Forensics," Internet Things Cyber Phys. Syst., pp. 23–50, Dec. 2022, doi: 10.1201/9781003283003-2.

[14] Sanghvi, H., Patel, V. J., Shah, R., Shukla, P., & Rathod, D. (2023, February). Web Browser Forensics: A Comparative Integrated Approach on Artefacts Acquisition, Evidence Collections and Analysis of Google Chrome, Firefox and Brave Browser. In *International Conference on Computing Science, Communication and Security* (pp. 204–218). Cham: Springer Nature Switzerland.

[15] K. Kaushik, S. Tayal, A. Bhardwaj, and M. Kumar, "Advanced Smart Computing Technologies in Cybersecurity and Forensics," Adv. Smart Comput. Technol. Cybersecurity Forensics, Nov. 2021, doi: 10.1201/9781003140023/ADVANCED-SMART-COMPUTING-TECHNOLOGIES-CYBERSECURITY-FORENSICS-KESHAV-KAUSHIK-SHUBHAM-TAYAL-AKASHDEEP-BHARDWAJ-MANOJ-KUMAR.

[16] Pizzolante, R., Castiglione, A., Mastroianni, M., & Palmieri, F. (2025, April). A Mobile Forensic Tool for Enhancing Cyber-Physical Security by Detecting XSS Attacks Through Web Server Access Log Analysis. In *International Conference on Advanced Information Networking and Applications* (pp. 119–129). Cham: Springer Nature Switzerland.

[17] K. Kaushik, M. Ouaissa, and A. Chaudhary, ADVANCED TECHNIQUES AND APPLICATIONS OF CYBERSECURITY AND FORENSICS. CHAPMAN & HALL CRC, 2024. Accessed: Sep. 16, 2024. [Online]. Available: https://www.routledge.com/Advanced-Techniques-and-Applications-of-Cybersecurity-and-Forensics/Kaushik-Ouaissa-Chaudhary/p/book/9781032479576.

CHAPTER 10

Leveraging Machine Learning for Mobile Forensics

Mobile digital investigations have assumed great importance in the past years because of the enhancement of the number of devices, especially mobile ones, in everyday life. Today, phones and tablet forms have become the means of communication, Internet, business, and social interaction. Therefore, these apparatuses bear extensive amounts of personal and business communications, such as text messages, emails, phone call records, GPS tracks, photographs, films, and information on the applications that are installed. Due to the increasing usage of mobile devices in everyday life activities and communications, they become the most valuable source of evidence during criminal investigations, cybersecurity incidents, or civil lawsuits, namely, regarding activity or communication tracks left by a suspect in a criminal investigation, cyberattack, or involved party in civil proceedings. However, this is an area of concern because the storage capacity on mobile devices is very large, thereby creating a challenge for forensic investigators. Current digital investigation techniques that involve mechanically copying or transferring data to a new environment are insufficient. Given the fact that thousands of mobiles churn out gigabytes of data each day, the process of knowledge extraction mandates a lot of time and can lead to the generation of numerous errors. However, mobile platforms, including Android and iOS, are revised more often, implying that new techniques of encryption, security elements, and operating system complications complicate the extraction and analysis of data. Tools that can be used for analysis of the cell phones affect the necessity of inventing more progressive technologies for analysis by forensic investigators due to constant development of mobile technologies and the growth of investigative complications.

CHAPTER 10   LEVERAGING MACHINE LEARNING FOR MOBILE FORENSICS

Machine learning is the technology that would change the "game" in digital forensics. It will then open opportunities for automation, speeding up, and accuracy improvement of mobile investigations. Machine learning is a subset of AI that enables computers to learn from data and better perform tasks without explicit programming. If applied to digital forensics, machine learning can automatically identify relevant evidence, suspicious patterns, and classify huge data sets. Using algorithms that could identify patterns in gigantic datasets, machine learning minimizes time and human effort in making manual analysis. For example, using machine learning to automatically identify anomalies or malicious behaviors in mobile network traffic, detect potential breaches, and classify files based on the likelihood of relevance for an investigation to prioritize evidence—for example, by ranking the most important critical information to highlight, such as suspicious messages, unusual app behaviors, or irregular login attempts. This reduces the burden that forensic investigators have to handle, thus enabling them to focus on the most crucial parts of the investigation. Additionally, machine learning can identify unknown patterns of threats that may not be possible to discern using traditional methods, thereby improving upon the accuracy and depth of analysis.

Beyond basic pattern recognition, the more advanced machine learning methods—supervised learning, unsupervised learning, semi-supervised learning, and reinforcement learning—can then be applied to specific forensic applications. Supervised learning models learn from labeled datasets, which could mean classifying mobile apps, messages, or network traffic as either benign or malicious. However, unsupervised learning methods can find anomalies or group similar data points, allowing investigators to identify numerous relationships or suspicious behaviors not immediately obvious. Semi-supervised learning is particularly helpful when labeled data is limited in number; the model can be trained using both labeled and unlabeled data, which ultimately gives rise to a more powerful model. Additionally, in reinforcement learning, real-time threat mitigation processes can be automated because systems learn about threats as they occur. It can further improve the overall flow of mobile digital investigations by incorporating machine learning into forensic tools and processes. These tools can automatically start extracting data from devices, analyze patterns, flag suspicious behavior for follow-up, and so on. More exposure to new datasets will even make the machine learning models better over time, hence enabling them to detect more elusive evolving threats, such as stealthy malware or zero-day vulnerabilities. This creates a continuous feedback loop in which machine learning not only improves ongoing investigations but also prepares forensic teams for future challenges.

Machine learning is the tool that has finally become an essential element in enhancing mobile digital investigations, based on the increasing complexity, scale, and speed required in modern forensic analysis. In other words, by automating recurrent tasks, such as the identification of critical evidence and the discovery of hidden patterns, machine learning helps forensic specialists dedicate their efforts to more complex analysis and decision-making processes, achieving faster, more reliable, and broader investigations. With the evolution of mobile devices, their role in digital forensics will continue to rise, and investigators' efforts will merely be working to keep pace with the continuously expanding mobile ecosystem.

Machine Learning (ML) Techniques for Mobile Digital Investigations

Machine learning is currently revolutionizing digital forensics; it provides powerful computing capabilities that support complex analytical tasks and large datasets. The application of machine learning in mobile digital investigations can significantly improve the efficiency and accuracy of evidence collection, analysis, and decision-making. This section outlines some of the basic machine learning techniques that can be applied to mobile digital investigations, including supervised, unsupervised, semi-supervised, and reinforcement learning techniques.

Supervised Learning

Supervised learning is a base learning technique that basically trains a model to classify or make predictions against labeled data. In the case of mobile digital investigations, this can be very helpful when data on smartphones, tablets, and other mobile devices needs to be analyzed. To start with, it all begins with the dataset—a collection of input data (features) and output data (labels or classes). This may include different types of attributes, including timestamps, application logs, location data, or even communication patterns; the labels represent categories such as "suspicious activity," "malicious app," or "normal behavior." The supervised learning model tries to learn the relationship between the features and the labels to which they correspond. It will continue to self-adjust during the training period by comparing its predictions with the actual labels and minimizing the error. Once trained, it creates accurate predictions or classifications of new, unseen data. This could involve analyzing logs on a mobile device and classifying whether certain behaviors indicate a possible security threat or benign activity. One of the major strengths of supervised learning is its ability to generalize from historical data: once the model learns from precedents, such as earlier security breaches, it can predict or even identify similar events in later investigations. The quality of prediction is always heavily dependent on the quality and volume of labeled training data. If the dataset lacks diversity or is small in size, then the model will not be able to generalize properly and thus fail to classify new cases correctly. Figure 10-1 [1] gives the basic understanding about the supervised learning model in a very precise way.

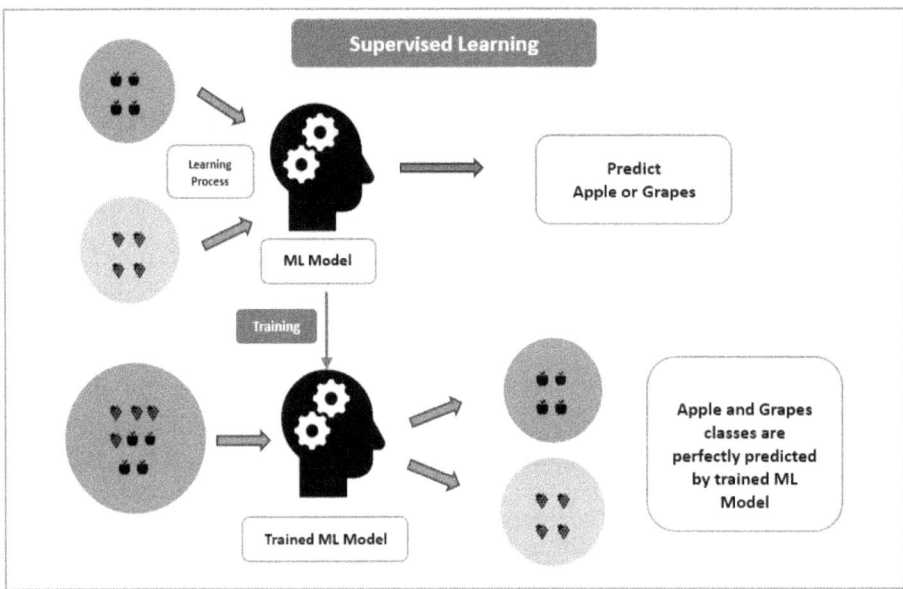

Figure 10-1. Basic concept of Supervised Learning

The fundamental advantage of the supervised learning model is its capability to generalize using historical data. Such a model learns from patterns of past cases—for example, known security breaches—and then predicts or identifies similar events in subsequent case studies. However, the quality of the prediction depends directly and solely on the quality and volume of the labeled training data. A minimal count or lack of diversity in a dataset will lead to poor generalization by the model and inadequate classification of new cases.

Unsupervised Learning

Unsupervised learning is a technique of machine learning whereby models learn on unlabeled data—this is to say, the dataset only includes input features without any corresponding output labels or predefined classes. Instead of explaining a specific outcome, as would happen with supervised learning, unsupervised learning mainly works on the identification of hidden patterns, structures, or relationships in the data. In other words, the model scans for inherent patterns in the input data and tries to group or organize them. This can be achieved either by using algorithms such as K-Means, Hierarchical Clustering, and DBSCAN to join different data points based on similar characteristics or dimensionality reduction methods such as PCA, transforming the data into a simpler and effective form while losing unimportant features. Unsupervised learning can be helpful especially in mobile forensics since the investigators are working with high volumes of raw, unlabeled data such as communication logs, app usage data or system logs without any prior knowledge of what is "normal" or "suspicious." This means unsupervised techniques are useful for exploratory data analysis—in other words, to discover unknown patterns or behaviors in the data—and anomaly detection, where the model is able to identify unusual or outlying data points, reflecting potential breaches in security or malicious activity. For example, by clustering device behavior, an unsupervised learning model can reveal a previously unknown pattern of user behavior that is indicative of unauthorized access or malware. It can classify potential security threats or vulnerabilities that have not been grouped or named before as part of identifying threats. For forensic investigators who sift through massive amounts of data to come up with actionable insights, it helps in automating the analysis process. Figure 10-2 [2] illustrates the basic concept of the unsupervised learning model in a clear and concise manner.

CHAPTER 10 LEVERAGING MACHINE LEARNING FOR MOBILE FORENSICS

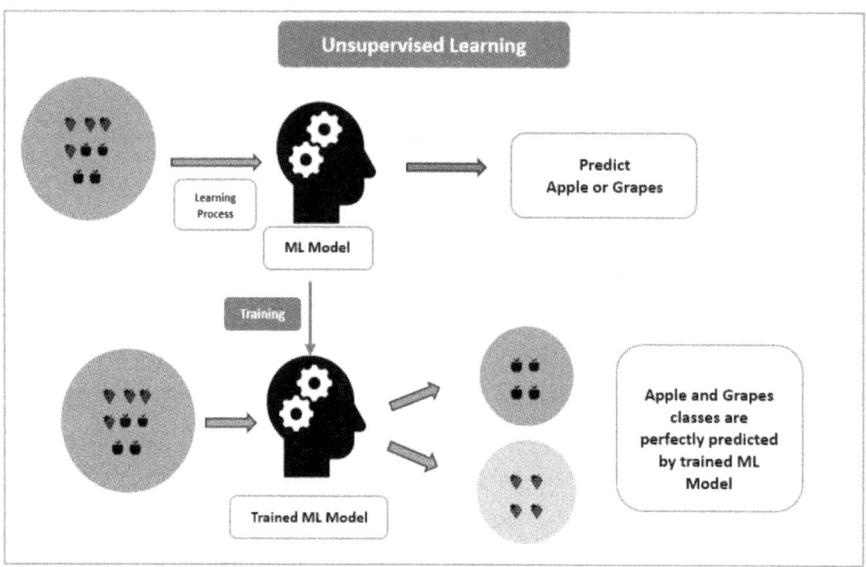

Figure 10-2. *Basic concept of Unsupervised Learning*

Semi-supervised Learning

Semi-supervised learning: It is a form of hybrid approach to train machines. It combines the best of both supervised and unsupervised learning, targeting those scenarios where it is costly and consequently nonsensical to acquire labeled data, but it is possible to obtain extensive datasets without labels. In technical terms, semi-supervised learning learns from a small amount of labeled data with much larger unlabeled datasets to better improve models' generalization and performance. The underlying idea here is that, by using unlabeled data, the model would learn more about the distribution and structure of the data and, therefore, predict even better on merely a few labeled samples. Often-applied techniques used for propagation of labels or labeling of similar, but not necessarily identical, unlabeled instances in some meaningful way include self-training, co-training, or graph-based methods. Semi-supervised learning is particularly useful for mobile forensics because, within a forensic investigation, labeled data—the truth about a particular piece of behavior, say, malicious or benign—is scarce and hard to obtain. It is difficult, for example, to annotate certain app usage patterns or certain kinds of communication behavior on mobile devices with expert knowledge. Figure 10-3 [3] illustrates the basic working functionality of semi supervised learning.

CHAPTER 10   LEVERAGING MACHINE LEARNING FOR MOBILE FORENSICS

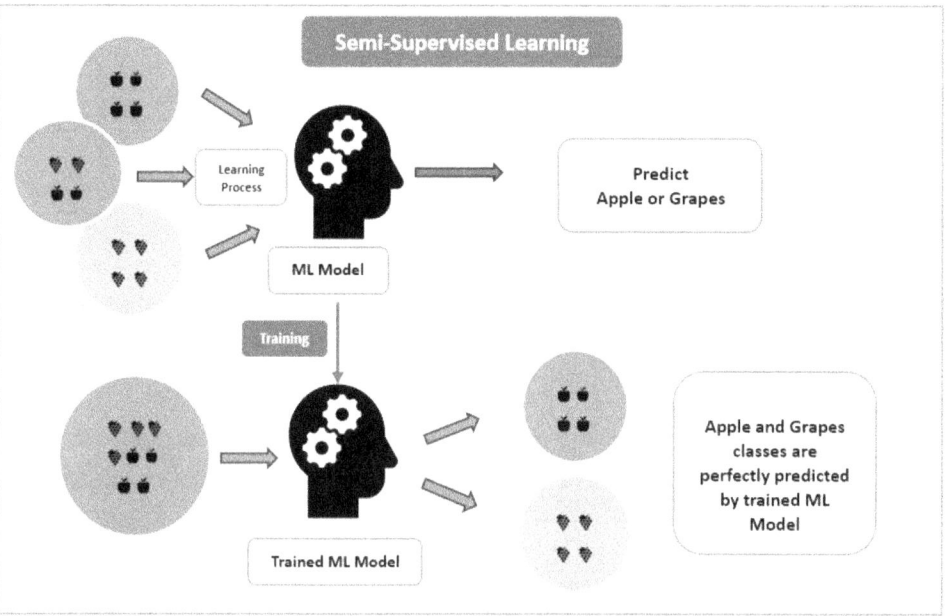

Figure 10-3. *Basic concept of Semi-Supervised Learning*

However, most mobile forensic investigators often find huge volumes of unlabeled data that might include information from system logs, network traffic, history of app usage, and records of calls or messaging. Such datasets are rich in information, but their nature makes the normal apart from suspicious behaviors unclear. Therefore, semi-supervised models are effective in learning patterns from few labeled examples—for instance, known malicious app interactions or communications flagged in previous cases—and yet they can use all the large amounts of unlabeled data for detection of as-yet-unknown threats or suspicious activity. For instance, a semi-supervised learning algorithm may learn new patterns of malware behavior by gathering knowledge from a few known malicious applications while processing large amounts of unlabeled app logs. Furthermore, semi-supervised methods like pseudo-labeling can iteratively label the unlabeled data. Pseudo-labels can then be utilized temporarily to further train the model on the unlabeled data by relying on the model's confident prediction based on the unlabeled data. Such a process would indeed enhance the capability of the model in detecting anomalies or even estimating the probability of security risk in mobile forensic investigations, based on the generalization ability beyond a small pool of labeled instances. Hence, semi-supervised learning strikes an excellent balance between having small amounts of expert-annotated data, on the one hand, and large-scale, often complex, unlabeled data typical of mobile forensics, on the other. Such a balance would result in more effective and accurate investigative efforts for identifying security threats, malicious apps, or unauthorized activity on devices.

Reinforcement Learning

A very powerful technique in machine learning is called Reinforcement Learning (RL), where an agent learns over time by trial and error through interaction with an environment that rewards or punishes it every time. Technically, the agent operates in a Markov Decision Process (MDP) framework, where at each time step it observes the state of the environment, takes an action, and moves according to the interaction to a new state. It gets feedback from the environment in the form of a scalar reward or penalty. Figure 10-4 [4] [6] gives a brief idea about the reinforcement learning. The objective for the RL agent should therefore be to learn an optimal policy—an appropriate mapping from states to actions for maximizing its cumulative reward over time, called expected return. Some of the core techniques used in RL are Q-learning, Deep Q-Networks (DQN), and policy gradient methods, which allow agents to learn even in the most complex and high-dimensional state spaces through the use of approximation techniques, such as neural networks. With traditional association being on more general application to robotics and gaming, and autonomous systems, reinforcement learning is slowly gaining prominence in the realm of digital forensics, in tasks involving autonomous decision-making and dynamic environments.

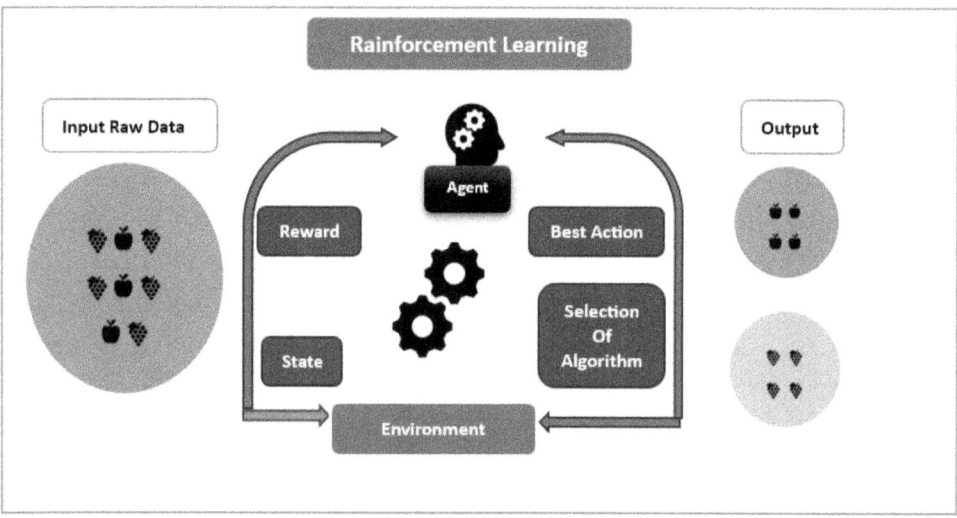

Figure 10-4. Basic concept of Reinforcement Learning

For instance, in mobile forensics, RL is applied to make threat detection and response real time, thereby making environments dynamic since there's a new threat every time, and devices and applications become dynamic day in and day out. For example, in real-time mobile threat detection, the RL agent might monitor network traffic, system logs, or application behaviors from a mobile device and learn how best to respond to potential threats. By interacting with the mobile system, rewards are given to the agent for actions decreasing threats—for instance, blocking rogue network requests or isolating unknown applications—and penalties are imposed on actions allowing threats to spread. This is useful for RL's adaptive learning property when the set of rules or model, as defined beforehand, is not strong enough because it is not fixed; the landscape of known threats will change. This means RL agents learn and detect security incidents, responding with real-time actions containing or mitigating threats without human input. For example, take mobile network security: the RL agent learns to react according to various types of attacks, like denial-of-service (DoS) attacks, by dynamically changing firewall rules, rerouting traffic, or temporarily disabling vulnerable services. Continuous learning and adaptation make RL a promising tool for managing complex, evolving environments like those found on mobile devices, with new vulnerabilities and attack vectors constantly emerging. The biggest challenge for using RL in mobile forensics is setting up the reward system. It needs to react quickly to threats without being so aggressive that it interferes with normal, legitimate phone use. Exploration-exploitation is another dilemma—exploring to find new strategies but exploiting known ones to maximize rewards. This balance in mobile forensics is crucial since overly aggressive exploration may compromise device performance or user experience, while overexploitation of current knowledge might miss evolving threats. However, despite this tension, RL's possibility to augment and enhance security measures in mobile environments makes for exciting frontiers in digital forensics.

Data Preparation and Feature Engineering

ML in mobile digital investigations highly relies on the quality and structure of the data to train the model. Thus, one way the characteristics of the data are prepared for the study is by preparing data and making the right feature engineering for a successful mobile forensic investigation. Data preparation and feature engineering are one of the most crucial stages that turn raw data into a usable format: extracting the essential features and the most powerful data points to be used for analysis, thereby training the models properly.

CHAPTER 10   LEVERAGING MACHINE LEARNING FOR MOBILE FORENSICS

Data Collection

Data collection is the first process in any mobile digital investigation—it gathers information from different data sources on the mobile devices. The variety of data that can be extracted on mobile devices and put to forensic use is broad and includes

a) **Mobile Apps Data:** The mobile apps store a lot of data that includes activities, log files, transaction history, and session information. This could be highly crucial during the investigation when it concerns messaging applications, financial applications, or social networking.

b) **Call Logs:** A mobile device's call log has information regarding incoming and outgoing calls, timestamps, and duration, along with phone numbers the calls were going to and coming from. Analysts can derive communication patterns and suspicious contacts from these records.

c) **SMS/MMS Messages:** Text messages contain very important data regarding communication that might be relevant to a case. The message content, information about the sender and receiver, and time stamp are therefore extracted from the cell phone for forensic investigation.

d) **Location Data:** GPS information, geotags, and other location-related information provide information about where the device or its owner physically is at a given time. This can be the most crucial information for establishing a suspect's or victim's movements.

e) **Multimedia Files:** Photographs, video clips, and audio files often contain meaningful evidence. Metadata appended to these files, in the form of timestamp and location, are often helpful in building a timeline or confirming incidents.

f) **Browser History and Cookies:** The mobile device retains all browsing history, cached data, and cookies. It provides knowledge of users' online actions, including which websites were accessed, their search queries, and downloads.

g) **App Usage Data:** Logs from mobile applications may expose information regarding usage behavior, including dates and times the app has been accessed and how often, serving as proof of behavioral patterns.

h) **Network Traffic:** Mobile devices generally send and receive data over different networks, like cellular and Wi-Fi. Such network traffic helps identify information regarding app communications, suspicious network activity, or downloaded malicious files.

Most of these types of data have large informative potential in the course of an investigation. However, raw data from mobile devices are mostly unstructured, noisy, and filled with irrelevant information, which makes preprocessing this data very crucial before it can be effectively used on a machine learning model.

Data Cleaning

Cleaning the data is the most critical stage in the pipeline for preparing raw data. This assures that successive machine learning processes are of quality and reliability. Raw data collected from a mobile device, such as system logs, usage statistics of applications, call records, or network traffic data, comes with different kinds of imperfections. Some of the main examples include: errors about values or timestamps, missing values such as incomplete logs or dropped packets, redundant information in terms of duplicate entries, repeated events, and noise due to noisy data points that are either irrelevant to the task or outlying data. Handling these cases improperly can severely degrade the performance of machine learning models, introducing various kinds of biases, inaccuracies, or inconsistencies into the training process.

Noise Handling Techniques

a) **Outlier detection and removal:** Data coming from mobile devices might have outliers or extreme values that are not following the normal data distribution. Outliers could result from corrupted data, malfunctioning devices, or false logs. The Z-score or Interquartile Range (IQR) can be used for outlier detection and removal purposes.

b) **Data Smoothing:** This is a technique to reduce noise in continuous data. Techniques such as moving averages or exponential smoothing help reduce fluctuations or spikes in data so that only the important trends are considered.

```python
from scipy import stats
import matplotlib.pyplot as plt
np.random.seed(42)
data = np.random.normal(loc=50, scale=5, size=100) # Generate normal distributed data
data[::15] += np.random.randint(20, 50, size=len(data[::15])) # Inject some artificial outliers
df = pd.DataFrame({'Value': data})
z_scores = np.abs(stats.zscore(df['Value']))
threshold = 3
df_z_removed = df[z_scores < threshold]
Q1 = df['Value'].quantile(0.25)
Q3 = df['Value'].quantile(0.75)
IQR = Q3 - Q1
lower_bound = Q1 - 1.5 * IQR
upper_bound = Q3 + 1.5 * IQR
df_iqr_removed = df[(df['Value'] >= lower_bound) & (df['Value'] <= upper_bound)]
df_iqr_removed['MA_Smoothed'] = df_iqr_removed['Value'].rolling(window=5).mean() # Moving Average (window size = 5)
df_iqr_removed['Exp_Smoothed'] = df_iqr_removed['Value'].ewm(alpha=0.3).mean() # Exponential Smoothing (alpha = 0.3)
plt.figure(figsize=(10, 6))
plt.plot(df['Value'], label='Original Data', linestyle='--', alpha=0.6)
plt.plot(df_iqr_removed['Value'].values, label='IQR Outlier Removed', marker='o')
plt.plot(df_iqr_removed['MA_Smoothed'].values, label='Moving Average', linewidth=2)
plt.plot(df_iqr_removed['Exp_Smoothed'].values, label='Exponential Smoothing', linewidth=2)
plt.title('Outlier Removal & Data Smoothing')
plt.xlabel('Index')
plt.ylabel('Value')
plt.legend()
plt.grid(True)
plt.show()
```

Figure 10-5. Sample code for outlier removal

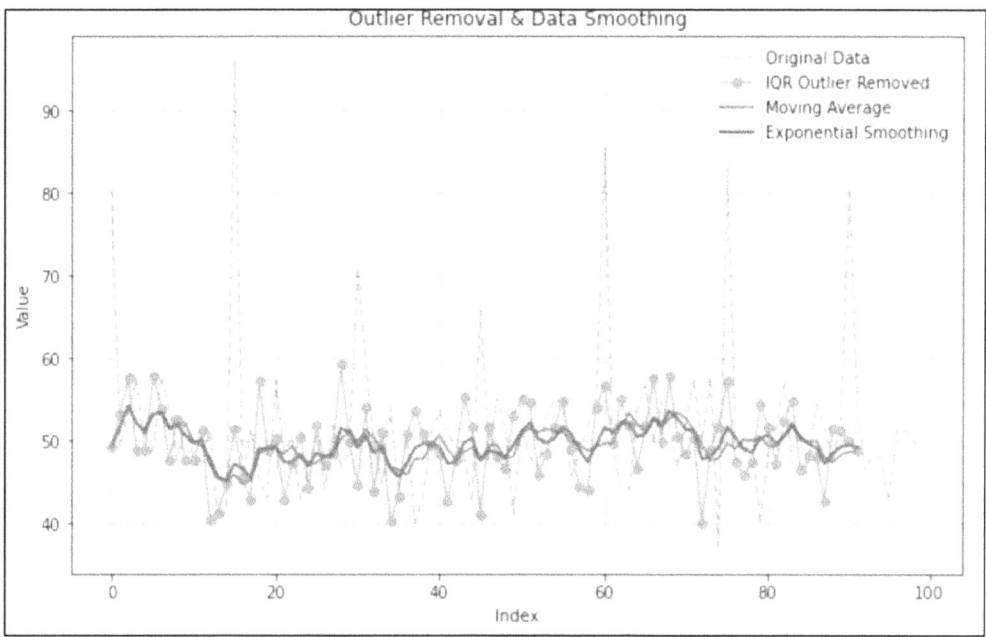

Figure 10-6. *Visualization of data after outlier removal*

The code mentioned in Figure 10-5 generates simulated sensor data with some artificial outliers, then detects and removes these outliers using two methods: Z-score (removing points far from the mean) and Interquartile Range (IQR) filtering. After cleaning, it applies two smoothing techniques. Moving Average and Exponential Smoothing to reduce noise and highlight trends in the data. Finally in Figure 10-6, it plots the original, cleaned, and smoothed data for easy visual comparison.

Handling Missing Values

a) **Imputation:** The problem of missing data occurs very frequently in mobile forensics—mostly in log or application data. Missing values can be filled by imputation techniques, mean, median, or mode imputation. Advanced methods include k-nearest neighbors (KNN) or multivariate imputation.

b) **Delete Partial Data Entries:** When imputation could change the data or if there is a large number of missing data, it might be best to delete partial data entries. This technique should be applied cautiously to avoid losing valuable information.

c) **Interpolation:** In time-series data, such as call records or network traffic, missing values can be interpolated by estimating missing values based on surrounding points.

Proper cleaning of data feeds the learning algorithms with accurate, relevant information, leading to better performance of machine learning-based pattern detection/anomaly discovery.

Feature Extraction

Feature extraction is the step of extracting useful information from raw data after data cleaning. It is a crucial part of the whole process because it transforms raw data formats into formats that can be used by machine learning models. Mobile forensics raw data are quite heavy; hence, the need to derive a concise and informative set of features.

Identification of Relevant Features from Raw Data Includes

a) **Timestamp Extraction:** Any data that is time-sensitive in mobile forensic datasets, like call logs or GPS data. Features extracted regarding time of day, day of the week, or frequency of specific events, such as calls or application usage, can be incredibly helpful in understanding user behavior patterns. In Figure 10-7, the sample code has been mentioned for timestamp extraction. This code takes a dataset with timestamps, like call logs or GPS records, and turns them into useful features.

It figures out what hour of the day and day of the week each event happened, and counts how many events each user had. These features help investigators spot patterns, such as unusual activity times or sudden spikes in user activity.

CHAPTER 10 LEVERAGING MACHINE LEARNING FOR MOBILE FORENSICS

```python
""" Timestamp Extraction for Mobile Forensic Data =>Demonstrates converting raw timestamps
(e.g., from call logs, GPS data) into features: 1. hour of the day 2. day of the week
3. event frequency per user"""
import pandas as pd
from datetime import datetime
# --- Sample forensic-style dataset ---
# In practice, these could be call logs, GPS pings, or app usage events
data = {
    'user_id': [101, 101, 102, 101, 103, 102, 103, 101],
    'event_type': ['call', 'gps', 'call', 'app_use', 'call', 'gps', 'app_use', 'call'],
    'timestamp': [
        '2025-07-10 08:15:27', '2025-07-10 09:30:45',
        '2025-06-11 21:10:05', '2025-07-11 14:22:15',
        '2025-05-12 23:59:59', '2025-07-12 06:45:00',
        '2025-03-12 11:00:00', '2025-06-12 08:05:15' ]}
df = pd.DataFrame(data)
# --- Convert timestamp column to pandas datetime ---
df['timestamp'] = pd.to_datetime(df['timestamp'])
# --- Feature: Hour of the day ---
df['hour_of_day'] = df['timestamp'].dt.hour
# --- Feature: Day of the week ---
# Monday = 0, Sunday = 6
df['day_of_week'] = df['timestamp'].dt.dayofweek
# --- Feature: Frequency of events per user ---
event_counts = df.groupby('user_id')['event_type'].transform('count')
df['event_frequency'] = event_counts
# --- Output result ---
print(df)
```

Figure 10-7. *Sample Python code for timestamp extraction*

b) **Text-Based Features:** NLP can be applied to messages or emails having text content to fetch relevant features in the form of tokenization, sentiment analysis, and identification of key or related entities for a text. Figure 10-6 illustrated simple NLP-based Python code for text-based feature extraction. This code highlights the usage of tokens, sentiments, and the entities.

CHAPTER 10   LEVERAGING MACHINE LEARNING FOR MOBILE FORENSICS

```python
def extract_text_features(text_series):
    """
    text_series: pandas Series or list of text messages/emails
    """
    features = []
    for i, text in enumerate(text_series):
        if not isinstance(text, str):
            text = ""

        # Sentiment Analysis
        blob = TextBlob(text)
        sentiment = blob.sentiment.polarity  # -1 to 1

        # Tokenization
        tokens = [word for word in re.findall(r'\w+', text.lower())]

        # Simple Entity Extraction (using regex: capitalized words)
        entities = re.findall(r'\b[A-Z][a-zA-Z]+\b', text)

        # Collect features
        feature_dict = {
            "tokens": tokens,
            "sentiment": sentiment,
            "entities": entities
        }
        features.append(feature_dict)

        # --- Display results ---
        print(f"\nText {i+1}: {text}")
        print(f"  Tokens    : {tokens}")
        print(f"  Sentiment : {sentiment}")
        print(f"  Entities  : {entities}")

    return features
```

Figure 10-8. *Sample Python code for text-based feature extraction*

    c) **Network Traffic Features:** Features of network data, such as packet sizes, destination IP addresses, and communication protocols, could be fetched to analyze network behavior and identify anomalies. In Figure 10-9, the sample Python code has been mentioned for the extracting network traffic-based features. This code takes a set of network traffic records and turns them into useful features for analysis. It calculates the average packet size for each session, counts how many unique destination IP addresses were contacted, and records how often different protocols like TCP or UDP were used. These features help investigators spot unusual network behavior or suspicious connections.

```python
[2]: """Network Traffic Feature Extraction for Mobile Forensics
     Extracts useful features from raw network log data: 1. Average packet size
     2. Number of unique destination IPs 3. Protocol usage counts"""

     import pandas as pd
     # --- Sample network traffic dataset ---
     # In real cases, these could be from pcap files or network monitoring logs
     data = {
         'session_id': [1, 1, 1, 2, 2, 3, 3, 3, 3],
         'packet_size': [500, 750, 600, 1500, 400, 300, 350, 320, 900],
         'dest_ip': [ '192.168.0.10', '192.168.0.10', '8.8.8.8','10.0.0.5', '10.0.0.5',
             '192.168.0.12', '192.168.0.13', '192.168.0.13', '8.8.4.4' ],
         'protocol': ['TCP', 'TCP', 'UDP', 'TCP', 'TCP', 'UDP', 'UDP', 'UDP', 'TCP']}
     df = pd.DataFrame(data)
     # --- Feature: Average packet size per session ---
     df_avg_packet = df.groupby('session_id')['packet_size'].mean().reset_index(name='avg_packet_size')
     # --- Feature: Unique destination IP count per session ---
     df_unique_ips = df.groupby('session_id')['dest_ip'].nunique().reset_index(name='unique_dest_ips')
     # --- Feature: Protocol usage counts per session ---
     df_protocol_counts = df.groupby(['session_id', 'protocol']).size().unstack(fill_value=0).reset_index()
     # --- Merge all features into one dataframe ---
     features = df_avg_packet.merge(df_unique_ips, on='session_id').merge(df_protocol_counts, on='session_id')
     print(features)
```

Figure 10-9. Sample Python code to extract network traffic features

d) **Multimedia Features:** For multimedia files like images or videos, metadata, such as file size, format, resolution, and geotags, could be used as features. Advanced techniques can even analyze the content of images using algorithms that detect images. Figure 10-10 demonstrates the basic and simple Python code to extract the metadata from the specific image.

```
def extract_image_metadata(image_path):
    """
    Extracts image metadata including size, format, resolution, geotags.
    """
    image = Image.open(image_path)
    metadata = {
        "file_size": os.path.getsize(image_path),
        "format": image.format,
        "resolution": image.size
    }

    # Extract EXIF data
    exif_data = image._getexif() or {}
    for tag_id, value in exif_data.items():
        tag = TAGS.get(tag_id, tag_id)
        if tag == "GPSInfo":
            gps_data = {}
            for t in value:
                sub_tag = GPSTAGS.get(t, t)
                gps_data[sub_tag] = value[t]
            metadata["geotags"] = gps_data
        else:
            metadata[tag] = value
    return metadata
```

Figure 10-10. Sample Python code for extracting metadata from the image

Dimensionality Reduction Techniques

This is a common characteristic of high-dimensional data in many mobile investigations. However, training models based on such data is computationally expensive and possibly leads to overfitting. Hence, the application of dimensionality reduction techniques reduces the number of present features while retaining all important information.

 a) **Principal Component Analysis (PCA):** PCA is a method of dimensionality reduction that transforms high-dimensional data into its low-dimensional forms by discovering the main components of the original data, that is, the directions of maximum variance. The features are reduced, and the data becomes more workable while still retaining the most relevant information.

 b) **Linear Discriminant Analysis (LDA):** LDA is a discriminant method for dimensionality reduction that maximizes the ratio between the between-class scatter and the within-class scatter

CHAPTER 10 LEVERAGING MACHINE LEARNING FOR MOBILE FORENSICS

of the data. It is particularly useful in mobile investigations, especially when dealing with classification problems, such as distinguishing good ware from malware.

c) **t-Distributed Stochastic Neighbor Embedding (t-SNE):** Another dimensional reduction technique is very effective in visualizing high-dimensional data by revealing relationships between different points and clusters or patterns not readily evident.

```python
[3]: """Example: Dimensionality Reduction Techniques (PCA, LDA, t-SNE)"""
import matplotlib.pyplot as plt from sklearn.datasets import make_classification
from sklearn.decomposition import PCA from sklearn.discriminant_analysis import LinearDiscriminantAnalysis as LDA
from sklearn.manifold import TSNE import pandas as pd
# --- 1) Create a sample classification dataset ---
X, y = make_classification(n_samples=300, n_features=10, n_informative=5, n_redundant=2,
                           n_classes=3, random_state=42)
# Put into a dataframe for easier handling
df = pd.DataFrame(X)
df['label'] = y
# --- 2) PCA (unsupervised) ---
pca = PCA(n_components=2) X_pca = pca.fit_transform(X)
# --- 3) LDA (supervised) ---
lda = LDA(n_components=2) X_lda = lda.fit_transform(X, y)
# --- 4) t-SNE (unsupervised, good for visualization) ---
tsne = TSNE(n_components=2, perplexity=30, random_state=42) X_tsne = tsne.fit_transform(X)
# --- 5) Plot results ---
fig, axes = plt.subplots(1, 3, figsize=(15, 4))
axes[0].scatter(X_pca[:, 0], X_pca[:, 1], c=y, cmap='viridis')
axes[0].set_title("PCA")
axes[1].scatter(X_lda[:, 0], X_lda[:, 1], c=y, cmap='viridis')
axes[1].set_title("LDA")
axes[2].scatter(X_tsne[:, 0], X_tsne[:, 1], c=y, cmap='viridis')
axes[2].set_title("t-SNE")
for ax in axes: ax.set_xlabel("Component 1") ax.set_ylabel("Component 2")
plt.tight_layout()
plt.show()
```

Figure 10-11. Sample Python code for dimensionality reduction PCA, LDA, and t-SNE

The code mentioned in Figure 10-11 creates a sample dataset and applies three popular dimensionality reduction techniques. PCA finds the main directions where the data varies the most, LDA reduces features while separating different classes as much as possible, and t-SNE creates a 2D map that shows patterns and clusters in complex data. It then plots the results so you can visually compare how each method represents the same high-dimensional dataset.

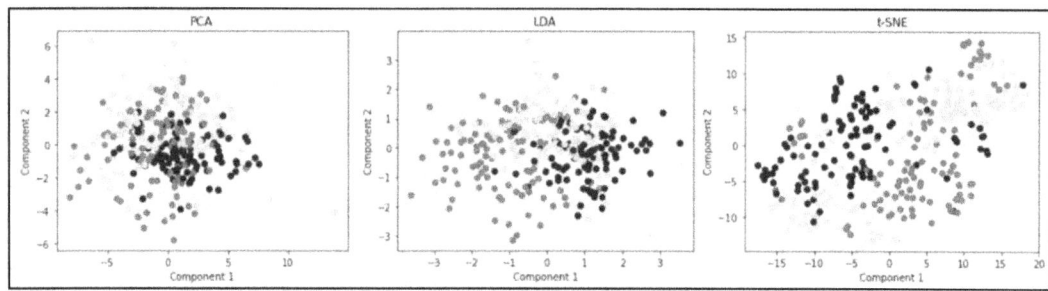

Figure 10-12. Comparison between PCA, LDA, and t-SNE

Figure 10-12 shows that PCA preserves the overall spread of the data but doesn't clearly separate the classes. LDA achieves better separation between the three groups since it's optimized for distinguishing classes. t-SNE reveals the most distinct clustering patterns, making it easier to spot natural groupings and relationships in the data.

Feature Selection

Feature selection is the process of selecting the most appropriate features, which should have the most significant impact on the results and performance of machine learning models. Therefore, after extracting all features from data, this process enhances the efficiency and accuracy of models by eliminating irrelevant or redundant features.

Methods of Feature Selection

a) **Recursive Feature Elimination (RFE):** RFE is a very popular feature selection technique that recursively removes the least important features. It works by fitting a model, ranking features according to how much they contribute to the model's performance, and removing those that contribute the least, repeating the process until the best set of features has been located.

b) **Correlation-Based Feature Selection (CFS):** CFS ranks features based on their correlation with the target variable and their pairwise correlations among themselves. Features with high correlations to the target but low intercorrelations are selected, thus minimizing redundant information.

c) **Mutual Information:** Mutual information calculates the amount of information shared between a feature and the target variable. Features with high mutual information are better and are included in a model.

d) **Lasso Regression:** Lasso regression is a linear regression variant with penalties for features that are far too numerous. The contribution of this variant is to shrink coefficients of less important features toward zero and eliminate those features in the model altogether.

In Figure 10-13, the code generates a sample dataset with several features and a target variable to mimic a real classification scenario, then applies four widely used feature selection techniques to identify the most important inputs for model building. Recursive Feature Elimination (RFE) fits a model, ranks the features by importance, and repeatedly removes the least useful ones until the desired number is reached. Correlation-based Feature Selection (CFS) selects features that are strongly related to the target variable while avoiding features that are highly correlated with each other, thus minimizing redundancy. Mutual Information evaluates how much knowing a feature reduces uncertainty about the target, keeping those with the highest scores. Lasso Regression uses L1 regularization to shrink the coefficients of less important features to zero, effectively removing them from the model. In Figure 10-14, the output lists the top features chosen by each method, clearly showing how different techniques may select different subsets based on their unique selection criteria.

CHAPTER 10 LEVERAGING MACHINE LEARNING FOR MOBILE FORENSICS

```
[4]: """Feature Selection Techniques:1. Recursive Feature Elimination (RFE)
     2. Correlation-based Feature Selection (CFS) 3. Mutual Information
     4. Lasso Regression (L1 regularization)"""
     import pandas as pd import numpy as np from sklearn.datasets import make_classification
     from sklearn.linear_model import LogisticRegression, LassoCV
     from sklearn.feature_selection import RFE, mutual_info_classif
     from sklearn.model_selection import train_test_split
     # --- Create sample dataset ---
     X, y = make_classification(n_samples=300, n_features=10, n_informative=5, n_redundant=2,
         n_classes=2, random_state=42)
     feature_names = [f"feature_{i}" for i in range(X.shape[1])]
     df = pd.DataFrame(X, columns=feature_names)
     df['target'] = y
     # --- 1) Recursive Feature Elimination (RFE) ---
     model = LogisticRegression(max_iter=2000, solver='lbfgs')
     rfe = RFE(model, n_features_to_select=5)
     rfe.fit(X, y)
     rfe_selected = [feature_names[i] for i in range(len(feature_names)) if rfe.support_[i]]
     print("RFE Selected Features:", rfe_selected)

     # --- 2) Correlation-based Feature Selection (CFS) ---
     corr_matrix = df.corr()
     target_corr = corr_matrix['target'].drop('target').abs()  # correlation with target
     # Select top 5 features with low intercorrelation
     sorted_features = target_corr.sort_values(ascending=False)
     selected_cfs = []
     for f in sorted_features.index:
         if all(abs(corr_matrix[f][sf]) < 0.7 for sf in selected_cfs):
             selected_cfs.append(f)
         if len(selected_cfs) == 5:
             break
     print("CFS Selected Features:", selected_cfs)
     # --- 3) Mutual Information ---
     mi_scores = mutual_info_classif(X, y, random_state=42)
     mi_ranking = sorted(zip(feature_names, mi_scores), key=lambda x: x[1], reverse=True)
     selected_mi = [f for f, score in mi_ranking[:5]]

     print("Mutual Information Selected Features:", selected_mi)

     # --- 4) Lasso Regression for feature selection ---
     lasso = LassoCV(cv=5, random_state=42).fit(X, y)
     lasso_selected = [feature_names[i] for i, coef in enumerate(lasso.coef_) if coef != 0]

     print("Lasso Selected Features:", lasso_selected)
```

Figure 10-13. Sample Python code for feature selection methods

```
RFE Selected Features: ['feature_0', 'feature_3', 'feature_5', 'feature_7', 'feature_8']
CFS Selected Features: ['feature_0', 'feature_1', 'feature_8', 'feature_7', 'feature_5']
Mutual Information Selected Features: ['feature_9', 'feature_0', 'feature_1', 'feature_7', 'feature_8']
Lasso Selected Features: ['feature_0', 'feature_1', 'feature_3', 'feature_5', 'feature_6', 'feature_7', 'feature_8']
```

Figure 10-14. Output of sample code mentioned in Figure 10-13

Feature selection thus focuses the machine learning model on the most impactful data, leading to increase in the accuracy of the model, a reduction in overfitting, and finally, an interpreted model. Through reducing the data's dimension and zeroing in on what is most relevant to the case, forensic investigators guarantee that machine learning algorithms are applied proficiently in mobile investigations.

Data preparation and feature engineering form part of any machine learning pipeline, but they are even more important in mobile digital investigations, where large, complex, and unstructured data sets abound. Proper collection, cleaning, feature extraction, and selection are critical aspects in developing an efficient and effectively accurate model for processing forensic data. These steps lay the foundation for effective analysis, so that meaningful patterns and insights are uncovered, furthering the investigation.

Machine Learning Models and Algorithms

Machine learning acts as a reservoir for mobile digital forensics regarding the discovery of trends, identification of anomalies, and automation of high-volume information analysis. Cell phones and mobile devices generate large amounts of data; therefore, it is possible for machine learning models to wade through such information to assist investigators. This section discusses the most common models applied in machine learning for mobile forensic investigations, identifies criteria applicable to model selection, describes training, and explores techniques to fine-tune and optimize model performance.

Overview of Common Models

The success of machine learning in mobile forensics relies heavily on the models and algorithms used. This section presents some common models for classification, regression, clustering, anomaly detection, among others. Each of them comes with specific advantages that depend on the type of data under analysis and the nature of the investigation to be conducted. It is, therefore, a difficult task to decide which one to use; yet, one needs to know what problem is to be analyzed, which data is available, and what outcome is desired.

CHAPTER 10   LEVERAGING MACHINE LEARNING FOR MOBILE FORENSICS

Neural Networks (NNs): Neural networks are a class of machine learning algorithms that mirror the human brain's structure, characterized by interconnecting nodes called neurons distributed in layers. In fact, NNs are pretty efficient in handling complex data patterns and non-linear relationships. NNs can be applied in mobile digital investigations, such as malware detection, pattern recognition, and anomaly detection.

```python
In [1]: import tensorflow as tf
        from tensorflow.keras.models import Sequential
        from tensorflow.keras.layers import Conv2D, MaxPooling2D, Flatten, Dense
        import numpy as np

        # Simulate image data (100 samples of 64x64 RGB images)
        X_images = np.random.rand(100, 64, 64, 3)
        y_images = np.random.randint(0, 2, 100)  # 0 = safe, 1 = suspicious

        # Define a simple CNN model
        cnn_model = Sequential([
            Conv2D(32, (3, 3), activation='relu', input_shape=(64, 64, 3)),
            MaxPooling2D(2, 2),
            Conv2D(64, (3, 3), activation='relu'),
            MaxPooling2D(2, 2),
            Flatten(),
            Dense(128, activation='relu'),
            Dense(1, activation='sigmoid')
        ])

        cnn_model.compile(optimizer='adam', loss='binary_crossentropy', metrics=['accuracy'])

        # Train the model
        cnn_model.fit(X_images, y_images, epochs=3, batch_size=10, verbose=1)
```

Figure 10-15. *Sample Python code of CNN Model to classify suspicious images*

- **Convolutional Neural Networks (CNNs):** CNNs are highly tuned to process images and videos, making them potentially useful for mobile forensics when dealing with multimedia content, for example, images or videos stored on the device. Figure 10-15 demonstrates the usage of CNN model in the classification of suspicious images recovered from a seized mobile device. In this process first we need to make a copy of the phone's storage, either partially or as a full bit-by-bit image, then all the images need to be extracted from locations like the gallery, messaging apps, and downloads. These images are then applied to CNN model to automatically spot any that might be suspicious or relevant to the case. Finally, the flagged images and their details such as file path, name, and other metadata are saved in a forensic report for investigators to review.

CHAPTER 10   LEVERAGING MACHINE LEARNING FOR MOBILE FORENSICS

- **Recurrent Neural Networks (RNNs)**: RNNs, especially Long Short-Term Memory (LSTM) networks, are highly optimized to deal with sequential data. They may therefore also be used for analyzing time-series data, such as call logs, app usage, or network traffic patterns. Figure 10-16 demonstrates the basic Python-based code which has capability to analyze the chat sequence using RNN based LSTM model. It first converts example text messages into numerical form, pads them to the same length, then feeds them into an LSTM network that learns to classify each as either normal or suspicious.

```python
from tensorflow.keras.preprocessing.sequence import pad_sequences
from tensorflow.keras.models import Sequential
from tensorflow.keras.layers import Embedding, LSTM, Dense
import numpy as np

# Simulated SMS/chat dataset
texts = [
    "Let's meet at the cafe",
    "He is planning something dangerous",
    "Lunch at 2 PM",
    "They are going to attack",
    "Call me when you reach",
    "The bomb will explode at midnight"
]
labels = [0, 1, 0, 1, 0, 1]  # 1 = suspicious, 0 = normal

# Tokenization and padding
tokenizer = Tokenizer(num_words=1000)
tokenizer.fit_on_texts(texts)
sequences = tokenizer.texts_to_sequences(texts)
X_chat = pad_sequences(sequences, maxlen=6)
y_chat = np.array(labels)

# Define LSTM model
rnn_model = Sequential([
    Embedding(input_dim=1000, output_dim=32, input_length=6),
    LSTM(64),
    Dense(1, activation='sigmoid')
```

Figure 10-16. Basic Python code of RNN-LSTM to analyze chat sequence

Decision Trees: A decision tree is an input data model that looks like a flowchart. In it, input data splits based on feature values, by which splitting in a decision tree leads to a decision. It's simple but really effective at classifying data in digital investigations. For example, in Figure 10-17, this shows how you can classify whether an app is potentially dangerous or not, relying on its permissions and network activities.

CHAPTER 10 LEVERAGING MACHINE LEARNING FOR MOBILE FORENSICS

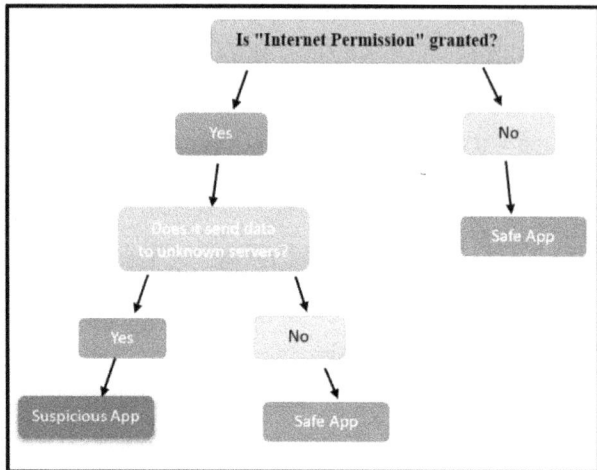

Figure 10-17. *Working style of decision tree*

Random Forest: Random forest is an ensemble learning method built using multiple decision trees combined to improve predictive results and avoid overfitting. It is widely applied in classification problems, for instance, benign vs. malicious apps classification or suspicious user behavior identification. Random forest accommodates big feature sets with useful feature importance metrics that fit the niche area of mobile forensics. Figure 10-18 presents simple Python code which can classify benign vs. malicious apps using a wider set of features (permissions, file access, resource usage) using random forest.

CHAPTER 10   LEVERAGING MACHINE LEARNING FOR MOBILE FORENSICS

```
In [2]: from sklearn.ensemble import RandomForestClassifier

        # Features: [CAMERA, INTERNET, READ_SMS, CPU_USAGE]
        X = [
            [1, 1, 1, 80],  # malware
            [0, 1, 0, 30],  # benign
            [1, 1, 0, 90],  # malware
            [0, 0, 0, 20],  # benign
        ]
        y = [1, 0, 1, 0]  # 1 = malware, 0 = benign

        rf_model = RandomForestClassifier(n_estimators=100)
        rf_model.fit(X, y)

        # New app analysis
        new_app = [[1, 1, 1, 70]]
        prediction = rf_model.predict(new_app)
        print("Malware" if prediction[0] else "Benign")

        # Feature importance
        importances = rf_model.feature_importances_
        print("Feature importances:", importances)

        Malware
        Feature importances: [0.40096618 0.07004831 0.14009662 0.38888889]
```

Figure 10-18. Classification of benign vs. malicious app using random forest

Support Vector Machine (SVM): They are well-suited supervised learning models for classification and high-dimensional data or data not linearly separable. As such, they are often used in mobile digital investigations when assuming anomalies for detecting unknown malicious mobile applications, web pages, or suspicious network behaviors. Figure 10-19 presents simple Python code which can detect unknown anomalies or malicious behavior in app/network activity (e.g., unusual data usage, access patterns) using SVM.

```
In [3]: from sklearn import svm

        # Features: [data_sent_MB, data_received_MB, connections_opened]
        X = [
            [1, 2, 3],     # normal
            [2, 2, 2],     # normal
            [10, 50, 25],  # anomaly
            [1, 1, 2],     # normal
        ]
        y = [0, 0, 1, 0]  # 1 = anomaly/malicious, 0 = normal

        svm_model = svm.SVC()
        svm_model.fit(X, y)

        # New behavior check
        new_behavior = [[8, 40, 20]]
        prediction = svm_model.predict(new_behavior)
        print("Anomaly" if prediction[0] else "Normal")

        Anomaly
```

Figure 10-19. Detection of anomalies in network/app activity using SVM

CHAPTER 10   LEVERAGING MACHINE LEARNING FOR MOBILE FORENSICS

k-Nearest Neighbors (k-NN): k-NN is a highly intuitive algorithm that can be used for both classification and regression tasks, as it makes predictions based on the majority class among its nearest neighbors closest to the chosen data point. In mobile investigations, the k-NN algorithm can be used to classify user behavior based on patterns of app usage.

Naive Bayes: Naive Bayes is a probabilistic classifier based on Bayes' theorem, assuming features are conditionally independent. So, it works well in text applications, such as SMS content analysis or emails found on mobile devices, detecting phishing attempts or spam messages. Figure 10-20 demonstrates a very simple Python code which can be used for text classification using Naive Bayes.

```
In [1]: from sklearn.feature_extraction.text import CountVectorizer
        from sklearn.naive_bayes import MultinomialNB

        # Sample SMS data
        texts = ["Meet me at 5", "There is a bomb", "Let's grab lunch", "Kill him now"]
        labels = [0, 1, 0, 1]   # 1 = relevant, 0 = irrelevant

        # Vectorization
        vectorizer = CountVectorizer()
        X = vectorizer.fit_transform(texts)

        # Training classifier
        model = MultinomialNB()
        model.fit(X, labels)

        # Predict new message
        new_text = ["Attack planned at night"]
        new_X = vectorizer.transform(new_text)
        prediction = model.predict(new_X)
        print("Relevant" if prediction[0] else "Irrelevant")

        Irrelevant
```

Figure 10-20. *Text classification using Naive Bayes*

K-Means Clustering: K-means is an unsupervised learning algorithm based on clustering data points into groups according to their similarity. This technique can help classify similar mobile network behaviors or certain usage patterns of a smartphone that require further investigation.

Model Selection Criteria

The selection of the appropriate machine learning model in mobile forensic investigations gives assurance of correctness and efficiency of the analysis procedure. Various factors influence the type of model that is required to be selected. Among them are the nature of data, the type of task, and especially the specific investigation goals, which have been explained in the next session.

Nature of Data

The type of data is of great importance when choosing the right machine learning model. These include

- **Text data:** Models like Naïve Bayes or Neural Networks with Natural Language Processing (NLP) techniques are suitable for data involving text. NLP helps analyze context, extract relevant keywords, and identify patterns in communication that may reveal intent or relationships.

- **Time-Series Data** (call logs, app usage): Models applicable for time-ordered data include Long Short-Term Memory (LSTM) networks or Support Vector Machines (SVMs). They can trace long dependency sequences in time for LSTMs and spot possible periodic trends in SVMs. These will be helpful in exploring recurrent patterns or anomalies in usage.

- **Multimedia Data** (images, videos): CNNs are also super effective in extracting and analyzing visual features automatically for image or video frame analysis. Thus, they are great for object recognition, face detection, or recognition of unusual patterns, normally important to determine visual evidence in mobile investigations.

Task Type

The kind of task to be addressed—whether classification, clustering, anomaly detection, or regression—also impacts model selection:

- **Classification Tasks:** Models used when the objective is categorizing data into different classes, such as flagging suspicious apps or users, are Random Forests, SVMs, and neural networks. They can work with large sets of complex patterns, and classifying objects yields reliable results in forensic investigations.

- **Clustering Tasks:** When the objective is to cluster similar data points, such as grouping similar user behaviors or device activities, K-means clustering or DBSCAN algorithms are used. These algorithms help in anomaly detection by finding natural groupings within the data, which is useful in cases where a dataset has no predefined classes—a scenario common in exploratory forensic analysis.

- **Anomaly Detection:** Applications looking for outliers or anomaly behavior, like detecting unauthorized access or unusual app usage, are best modeled using Isolation Forest or One-Class SVMs. These types of models are particularly designed to determine anomalies in huge datasets, useful for identifying suspicious activities that deviate from normal usage patterns

Other Factors

Other factors considered when making a model choice include

Model Complexity vs. Interpretability: Simpler models, such as Decision Trees or Naïve Bayes, have better interpretability, particularly for investigators who must discuss findings in a court or legal setting. More complex models, such as deep learning networks, would typically offer more accuracy but might be harder to interpret.

Scalability and Speed: Random Forest, CNN, or other scalable models would be more effective in larger data sets since they are suited to high-dimensional space, but at the cost of more computational cycles. Therefore, the focus for real-time investigation or large data sets should lie on efficiency and performance.

Labeled Data Size: In scenarios where, labeled data may be scant, it may even be necessary to develop models for unsupervised or semi-supervised learning. Semi-supervised learning approaches can leverage a few pieces of labeled data and a much larger pool of unlabeled data, so the model can generalize much better in scenarios where labeling is difficult. Considering these facts, with respect to data nature, task kind, and operational-level constraints, investigators can then choose an appropriate machine learning model that suits the need for investigating a mobile case in the most accurate and efficient way possible.

Training Machine Learning Models

Once a model has been selected, it must be trained on the available data. The proper training of a model will decide whether the model generalizes well to new data and whether it will work correctly in real forensic investigations.

Data Splitting

- **Training Set:** This is the part of the dataset that trains the machine learning model itself. From this set of data, the model learns patterns, relationships, and rules.

- **Validation Set:** During training, a validation set comes into play. The model's parameters are fine-tuned using this validation set to prevent overfitting. Therefore, it provides an unbiased evaluation of the model during training.

- **Test Set:** After the model is trained and validated, the test set is used to check how the model performs on data it has never seen before. This ensures the model will have good generalization behavior rather than memorizing the training data. For instance, 60–70% of the data may be used for training, 15–20% for validation, and the remaining 10–20% for testing. The ratio may vary with the dataset size and model complexity.

Training Methodologies

- **Cross-Validation**: The most widely used training methodology is probably k-fold cross-validation. Here, data are divided into k subsets. For validation, the model will be trained on k-1 subsets. This process is repeated k times, such that each subset acts as a validation set once. Cross-validation ensures the model is robust and less prone to overfitting.

```python
# Create an imbalanced classification dataset
X, y = make_classification(
    n_samples=500, n_features=10, n_classes=2,
    weights=[0.85, 0.15],  # imbalanced
    random_state=42)
# ---- Stratified Sampling ----
# Keeps class distribution similar in train/test sets
X_train, X_test, y_train, y_test = train_test_split(
    X, y,
    test_size=0.2,
    stratify=y,   # ensures same class proportions
    random_state=42)
print("Class distribution in full dataset:", np.bincount(y))
print("Class distribution in training set:", np.bincount(y_train))
print("Class distribution in test set:", np.bincount(y_test))
# ---- k-Fold Cross-Validation ----
print("\n--- Regular K-Fold Cross-Validation ---")
kf = KFold(n_splits=5, shuffle=True, random_state=42)
model = LogisticRegression(max_iter=1000)
scores = cross_val_score(model, X_train, y_train, cv=kf, scoring='accuracy')
print("Accuracy scores:", scores)
print("Mean accuracy:", scores.mean())
# ---- Stratified k-Fold Cross-Validation ----
print("\n--- Stratified K-Fold Cross-Validation ---")
skf = StratifiedKFold(n_splits=5, shuffle=True, random_state=42)
strat_scores = cross_val_score(model, X_train, y_train, cv=skf, scoring='accuracy')
print("Accuracy scores:", strat_scores)
print("Mean accuracy:", strat_scores.mean())
```

Figure 10-21. Stratified K-fold validation technique

- **Stratified Sampling:** Stratified sampling is particularly useful for classification problems because it establishes the same or similar natural proportions of classes (e.g., malicious to benign apps) in the training, validation, and testing sets. Stratified sampling makes sure that the training and test sets have the same mix of classes as the original data, so nothing is over- or under-represented. In k-fold cross-validation, you split the training data into *k* parts (folds) and train the model *k* times, each time using a different part for testing and the rest for training. Stratified k-fold cross-validation works the same way, but it also keeps the class balance in every fold. This is especially important where imbalanced datasets prevail, as is very common in mobile forensics.

Overfitting and Underfitting

Overfitting and underfitting are two significant issues that come into play during the development of a model, particularly in forensic applications where precision and generalization abilities are of utmost importance. This makes it a key issue in developing effective machine learning models for mobile digital investigations.

Overfitting

Overfitting is said to occur when a model learns the noise and minor details of the underlying patterns in the training data set, meaning it performs very well on the training data but performs very poorly on new, unseen data. In mobile forensic investigations, an overfitted model might perfectly identify every detail in some particular training dataset but fails to generalize to other datasets, which can lead to possible wrong conclusions. Other indicators of overfitting include performance that is very good on the training set but much poorer performance on validation or test sets, and an immense gap between the training and validation errors. Overfitting can be addressed through the implementation of some solutions.

The regularization techniques used to penalize large coefficients in the model, so as not to focus too much on minor differences within data, are L1 (Lasso) and L2 (ridge) regularization. Also, methods such as k-fold cross-validation may be applied to catch overfitting by compelling a model to perform well on more than one subset of the data. Simplifying the model can also be effective; for example, the application of pruning decision trees or simply reducing layers effectively prevents overfitting.

Underfitting

This is underfitting, where the model is too simple to capture the underlying structure of data, thereby performing poorly on both the training set and new data. Underfit models in mobile forensic investigations may fail to identify critical patterns; therefore, they will miss key indicators of malicious activity.

CHAPTER 10   LEVERAGING MACHINE LEARNING FOR MOBILE FORENSICS

In case of underfitting, the model is too simple to capture the underlying structure of data, thereby performing poorly on both the training set and new data. Underfit models in mobile forensic investigations may fail to identify critical patterns; therefore, they will miss key indicators of malicious activity. To address underfitting, more complex models can be used, relevant features can be added, and training time can be increased so the model learns richer representations of the data.

Balancing overfitting and underfitting is a challenging task when developing models for real-world forensic environments. With the right application of validation techniques and careful tuning of the model by forensic investigators, machine learning models assure reliability and effectiveness in digital evidence discovery.

Fine-Tuning and Optimization

Once the model is trained, fine-tuning and optimization can be performed on it. Fine-tuning refers to the fine adjustment of hyperparameters or the parameters set before models are trained. These are not learned by the models.

Hyperparameter Tuning Techniques

Hyperparameter tuning techniques include varying parameters, such as learning rate, strength of regularization, or number of layers of a neural network, to optimize machine learning model performance. The following are popular techniques that have been applied:

a) **Grid Search**: A grid search is an exhaustive search that tries all possible combinations of hyperparameter values and selects the one that gives the best performance. Though it guarantees an optimal solution, it might become computationally costly for models with many hyperparameters, such as neural networks, where hyperparameters like learning rate, number of layers, batch size, and activation functions are tuned to produce an optimal performance model. An important task in hyperparameter tuning is fine-tuning random forests, including the number of trees, maximum depth, and split criteria.

CHAPTER 10 LEVERAGING MACHINE LEARNING FOR MOBILE FORENSICS

```python
def create_model(learning_rate=0.01, layers=1, activation='relu'):
    model = Sequential()
    model.add(Dense(32, input_dim=10, activation=activation))
    for _ in range(layers - 1):
        model.add(Dense(32, activation=activation))
    model.add(Dense(1, activation='sigmoid'))
    model.compile(optimizer='adam', loss='binary_crossentropy', metrics=['accuracy'])
    return model

model = KerasClassifier(build_fn=create_model, verbose=0)

param_grid = {
    'learning_rate': [0.001, 0.01],
    'layers': [1, 2, 3],
    'activation': ['relu', 'tanh'],
    'batch_size': [16, 32]
}
```

Figure 10-22. Basic Python code of grid search-based fine tuning

The code mentioned in Figure 10-22 builds a customizable neural network that can change its learning rate, number of layers, activation function, and batch size. The *create_model* function sets up the network's structure and training method, while *KerasClassifier* makes it compatible with scikit-learn tools. The *param_grid* lists all the possible combinations of these settings so a Grid Search can test them and find the best one for performance.

b) **Random Search**: Instead of trying all permutations of hyperparameters, random search takes random permutations within a prescribed range. Rather than being time-consuming, as with grid search, random search is much more efficient because it has the potential to perform as well as or even better than grid search, especially when the hyperparameter space is large.

c) **Bayesian Optimization**: This is a more complex approach that uses probability to predict which set of hyperparameters will perform better. This method iteratively searches and is generally more efficient than grid search or random search if the number of available hyperparameters is large.

Early Stopping

Among the various techniques widely used to prevent overfitting in deep learning models, such as neural networks, is early stopping. Overfitting occurs when the model learns not only the underlying pattern of the training data but also the noise and random fluctuations; therefore, it leads to poor generalization to unseen data.

Early stopping is an approach that monitors the performance of the model on the validation set while training, rather than focusing merely on the loss or accuracy of the training set. The model's weights are updated over multiple iterations or epochs. With each epoch, the model does its best to minimize a loss function. Therefore, the training data will continue to see a decline in loss for more epochs, but the validation set will most probably increase in loss after a certain number of epochs, indicating that the model is overfitting to the training data.

The main concept of early stopping is to stop training before the overfitting process takes place. It works as follows:

a) **Validation Monitoring**: The model's performance is monitored at the end of every epoch for validation loss or accuracy. This is done using a separate validation set that isn't used for training; instead, it serves as a proxy for how well the model will perform on unseen data.

b) **Patience Parameter**: A patience parameter is defined, referring to how many epochs the model is allowed to continue training before improvements stop. For instance, if there is no improvement in the validation performance of the model after a specified number of epochs (e.g., 5 or 10 epochs), training stops.

c) **Stopping Criterion**: The training process stops when it fails to improve on the validation set for the duration of the patience period. This avoids overfitting the model to the training data while preventing it from becoming too complex or overfitting to the noise in the training data.

d) **Best Weights Retained**: In most cases, the weights corresponding to the epoch with the best validation performance are retained. When early stopping is triggered, it reverts to the saved weights and continues with the version of the model that best generalizes.

Using early stopping provides a balance between overfitting when the model is too simple and cannot grasp the underlying patterns, and when the model is too complex and learns to memorize the data provided. This technique is helpful in developing a model that generalizes well to new, unseen data, which is critical in applications such as mobile forensics or any predictive task requiring robustness to new data.

Regularization Techniques

Overfitting happens when a model becomes too complex and learns not only the underlying patterns but also noise and outliers present in data, hence making poor performance on unseen data. Regularization techniques are methods in the machine learning domain that prevent the overfitting of complex models, such as neural networks, during training. Regularization includes various kinds of constraints or penalties added during learning to make the model simpler and more generalizable.

Common types of regularization include

a) **L1 Regularization (Lasso)**: In this technique, the absolute value of weights is added as a penalty to loss. This encourages sparsity—some weights are driven to zero, hence it selects only the most important features. This comes in handy in situations where feature selection is required.

b) **L2 Regularization (Ridge)**: This method adds the square of the weights to the loss function. However, unlike L1, it is less likely to send larger weights toward zero; rather, it spreads out the contribution of all features, reducing the risk of overfitting.

c) **Elastic Net**: It combines both L1 and L2 regularization. The advantage of this is that it ensures a balance between feature selection (the L1 aspect) and weight shrinkage (the L2 aspect). It turns out to be quite useful when features are correlated.

CHAPTER 10   LEVERAGING MACHINE LEARNING FOR MOBILE FORENSICS

```
from sklearn.linear_model import Lasso, Ridge, ElasticNet
from sklearn.model_selection import train_test_split
from tensorflow.keras.models import Sequential
from tensorflow.keras.layers import Dense, Dropout, BatchNormalization
from tensorflow.keras.preprocessing.image import ImageDataGenerator
import numpy as np
X, y = load_boston(return_X_y=True)
X_train, X_test, y_train, y_test = train_test_split(X, y, test_size=0.2)

# L1 Regularization (Lasso)
lasso = Lasso(alpha=0.1)
lasso.fit(X_train, y_train)

# L2 Regularization (Ridge)
ridge = Ridge(alpha=1.0)
ridge.fit(X_train, y_train)

# Elastic Net
elastic = ElasticNet(alpha=0.1, l1_ratio=0.5)
elastic.fit(X_train, y_train)

print("Lasso score:", lasso.score(X_test, y_test))
print("Ridge score:", ridge.score(X_test, y_test))
print("Elastic Net score:", elastic.score(X_test, y_test))

Lasso score: 0.7532812785091995
Ridge score: 0.7583622831961011
Elastic Net score: 0.7443550301603055
```

Figure 10-23. Implementation of L1, L2, and Elastic Net

The code mentioned in Figure 10-23 shows the basic implementation of three regularization methods called Lasso (L1), Ridge (L2), and Elastic Net on the Boston housing dataset to prevent overfitting. The parameter alpha indicates the regularization strength. It controls how much penalty is added to the model's loss function for large weights. The output shows Ridge performing slightly better (0.758) than Lasso (0.753) and Elastic Net (0.744), meaning Ridge generalized best for the dataset used in Figure 10-23.

d) **Dropout**: Dropout is a regularization technique applied in deep learning, whereby all neurons during training become "dropped" (i.e., ignored). This prevents the network from relying too much on any neuron or feature; it compels the model to learn more robust patterns, which generalize better to new data.

e) **Data Augmentation (in Deep Learning)**: While not an overt form of regularization, data augmentation artificially increases the size and variability of the training set. For instance, when dealing with images, transforms may be applied in different orientations—rotate, flip, and scale for better generalization. By doing so, the model is exposed to much more varied data than it otherwise might have been exposed to by the training set alone, curbing the possibility of overfitting.

f) **Batch Normalization**: It normalizes the input of every layer within a neural network to always have a consistent mean and variance. It stabilizes and accelerates training and has a slight regularizing effect by reducing sensitivity toward initial weights and learning rates.

```
model = Sequential([
    Dense(64, activation='relu', input_shape=(X_train.shape[1],)),
    BatchNormalization(),
    Dropout(0.5),  # Drop 50% neurons during training
    Dense(32, activation='relu'),
    Dense(1)
])
model.compile(optimizer='adam', loss='mse')

# Fake image data augmentation
datagen = ImageDataGenerator(rotation_range=20, horizontal_flip=True, zoom_range=0.2)

fake_images = np.random.rand(10, 64, 64, 3)
aug_iter = datagen.flow(fake_images, batch_size=2)
```

Figure 10-24. *Batch normalization, Dropout, and Augmentation*

The code mentioned in Figure 10-24 builds a neural network with batch normalization and dropout (removing 50% of neurons during training to reduce overfitting), then compiles it with the Adam optimizer and mean squared error loss. It also sets up fake image data and applies augmentation like rotation, flipping, and zooming to simulate varied training images. Regularization techniques are important for building robust models capable of generalizing well on unseen data. They avoid overfitting to some features, allowing proper generalization by penalizing complexity to improve model performance and reliability in most machine learning applications.

Evaluation and Validation of Machine Learning Models

Once the model has been developed, evaluation and validation come into operation; these are good checks that help ensure the reliability of the machine learning models by checking how well a model generalizes to unseen data. Several metrics—accuracy, precision, recall, F1-score, and area under the curve (AUC)—are used to measure the quality of the model's performance. Techniques used for cross-validation, particularly k-fold cross-validation, are used to prevent overfitting because one needs to split the dataset into numerous subsets, while the model must be tested on different parts of the data. Another very important subtask of the cross-validation technique is using a dataset to validate the model's ability to make predictions, so as not to predict on new instances. This means it tests the model under realistic conditions, such that it solves the problem and satisfies deployment practicality standards.

Applications of Machine Learning in Mobile Digital Investigations

The most distinguishable applications of ML in mobile digital investigations handle big volumes of data for critical insights. Among its primary applications is the automated classification of data, wherein large volumes of mobile device data, whether in the form of messages, call logs, or app usage, are categorized by ML models into predefined categories at a much faster rate than traditional methods for investigations. Another very important application is anomaly detection, where ML algorithms utilize anomalies in mobile data to detect potential security breaches or malicious access or behavior.

Another application is in the analysis of app behavior, where machine learning models point out pattern characteristic of malware or suspicious activity on apps based on features like permissions requested, background processes, and data transmission patterns. Pattern recognition and clustering techniques also allow investigators to group related data points together, which helps spot connections between activities or individuals when conducting investigations.

Furthermore, file recovery operations use ML models to restore deleted or hidden files from mobile devices. NLP can be applied, for example, to analyze text messages, social network posts, or emails for the presence of words or phrases indicating a particular sentiment, tone, or pattern of communication relevant to the case. Overall, machine learning significantly boosts efficiency and accuracy in mobile digital investigations.

In the next chapter, the various applications of ML in mobile digital investigation are explained in detail.

Integration of Machine Learning with Forensic Tools and Processes

Machine learning (ML) today transforms the face of mobile digital investigation by automating extractions of various patterns from big data and analyses that would be difficult to do manually. The introduction of ML models into forensic tools and workflows facilitates the investigation process, increases accuracy, and makes the process reliable. This chapter discusses how to incorporate ML into existing forensic processes, and touches on the pains and limitations investigators may experience after deploying ML in mobile forensic investigations.

Workflow Integration

The integration of machine learning into already-established forensic workflows calls for a highly structured approach that brings together the power of ML with the objectives and procedures characteristic of the investigation process. This integration can be broken down into a series of main phases:

a) **Preprocessing and Data Collection:** Mobile digital investigations include the collection of unprocessed data on mobile devices, such as application data, logs, and geolocation information, and corresponding communication records. It integrates data extraction and filtering tools with machine learning models, in which the preprocessing process can be enhanced automatically by ML, based on identifying relevant data, tagging suspicious patterns, or filtering out unwanted data.

b) **Analyzing Data and Detecting Patterns**: Once data is pre-processed, ML algorithms can be used to identify patterns indicating malicious behavior, such as the presence of malware, unauthorized access, or exfiltration of data. ML models can analyze data sets much larger than those that humans could process in far less time, surfacing relevant evidence and correlations. For instance, anomaly detection algorithms may be used to detect unusual user behavior or access patterns.

c) **Report Generation and Visualization:** Forensic tools will assist in integrating ML to automate report generation and visualize findings. Machine learning algorithms can group related evidence or show correlations between activities, enabling investigators to understand complex data better. A tool can generate insights and recommendations automatically based on the analysis presented above, providing a clearer path for solving a case.

d) **Automation of Repetitive Processes:** ML models are well-suited to automate repetitive processes within a forensic workflow, such as reviewing thousands of messages or duplicate file identification. This allows investigators to focus on more technical areas of the incident. Moreover, ML algorithms can be scheduled to scan continuously for signs of digital system compromise or anomalies without manual intervention, alerting investigators when potential evidence is found.

e) **Predictive Analytics and Decision Support:** Machine learning algorithms can predict the likelihood of future events or behaviors based on patterns uncovered in historical data, making them extremely useful for predicting security breaches or investigation outcomes. ML-based decision support systems may assist forensic investigators in decision-making at different investigation levels by providing further actions based on predictive insights.

Challenges and Limitations

Even though numerous advantages are associated with integrating machine learning into forensic workflows, challenges and limitations accompany them. To successfully address the gap in real-world forensic activities, integration and deployment can be successful and run properly only if these challenges are overcome.

a) **Data Availability and Quality:** ML models need massive, high-quality training data, as well as testing. Relevant and labeled data may be hard to access in mobile digital investigations, where most information is sensitive or encrypted. Furthermore, data obtained from mobile devices might not be complete or may be corrupted, which challenges ML models requiring clean, structured datasets. Data quality issues must be resolved for effective ML integration.

b) **Model Interpretability:** One of the biggest issues in applying machine learning to forensic investigation is that complex models, such as deep learning networks, lack interpretability. Investigators need to understand how models use rules and inputs to reach certain conclusions, so they can apply those findings in court. Investigators, legal teams, or judges are unlikely to trust machine learning models functioning as "black boxes" where nobody knows what happens within them. Research is ongoing on more explainable models and tools that can explain their decision-making processes.

c) **Legal and Ethical Issues:** In forensic investigations involving machine learning tools, legal and ethical standards must be met, mainly revolving around sensitive data handling. With myriad laws governing data privacy, such as GDPR, imposing tight restrictions on personal data collection, processing, or analysis, one must stay up-to-date. Machine learning use should not violate these rules, and forensic teams must obtain proper authorization when accessing and analyzing data. Additionally, there is a responsibility to carefully consider the ethical implications of conducting surveillance or data collection investigations with automated tools. If ML models are shared publicly or within the DFIR community, it's important to mask or anonymize personal data so that no private details are exposed. Without these protections, the models could be misused to uncover information about individuals (for example, through hacking or unrestricted access), which might later be used to target them.

d) **Resource and Expertise Constraints:** Applying machine learning models in forensic tools involves massive computation and deep expertise in data science and digital forensics. Most forensic teams lack the necessary infrastructure or personnel with the technical acumen to design, train, and deploy machine learning models. This makes it difficult for ML-driven tools to be adopted in smaller forensic labs or law enforcement agencies. It requires training forensic practitioners in machine learning algorithms and availability of adequate computational resources.

e) **Model Maintenance and Adaptability:** Machine learning models must be updated and retrained constantly to avoid becoming outdated in the rapidly changing world of mobile digital investigations. New malware varieties, changes in attack patterns, and mobile technology improvements require ML model adaptability. Incoherent systems cannot identify newer threats or handle new data types if models are not maintained and upgraded.

Summary

The integration of machine learning in the investigation of mobile digital potentially benefits with quicker, more accurate, and scalable forensic processes. Tasks automation, pattern identification, and predictive insights from machine learning make it a highly valuable tool to be used during investigation processes. However, teams have the task of overcoming the challenges that are arising in terms of data quality, model interpretability, legal compliance, and resource constraints to take full advantage of the capabilities of ML. The success achieved in the case studies is in moving forward to revolutionize this world of mobile forensics while becoming more efficient and reliable in terms of investigation as there is evidence in the multilevel digital landscape.

References

[1] https://medium.datadriveninvestor.com/exploring-the-basics-of-supervised-learning-bfc18fe5ce72

[2] https://medium.com/@christophe.atten/exploring-the-basics-of-unsupervised-learning-6b35df79064e

[3] https://medium.datadriveninvestor.com/the-ultimate-beginner-guide-of-semi-supervised-learning-3bd11cb19835

[4] https://medium.com/@qjbqvwzmg/understanding-the-three-pillars-of-machine-learning-supervised-unsupervised-and-reinforcement-b3a5317b5612

[5] All codes are available at: https://github.com/raviesheth2608/Android-and-IOS-Mobile-Forensics-A-Press-/tree/main

[6] https://pythongeeks.org/

CHAPTER 11

Applications of Machine Learning in Mobile Forensics

Mobile forensics has emerged as one of the most important areas within the field of digital forensics, given that mobile devices are ubiquitous and offer gigabytes of data. Mobile forensics is a process for recovering data from mobile devices, such as smartphones and tablets, and then analyzing it for purposes of aiding criminal investigations, legal proceedings, or cybersecurity efforts. As the complexity of mobile phones increases, so does the data; it is surprising to find that traditional forensic techniques often fall short in efficiently and effectively analyzing that amount of information. Machine learning (ML) has recently been proven to be one of the subsets of AI that is a very important tool in solving the problems and issues of mobile forensics. Its algorithms enable systems that can learn from data, which improves their performance over time in identifying patterns, anomalous patterns, and classifying data in ways humans are not capable of performing. The integration of ML helps improve the detection of digital artifacts, automate most tedious actions, and provide predictive insights that can be used proactively during investigations.

Applications of Machine Learning in Mobile Forensics include identifying different types of digital artifacts, such as text messages, images, and even application data; anomaly detection, which may show malicious activity; predictive analytics for forecasting criminal activities and behaviors. Machine learning is also instrumental in financial fraud detection, social media analysis, threat intelligence, and email classification. Each of these applications uses specific ML models and techniques adapted to the peculiar characteristics of mobile data. In this new trend of evolution

© Ravi Sheth, Keshav Kaushik, Chandresh Parekha, Narendrakumar Chayal 2025
R. Sheth et al., *Android and IOS Mobile Forensics*, https://doi.org/10.1007/979-8-8688-1748-9_11

in mobile forensics, the use of machine learning is bound to increase even more, thus opening up new avenues for improving the outcomes of investigations while increasing the accuracy and efficiency of forensic analyses.

This chapter will discuss the key applications of machine learning in mobile forensics. It will discuss the possible transformation that the field may undergo with the inclusion of new technologies to match the need for a higher complexity level in digital investigations.

Machine Learning Models for Anomaly Detection in Mobile Phones

Anomaly detection can be defined as the process of identifying patterns that do not comply with expected or normal behaviors. These anomalies are also commonly known as outliers and can suggest unusual events of interest in investigations made in forensic analysis. Generally, anomaly detection plays a crucial role in finding abnormal usage behaviors, unauthorized accesses, malware activity, fraud, and all other security breaches in mobile forensics. Because anomalies represent data points or events that do not occur within a defined set of norms, it can be beneficial to detect those anomalies as early as possible to prevent subsequent damage or loss. Anomaly detection is really useful in mobile forensics. Indeed, due to the large quantity of data generated by mobile phones, from call log information to app usage, messaging, geolocation data, and network traffic, discovering unusual patterns in data can reveal critical information relating to threats or malicious activity. Anomalies in mobile data could include any peculiar login times, sudden increases in data usage, unexpected application behaviors, or irregular communication patterns. Forensic analysis detects such anomalies to trace suspicious activity, compromised devices, or identify insider threats. Anomaly detection is just one application in cybersecurity, but it could be utilized in many fields, including finding fraud in financial systems, medical diagnostics, and monitoring industries. It is especially useful in mobile forensics because it helps identify activities that might indicate criminal behavior or fraud, or incriminate network breaches.

There are three types of anomalies commonly identified by investigators in forensics:

- **Point anomalies:** This refers to a single data point that is significantly different from other data. For example, an abnormal login attempt from an unusual location can be classified as a point anomaly.

- **Contextual anomalies:** These are anomalies only in specific contexts. Executing several data transfers could be normal for a user in a work environment during working hours but an anomaly if executed late at night.

- **Collective anomalies:** This refers to a collection of correlated observations that, when taken as a whole, demonstrate abnormal behaviors. For instance, an abnormal increase in usage data coinciding with several login attempts from other IP addresses could be a symptom of a system breach.

Common Mobile Data Anomalies

Machine learning (ML) has come to be known as one of the more important tools used in anomaly detection. It can process vast amounts of data and find complex patterns that would otherwise go unnoticed with other, more traditional statistical methods. Some machine learning algorithms, specifically trained on historical data by first establishing what "normal" behaviors are, can automatically locate any deviation from this norm in mobile forensics. Following is a number of anomalies that indicate malicious activities, unauthorized access, or abnormal user behavior in mobile forensics. A few examples of common mobile data anomalies include the following:

- **Anomalous locations of login**: If logins come from a distant location from the expected geo-location of the user, this is in bad faith and usually indicates unauthorized access. For example, a user might log in from a particular city most of the time, but log in from another continent one day; such a login might turn out to be a breach.

- **Unexpected binge on data**: Sudden spikes in data usage could point toward the presence of malware or some other unauthorized application running in the background. Machine learning models monitor patterns of data consumption and send an alert should there be an unusual, spiky increase in data usage.

- **Irregular call patterns**: Some anomalies reveal themselves within call logs—increased calls to a few specific international numbers or calls placed at irregular times. Analyzing the pattern of calls over time can help identify some fraudulent activity, such as scams or social engineering.

- **Suspicious application installations:** Unusual or unauthorized applications installed when it seemingly deviates from the normal behavior of the user's app installation can be a sign of malware or hacking. Models for anomaly detection observe app installation activity to be alerted of deviation from typical behavior.

- **Inconsistent GPS data:** It might imply spoofing or unauthorized tracking of a location if the data from the user's GPS shows the user to be in two places far apart within an impossible time frame. Forensic investigators use anomaly detection to identify these inconsistencies and then validate the location data authenticity.

- **Unusual device activity late at night:** Smart devices operate in trends as seen by the pattern of a device user's activities. If a device suddenly starts transmitting enormous quantities of data at odd hours late at night, it could be an indication of an intrusion. Anomaly-detecting software checks time-based usage patterns of devices to discover such unusual activities.

An outline of some of the most common approaches applied in the context of anomaly detection within forensic investigations follows:

Supervised Learning for Anomaly Detection

In supervised learning, the model is trained with labeled data; that is, both normal and anomalous examples are provided during training. Subsequently, it learns the way to distinguish between normal and anomalous instances based on the features within the data.

- **Classification Algorithms:** Algorithms include decision trees, random forests, support vector machines, etc. These algorithms classify data as normal or anomalous. For example, an SVM can be trained with known normal and anomalous patterns of mobile app usage to classify data points later on.

- **Neural Networks:** Neural networks, especially deep learning models, are useful in identifying complex patterns indicating anomalies. In mobile forensics, they can be applied for scrutinizing logs from app usage, network activity, or patterns in user behavior to identify barely noticeable anomalies signifying unauthorized access or malware.

Unsupervised Learning for Anomaly Detection

Unsupervised learning is used in approaches that lack labeled data. In such cases, the algorithm is trained only on normal data, and all data points with significant differences from those learned normal behaviors are flagged as anomalies.

- **Clustering Algorithms**: Clustering algorithms, such as K-means and DBSCAN (Density-Based Spatial Clustering of Applications with Noise), are commonly used for anomaly detection. They group similar data points into clusters and identify points that do not belong to any cluster or are far from the cluster centers, which then become anomalies. In the context of mobile forensics, anomalies could be patterns that appear unusual during app installation, communication behavior, and sometimes geolocation data.

- **Isolation Forests**: Isolation Forests are one of the popular unsupervised machine learning algorithms for anomaly detection. Their principle is that they can arbitrarily choose any feature and then arbitrarily split data. Anomalies are isolated when there are fewer records and distinct characteristics. Isolation Forests can be used in mobile forensics to recognize abnormal user behavior, such as sudden changes in app usage or network activity.

- **Autoencoders**: Autoencoders are neural networks typically used for anomaly detection in unsupervised learning. They are trained to compress input information into a smaller representation and then reconstruct it. Where the reconstruction error is significantly higher than expected—meaning the input does not fit normal behavior—anomalies are flagged. For example, an autoencoder trained on normal mobile network traffic patterns might flag spikes in data transfer as anomalous, potentially indicating malware activity.

CHAPTER 11   APPLICATIONS OF MACHINE LEARNING IN MOBILE FORENSICS

Semi-supervised Learning for Anomaly Detection

Semi-supervised learning involves training your model on a small amount of labeled data and a large amount of unlabeled data, which can be very useful for mobile forensics, where there are limited examples of anomalies but plenty of normal data.

- **Hybrid Models**: Hybrid models are a mix of supervised and unsupervised learning techniques. It can be hard to clearly define anomalous behavior or label it, which is one reason why mobile forensics is not usually done in this fashion. These models can use known anomalous data points and implement unsupervised methods to find newer ones that have been unseen thus far.

Since data captured by a mobile device is largely time-dependent—for instance, call logs and GPS data time-series analysis is very crucial. Machine learning algorithms in this case analyze time-based data and create patterns to find anomalies different from the expected sequence or trends.

- **Long Short-Term Memory (LSTM Networks)**: LSTMs are a form of recurrent network (RNN) specially designed for sequential use and capable of memorizing patterns over time. This algorithm will be very handy when it comes to detecting anomalies in time-series data, such as irregular call patterns or unusual app usage over time.

- **Seasonal Hybrid Extreme Studentized Deviate (S-H-ESD)**: It is applied to time-series anomaly detection when the data possesses a seasonality characteristic. Here, it detects unusual patterns in call frequencies, application usage, or network access deviating from a normal seasonal pattern of a user in mobile forensics.

One of the most important tools used in mobile forensic investigations is anomaly detection, which makes it possible to find abnormal patterns of activity that might point to security breaches, fraud, or unauthorized access. Utilizing machine learning approaches—supervised, unsupervised, and time-series analyses—allows investigators to automate much of the work when dealing with large portions of mobile data; this makes the entire process more efficient and, above all, more accurate. Detecting anomalies at an early stage prevents further damage and helps uncover digital evidence that would otherwise remain hidden.

Machine Learning Approaches Predictive Analysis

Predictive analysis is a statistical technique that combines machine learning algorithms and data mining to use historical data to make forecasts for later events, behaviors, or trends. In other words, it is an attempt to gain insightful knowledge that enables right decision-making, foresight, and prevention of emergent threats before they actually happen. A good predictive analysis should be essential in mobile forensic analysis, as it would let investigators predict actions, forecast outcomes, and find connections in digital data that may imply criminal activity, fraud, or system vulnerabilities. At the core of predictive analytics, it revolves around the concept of modeling futures according to patterns of past behaviors and data. For example, in mobile forensics, predictions might include the probability that a device has been compromised, possible user behavior based on communication patterns, or predictive data usage trends that likely point to unauthorized access or data exfiltration.

Use of Predictive Analysis in Mobile Forensics

The primary use of predictive analysis in mobile forensics encompasses extensively aiding investigators to track and anticipate unlawful activity, fraud, or unauthorized access. Some major applications are discussed below:

- **Prediction of User Behavior**: Predictive analysis finds one of its prime applications in mobile forensics, predicting user behavior. Based on historical data analysis of user communication patterns, application usage, and location history, these tools can predict future behavior patterns. For example, if a user tends to use a certain set of applications and communicate with a specific group of contacts at a particular time, then deviations from this norm will appear as possible anomalies or even indications of malicious activity.

- **Risky activity detection:** Machine learning algorithms can be used to predict when a user is more likely to engage in potentially risky activities. These may include installing unverified applications from untrusted sources, visiting suspicious or phishing websites, clicking on malicious links in emails or messages, disabling security settings, or sharing sensitive information on unsecured platforms. It is important to note, however, that some activities, such as downloading applications outside official app stores, may be harmless for experienced users (e.g., power users or developers). The distinction between legitimate and risky behavior must therefore be made carefully to avoid false alarms. With such predictions, forensic investigators can prepare ahead of time to monitor and investigate risky users or devices.

- **Fraud detection:** The fraud case of mobile financial involves predictive analysis in cases where its capabilities help identify fraudulent transactions or behaviors. Based on historical transaction data, machine learning models check patterns and identify those related to legitimate and fraudulent transactions. Predictive models from transaction amount analysis, device location, and user behavior can pinpoint abnormal activities that might involve fraud, such as fast transfers of large amounts of money or transactions starting elsewhere, unfamiliar to the user. Predictive analytics can also prevent fraud by identifying at-risk users or devices based on past behaviors. For example, if a device exhibits behavioral characteristics similar to those in past fraud events, forensic tools can raise alarms for further scrutiny or preventively block access to sensitive services. It is explained in detail in the next section.

- **Malware Detection:** Predictive analytics can detect and prevent malware attacks on mobile devices. By examining historical data on application usage, network activity, or system logs, machine learning models identify patterns descriptive of malware. For example, if a device acts erratically at odd hours by downloading or uploading data, or if an application behaves contrary to its intended purpose, predictive models flag this as anomalous behavior indicating potential malware activity. Besides detecting malware, predictive models forecast which devices or networks are at risk of infection, helping forensic investigators prioritize at-risk devices for inspection. Predictive analysis identifies devices vulnerable to malware propagation in the past.

- **Data Exfiltration Prediction:** Data exfiltration, the unauthorized transfer of data from mobile devices to external sources, is a key concern in mobile forensic investigations. Predictive analysis detects and prevents data exfiltration by analyzing network traffic and device usage patterns. For instance, if a device transmits large amounts of data to unknown servers. Machine learning models detect devices with a greater tendency toward exfiltration based on historical behavior. If a device was associated with suspicious network activities in the past, it may engage in exfiltration activities. Predictive analysis helps investigators focus on high-risk devices, making forensic investigations more efficient.

- **Unauthorized access:** Predictive analysis can be used for anticipating and preventing unauthorized access in mobile devices. Machine learning models identify patterns in historic data of login attempts, device usage, and network activity that mark them as unauthorized access. For example, if a device has been logged in from an unfamiliar location or at an unusual time, the predictive model can flag it as potential unauthorized access. Besides detecting unauthorized access, predictive models can prevent it by predicting which machines will likely present a problem. For example, if a machine was previously used in failed login attempts or suspicious network activities, it is likely that such things could happen again in the future. Predictive analysis enables investigators to take pre-emptive action, such as increasing security around high-risk devices.

- **Predictive Analysis of Insider Threats:** Insider threats, resulting from mishandling access provided to employees or trusted individuals, are a major concern in mobile forensic investigations. Predictive analysis helps identify potential cases of privilege misuse by analyzing user behavior and tracing abnormal patterns. For example, if an employee accesses sensitive data at unusual times or unfamiliar locations, predictive models flag them as suspicious behaviors.

Based on past behaviors and contextual data, machine learning algorithms predict who poses the highest risk as insider threats. A worker recently disciplined or showing signs of disgruntlement may be suspected as a malicious actor. Predictive analysis thwarts insider threats in real time by identifying prospective actors.

CHAPTER 11   APPLICATIONS OF MACHINE LEARNING IN MOBILE FORENSICS

Appropriate Machine Learning Model for Predictive Analysis

Machine learning plays a primary role in predictive analysis, using vast datasets to realize trends and behavior. In mobile forensic investigations, machine learning models can be built from historical data, such as user activity logs, call records, app usage patterns, and geolocation data, to predict future actions or detect possible anomalies. Some common machine learning algorithms in predictive analysis include

- **Linear Regression:** This algorithm predicts a continuous outcome variable based on independent variables or predictors with a dependent variable or outcome. For example, linear regression can predict the likelihood of an event, such as how data usage increases over time, based on historical data in mobile forensics.

- **Logistic Regression:** Logistic regression is useful in situations involving binary classifications, allowing the outcome to be a categorical variable. This model is helpful when predicting whether a device possesses indicators of unauthorized access. It is particularly handy during forensic investigations where predictions of "malicious activity" vs. "normal behavior" are needed.

- **Decision Trees and Random Forests:** Decision trees represent the decision-making process by building a tree-like structure, where each branch represents the rule of the decision based on feature values. Random forests constitute an ensemble of several decision trees, providing more robust predictions by averaging the outcomes of several trees. These algorithms can predict user behaviors, such as app use patterns, and find patterns in communication data signaling looming criminal activity

- **Support Vector Machine (SVM):** SVMs in mobile forensics enable data to be classified into different categories. For instance, the application identifies whether some app interactions or network behaviors reveal fraudulent behavior or are normal.

- **Neural Networks:** Neural networks, which include deep learning models, basically learn by observing patterns in large datasets. Neural networks in mobile forensics can analyze massive data from mobile devices to predict user behaviors or trends of network usage that may indicate security threats or some forms of suspicious activity. The recent developments in RNNs and LSTMs, which are perfectly tailored to the task of handling sequential data, also indicate the possibility of making predictions about future behaviors based on historical sequences of activities, such as app usage or patterns of communication.

- **Clustering Algorithms:** Clustering algorithms, such as K-means, group data points into clusters based on likeness. Such information can often come in handy for mobile forensic investigators to determine groups of users who have similar types of behavior or detect anomalous behavior by detecting outliers from a dataset.

Predictive analytics can be an effective tool in mobile forensic investigations to predict a host of threats, whether it is access by others to a mobile device without permission, fraud, or malware attacks. Machine learning models, as well as historical data collection, allow forensic analysts to make accurate predictions based on user behaviors and such other things as data exfiltration and insider threats. In that case, being able to foresee and deter criminal activity will be very important in maintaining integrity while conducting a forensic investigation of mobile devices, given the rise of mobile devices in our lives.

Financial Fraud Detection

Increased mobile usage in banking, digital payment services, and e-commerce has drastically increased the concern factors toward mobile-related financial fraud. Due to their portability and increasing reliance for general daily usage, mobile devices are becoming the most pursued options by cyber hackers who seek to use them as centers for financial transactions, stealing personal information, and committing fraud. Financial fraud in mobile contexts is quite diversified and ranges from identity theft and phishing to credit card fraud and scams concerning mobile payment services. Financial fraud in the mobile domain can take many forms:

- **Unauthorized Transactions:** Cybercriminals can access a user's mobile banking or payment application and begin conducting fraudulent transactions. This is most often accomplished through phishing attacks or malware harvesting login credentials.

- **Mobile Payment Fraud:** This includes popular mobile payment services, such as Apple Pay and Google Pay, and many other digital wallets. Fraudulent use has gained much momentum recently, mainly by exploiting vulnerabilities in the mobile payments process, stealing card information, and fraudulent manipulation of transaction data.

- **SIM Swap Fraud:** Since hackers can convince telecom companies to swap a victim's SIM card for a new one in the hacker's possession, fraudsters receive access to two-factor authentication codes, thereby bypassing security measures that may lock access to financial accounts.

- **Credential Stuffing:** It involves using previously stolen login credentials from other applications to gain unauthorized access to mobile banking apps. Most people reuse passwords across different applications, making it a widely spread attack vector.

With the high prevalence of these mobile-based threats today, a well-developed detection system that is able to provide capabilities for detecting fraudulent behavior before serious damage can be dealt is required. Nowadays, fraudsters have changed their tactics so much that traditional rule-based systems cannot match them and are not effective for this purpose. In the domain of ML models, it has been found to be more effective.

Machine Learning Models for Fraud Detection

Machine learning is a technique that comes with many benefits over traditional detection methods. It can learn continuously from huge data, detect latent patterns, and modify itself according to new forms of fraudulent activity, unlike static rule-based systems.

The machine learning model, with a database of so-labeled cases where fraud cases are easily identifiable, is trained on a database of legitimate and fraudulent transactions. The model has the capability to learn from legitimate and fraudulent transactions, and from similar characteristics, it can predict which transactions would appear fraudulent based on future transactions. Table 11-1 lists the primary machine learning techniques used in financial fraud detection inside mobile environments.

Table 11-1. Machine Learning Techniques with Descriptions and Use Cases

Technique	Description	Example use case
Logistic regression	A simple yet powerful algorithm for binary classification problems. It analyzes features such as transaction amount, frequency, location, and device used.	Predicting whether a transaction is fraudulent or not
Decision trees	Splits data into smaller groups based on features (e.g., amount, device used). Each branch represents a decision, and the leaf nodes give the prediction.	Classifying a transaction as fraud or non-fraud
Random forest (ensemble)	An ensemble of multiple decision trees whose results are aggregated to improve accuracy and reduce overfitting	Detecting fraud by analyzing patterns like transaction time and user behavior
Support vector machines (SVMs)	Finds the optimal separating boundary (hyperplane) between fraud and non-fraud transactions. Works best when classes have a clear margin	Identifying fraudulent transactions with clear distinguishing features

In most cases, it is hard to find labeled fraud data because fraudsters always change their tactics. Since unsupervised learning models do not rely on labeled data, they are well-suited for discovering unknown or previously unseen fraud patterns. In some cases, it has a few labeled data, such as known fraudulent cases, and bulk unlabeled data, such as normal transactions, to discover new fraud patterns.

- **Clustering Algorithms (K-Means, DBSCAN):** These algorithms group similar data points together and flag the remaining transactions as anomalous. For example, K-Means could cluster transactions from a normal group of spending habits; therefore, any new transaction not a member of a particular cluster would be marked anomalous.

- **Autoencoders:** These neural networks learn compressed representations of normal transactions and can thereby detect anomalies. So, if a new transaction is processed in such a way that it significantly differs from the learned representation, it flags that as the probable fraud. Autoencoders are very useful for finding new types of fraud for which signatures have not been learned previously.

- **Self-Training Models:** Starting with only a small set of labeled fraud cases, they classify new transactions using unsupervised methods. As they continue to find more potential fraud cases, they refine their classifications over time.

Financial fraud is very often anomalous in the sense that some transactions exhibit behavior quite different from normal user behavior. Algorithms designed to detect anomalies are particularly good at identifying such outliers.

- **Isolation Forest:** This algorithm works on the basis of selecting a random feature, after which they split the data. The fewer and more disparate fraudulent transactions, they separate easily and can hence be readily identified.

- **One-Class SVMs:** One-class SVMs are useful when there are only a few fraud cases. It learns from all the "normal" transaction data and marks anything that is not modeled as likely fraudulent.

The same is applicable to financial fraud detection in mobile contexts: machine learning has transformed the way organizations detect and prevent fraudulent activities. This is because machine learning models are very effective in analyzing vast amounts of transactional data and discovering anomalies, thereby adapting to new fraud patterns that could be hidden in transactions. Applications include simple unsupervised clustering techniques to deep learning models. These algorithms find immense application in protecting mobile users against unauthorized transactions, SIM swap fraud, and payment scams.

Social Media Forensics in Mobile Phones

Social media applications are some of the most potent vehicles for communication, information exchange, and interaction in the mobile phone ecosystem. Since social media contains vast amounts of personal data, in many cases, social media data often has a significant role to play in the forensic value of mobile investigations. Forensic social media analysis refers to the collection, preservation, and analysis of information from these sources to unearth digital evidence related to criminal activities or incidents of cybersecurity, as well as personal fights. Mobile devices include social media activity, which is very crucial in forensic investigations due to the following reasons:

- **Communication Evidence:** Information about social media applications in mobile phones carries messages, photos, videos, location, and contact information. This information may be used as pivotal evidence in harassment and cyberstalking cases or in crimes planned.

- **Geolocation Information:** The posts or photos often geotag in social media applications. This will help trace out the user's location and their movements.

- **Content Analysis:** Online content can include uploading, commenting, and sharing multimedia files that will give proof of intent or conduct, such as online radicalization crimes, hate crimes, or fraudulent digital crimes.

- **Timeline Reconstruction:** Many social media applications maintain a record of interactions with their timestamp, along with a history of interaction. This will enable forensic experts to build a timeline of activity, which can be very useful in fraud, misconduct, or cybercrime cases.

Social media applications installed on the mobile devices of suspects or victims can be very useful for mobile forensic investigators when building evidence and verifying claims. As a result, analyzing social media data has become an essential part of modern mobile forensic investigations.

Machine Learning Methods for Analyzing Social Media Data

The volume of unstructured data on social media platforms is so high that manual analysis is not possible. It mainly relies on machine learning to analyze social media data automatically to quickly identify important patterns, trends, or anomalies for forensic investigators.

- **Natural Language Processing for Text Analysis:** NLP happens to be one of the most important machine learning techniques applied when assessing social media data. Social media content is rich with text information—from posts and comments to messages and hashtags. The NLP model, like BERT or GPT, analyzes text content for sentiment, intent, or specific keywords that could establish malicious activity.

- **Sentiment Analysis:** It is the application of machine learning in determining the emotional content of a social media post, which helps identify whether the message is abusive, threatening, or dangerous. For instance, while harassing someone, it identifies the word or statement that is either offensive or threatening and is written in a post or message.

- **Keyword and Topic Detection:** NLP algorithms can pick out specific keywords or phrases relevant to investigating certain illegal activities or extremist views. For example, NLP models can filter thousands of messages exchanged between the victim and suspect and flag pieces containing threatening language or aggressive behavior in a cyberstalking case.

- **Multimedia Sharing or Posting:** Multimedia applications loaded on phones can share images and videos. Machine learning models, in such scenarios, include computer vision algorithms that analyze multimedia files for forensic purposes. Techniques applied include object detection, facial recognition, and scene analysis to identify people, locations, or events captured on multimedia.

- **Facial Recognition:** Through training machine learning models on facial recognition, it easily matches individuals in social media images with known suspects or individuals in a criminal investigation.

- **Content Classification:** Algorithms can classify media based on content. For example, identifying explicit or violent content, such as firearms, knives, or CSAM (Child Sexual Abuse Material), might point toward criminal intent or evidence of wrongdoing. In a drug trafficking case, computer vision algorithms can identify drug paraphernalia or locations associated with previous crimes from posts shared on mobile devices via social media.

- **Network Analysis:** Social media applications have large relational data, including friend lists, followers, and group memberships. With machine learning models, such relations can be analyzed to bring out associate networks engaged in illicit activities.

- **Graph Analysis:** Graph-based learning techniques map and analyze ties formed between users. Through these structures and strengths, which characterize ties between users, investigators can identify influential figures within the criminal network.

- **Community Detection:** In a social media set, ML models identify communities or clusters across the dataset whereby forensic analysts can easily identify groups of users whose activities appear similar and suspicious. For example, in an investigation into organized crime, graph analysis can identify hidden links among various users of multiple mobile social media accounts to facilitate secret communication.

- **Anomaly Detection:** The anomaly detection machine learning approach can identify user behavior not similar to or suspicious in social media. Algorithms flag anomalies when major deviations with activities that do not characterize a particular user are observed; this includes sudden surges in posts, the appearance of incongruent interaction, or account activity. For example, applying machine learning models can flag anomalies, including login attempts from anomalous locations or posts that don't resemble the user.

- **Emotion and Behavioral Profiling:** Machine learning profiles user behavior and emotions. It would be plausible for forensic experts to infer the emotional and psychological state of a user through analyzing posting activity, frequency, topics, or interactions' tone while using social media, questioning possible involvement in crime or susceptibility to manipulation. For example, changes in posting patterns or tone of communication may help determine if a social network user is being coerced to upload unlawful updates through behavioral profiling based on machine learning.

Insights and Value from Social Media Analytics

The mobile phone-based social media analytics contains a plethora of insights that can be used to hone forensic examinations. Some relevant insights include

- **Detection of Criminal Intent:** This can be done by analyzing text, images, and video data using machine learning, thereby breaking down the underlying intent or expression of criminal intent inherent in interactions on social media. For example, with sentiment and keyword analyses, messages intending illegal activity planning or posts celebrating criminal acts can be detected in time to prevent such crimes before they happen.

- **Tracking Movement and Locations:** Most posts on social media contain geotags or implicit location data that machine learning-based algorithms can use to reconstruct all locations a person has visited. This is convenient for alibi purposes and checking suspicious stories created by suspects. Access to geolocation metadata in a suspect's Instagram posts will reveal whether they were at the crime scene location and at that time.

- **Relationships and Networks:** Network Analysis with Machine Learning defines the relationships and associations of the suspect. You will know how to draw interactions between and among social networks, mapping co-conspirators, enablers, or other groups committing crimes. For example, analyzing graph structures of social media followers of a drug dealer reveals previously unknown associates involved in trafficking.

- **Patterns of Deception Detection:** Social media is also used for fraudulent schemes or scams. Through social media posts and interactions, machine learning models develop patterns for deception detection. For example, machine learning identifies fraudulent offers or phishing attempts on social media platforms targeting mobile users by analyzing historical conversations and transactions.

- **Psychological and Behavioral Insights:** Machine learning extracts psychological profiles from a user's social media activity, giving clues about their emotional state, motivations, or mental health. This is critical in cases involving cyberbullying, suicide threats, or radicalization. For example, an examination of the suspect's posts through forensic analysis discloses patterns of increasing aggression and violence built up over time, indicating aggravating criminal behavior.

The analysis of social media-based data with machine learning would be significantly important in mobile forensic investigations, since it enables the use of powerful tools to identify otherwise hidden patterns, criminal intent, and sequences of events. Social media data is massive in volume and normally unstructured; hence, it calls for more advanced ML techniques, such as network analysis, NLP, and computer vision, to be efficiently analyzed and derive insights from. This can be applied to social media data collected from a mobile phone, which would help illuminate persons, their actions, and relationships, because these are fundamental in modern digital forensics.

Threat Intelligence in Mobile Phones

The role of threat intelligence in the mobile ecosystem is to identify, analyze, and mitigate risks related to mobile devices, applications, and networks. With mobile phones increasingly replacing traditional computing devices such as PCs for accessing bank accounts, emails, and storing personal data, they have become prime targets for cybercriminals. Threat intelligence in this context relies heavily on telemetry sources such as Endpoint Detection and Response (EDR) or Mobile Device Management (MDM) agents that collect logs and activity data from devices, enabling deeper analysis of potential threats. In addition, forensic acquisition of mobile device images can provide critical insights for identifying malicious activity and understanding attack vectors. By leveraging these data sources, threat intelligence focuses on gathering actionable information to safeguard devices against malicious attacks, unauthorized access, and data breaches. Mobile threat intelligence tracking involves monitoring a range of sources, including

- **Mobile Malware:** Analysis of recently emerged malicious codes on mobile, including spyware, ransomware, and trojans, which target specific vulnerabilities within the mobile OS. The exploits may use rootkits to gain escalated privileges, steal sensitive information with a keylogger, and malware exploiting zero-day vulnerability or bypassing security features in app sandboxing and code signing. In addition, monitoring malware delivery channels, such as compromised application stores, side-loading, or malicious web links, can help track the pace of changing threats.

- **Phishing Campaigns:** These campaigns utilize email applications, messaging services, or mobile browsers to target users. Such campaigns may adopt a legitimate appearance but use tactics such as dynamic DNS, shortened URLs, and fake application overlays to steal user data. This includes domain squatting, URL redirection, and malicious QR codes, among others, to bypass traditional security checks on mobile. Also included in this group is phishing through rogue apps that may use social engineering to surreptitiously gain unauthorized access to sensitive information stored in mobile applications.

- **Endpoint Vulnerabilities:** The nitty-gritty of some unpatched security vulnerabilities in mobile OS applications lies in, for instance, Android fragmentation, because different OEMs customize it differently and have different patch schedules; or iOS zero-click exploits. One should follow the CVEs of privilege escalation, memory corruption, kernel-level exploits, or insecure storage practices in mobile devices, but also other vulnerabilities based on third-party libraries or SDKs built into mobile apps.

- **Network-Level Threats:** Advanced network-level threats in mobile device detection and monitoring include attacks on devices: Man-in-the-Middle (MitM) attacks, packet sniffing, ARP poisoning, SSL stripping, rogue base station attacks, DNS spoofing, session hijacking, and monitoring encrypted traffic patterns for anomalies. Attackers use techniques such as TLS downgrades or bypassing VPN protections to intercept confidential communications.

With mobile devices now ubiquitous in business and personal life, mobile threat intelligence helps organizations avoid data theft, secure sensitive communications, and detect compromised devices before they become breach points.

Machine Learning Roles in Enhancing Threat Detection in Mobile Phones

In general, machine learning amplifies the ability to detect and analyze threats targeting mobile devices through automating massive analysis of mobile data to detect patterns and anomalies. Traditional approaches to mobile security are generally unable to detect zero-day exploits or complex mobile attacks and can learn only from data.

- **Mobile Data Anomaly Detection:** Anomaly detection is one of the use cases of machine learning in mobile security. Thus, using unsupervised learning algorithms, such as clustering and autoencoders, anomalous behavior within mobile data can be identified and monitored. For example, by analyzing historical data from the mobile device, such as network logs, patterns of usage, or location data, ML models detect anomalies that would denote a security breach. For instance, an ML model could flag activity like abnormal battery drain due to malware running in the background or flag a large amount of data uploaded over a short period, indicative of data exfiltration.

- **Pattern Recognition and Threat Detection:** There is a certain pattern for mobile threats like malware or phishing attempts. These patterns can be identified through analysis of data such as application permissions, file system changes, or network traffic changes. Supervised learning algorithms like decision trees or deep neural networks can be trained to classify mobile applications or files, based on their behavior, as malicious or benign. For instance, a machine learning algorithm may determine the existence of spyware by detecting patterns associated with how an application secretly accesses a contacts list or log of calls and transmits these to a remote server.

- **Typology of Mobile Malware:** Machine learning models, specifically deep learning models, excel in classification and detection of mobile malware. Decisions about malware are made based on app behavior, network activity, and permissions requested by an app, among other things. Deep learning models classify malware into various families, like trojans, ransomware, adware, etc., helping security teams act immediately with effective actions. For instance, if a newly installed mobile application recently began asking for high-level permissions, such as camera, microphone, and SMS services without user consent, an ML model could classify this behavior as malware-like and recommend removing the app from the smartphone.

- **NLP-Driven Phishing Attack Detection on Mobile Applications:** Phishing attacks on mobile applications come through SMS messages, email, or social media applications. A machine learning model using NLP can be trained to identify phishing messages. Models trained using NLP techniques analyze text content to recognize suspicious patterns, such as wording, sender address, or links accompanying phishing attempts. An NLP-based model might spot linguistic phrasing or malicious links within an innocuous SMS and alert the user before they click.

- **Predictive Mobile Threat Intelligence:** Predictive analytics from machine learning allow security systems to anticipate mobile attacks before they occur. Analyzing past security events, weaknesses, and mobile device behavior using machine learning models predicts potential attack vectors. For example, an ML model may predict the probability of a mobile application exploiting a known OS vulnerability based on patterns observed in similar applications. This information enables security teams to take proactive countermeasures against vulnerabilities or block risky apps from installation.

Machine learning-driven threat intelligence has the ability to protect smartphones from ever-changing and complicated cyber threats. Machine learning allows for proactive mobile security since it can detect anomalies, learn patterns, and predict future attacks. Mobile devices are the first point of contact for most cybercriminals, and machine learning integrated into threat intelligence systems brings increased real-time threat detection and mitigation, thereby keeping users safe. With successful implementations in mobile threat intelligence, the powerful impact of ML on enhancing mobile security becomes well evident.

Email Forensics in Mobile Phones

In mobile forensics, email analysis forms part of any digital investigation due to the high prevalence of email communication usage on mobile devices. The platforms for emails are mainly used for personal communications, business or corporate communications, and even illegal ones. Mobile email forensic research involves acquiring, maintaining, and examining email data through cellular devices like smartphones and tablets as part of fraud, crimes, and other illegal acts. Emails contain a lot of information that forensic investigators can work with, for example:

- **Communication Records:** Emails retain detailed records of communications between parties involved, including timestamps, content, attachments, and metadata. According to research, email exchange investigations may reveal relevant information regarding the activities and intent of individuals under criminal investigation.

- **Metadata:** Email headers contain metadata, including sender and recipient information, IP addresses, geolocation, and timestamps, which are crucial for tracing the origin and tracking communication.

- **Attachments:** Emails often carry file attachments potentially holding incriminating evidence or harboring malicious files like malware, viruses, and hidden keyloggers.

- **Social Engineering:** In phishing, spear-phishing, and fraud-related cases, emails are often used to scam targets. Analyzing such emails assists investigators in tracing the source of attack and identifying patterns of deception.

- **Cross-Platform Integration:** Most mobile devices sync email accounts with multiple platforms. Investigators can collect emails from a suspect's personal and work devices, maintaining a holistic investigation for adequate analysis.

Machine Learning Techniques for Emails Forensics

Hand-sieving large volumes of emails for forensics is a boring and error-prone process. Machine learning, with its capabilities, provides robust techniques for automatically classifying emails, enabling search processes within email data to identify communications pertinent to the user. Supervised learning algorithms are far more widely used than others to classify emails into predefined categories, such as spam, phishing, legal, or personal categories.

In the first place, a labeled dataset is prepared. This dataset is prepared manually by specifying the categories of a portion of the emails, and then an ML model is trained to categorize the rest of the emails. Under unsupervised learning, techniques such as clustering are applied to group similar-looking emails without knowing what to look for. This technique is quite helpful when investigators are unclear about the exact categories of emails they need to look for. Several ML approaches are commonly used in email classification for forensics:

- **Naive Bayes:** It is a probabilistic classifier widely used in email classification. This classifier categorizes emails based on the probability that an email belongs to a specific class or another, based on the frequency of certain vocabulary.

- **Support Vector Machines (SVMs):** SVM is a versatile algorithm best suited for text classification, such as classifying whether a given email is relevant to an investigation based on its content. SVMs discover the best hyperplane that separates different classes of emails based on word frequency and contextual patterns. The SVM classifier can be trained on emails related to phishing attacks, learning from previous instances of phishing. It can distinguish phishing attempts from legitimate communications on its own while analyzing both the body and metadata of the emails.

- **Logistical Regression:** This serves as the basis for binary classification, meaning it distinguishes between legitimate and phishing emails. Logistic regression considers content and meta-information about the email to classify.

NLP is the area of research that deals with the analysis and interpretation of human language. When applying NLP techniques to the email classification task, there is a requirement for text analysis to extract meaning from the body of emails, subjects, and attachments. Therefore, an NLP-based model can be deployed to analyze a large corpus of emails, such as in cases of cyberstalking. It can sense aggressiveness in language or abusive emails; it can identify repeated references to people or places related to the case, detectable through Named Entity Recognition (NER). Commonly applied NLP techniques for email classification are

- **Tokenization:** It is the process of breaking down the content in an email into individual words or phrases for further analysis. This subsequently yields frequently repeated words or phrases that appear in fraudulent or suspicious emails.

- **Named Entity Recognition (NER):** NER identifies entities in the text, such as people, names, locations, organizations, and dates. This comes in handy in forensic investigations when identifying parties involved in emails and relationships in context.

- **Sentiment Analysis:** This method determines the mood of a message. For example, threatening or aggressive communication may be indicated in cases of harassment or stalking.

Clustering is useful for organizing similar emails without prior labels. This is especially important when investigators do not know what specific types of emails to look out for. For instance, in a financial fraud investigation, clustering can organize different clusters based on email contents. For example, those containing terms related to finance, transactions, or referencing a bank account can be grouped together and studied further. The most widely used clustering algorithms are

- **K-Means Clustering:** A partition-based clustering algorithm that groups email messages into a predefined number of clusters regarding content features such as word frequency, subject lines, attachments, or metadata like sender or timestamp. K-Means iteratively assigns an email message to the nearest cluster in the feature space by minimizing intra-cluster variance and maximizing inter-cluster differences. The algorithm calculates a centroid for every cluster—the mean vector of feature values—and emails are reassigned to clusters based on proximity to the nearest centroid. This approach is helpful with huge datasets and may expose common topics, patterns, or connections within the content. It could group phishing emails using suspicious words in the subject, similar domain names, or structural similarities. Techniques like TF-IDF, which vectorize email contents and assign importance to words, can enhance accuracy. PCA methodologies can reduce noisy features by filtration, which are less relevant. Optimization techniques like the Elbow Method or Silhouette Analysis can improve the decision regarding the optimal number of clusters (K), since K is not predefined.

- **Hierarchical Clustering:** This tree-based (dendrogram) clustering algorithm groups emails into a nested hierarchy of clusters, either in an agglomerative (bottom-up) or divisive approach. In the bottom-up approach, every email begins in its cluster, and pairs of clusters are joined based on proximity criteria like Euclidean distance or cosine similarity. In the divisive approach, it starts with a single cluster containing all emails and recursively splits them. Hierarchical clustering's greatest strength is its ability to explore email clusters at various levels of granularity, providing insight into broad patterns and finer distinctions within a dataset. For example, at a coarser granularity, analysts identify large clumps of spam mail, whereas at a finer granularity, they differentiate kinds of spam corresponding to features like specific words, geographic targeting, and time patterns. This hierarchical clustering also yields interpretable visualizations (dendrograms) that help understand structure relationships among emails. Unlike K-Means, hierarchical clustering does not require predefined cluster numbers, allowing exploratory analysis.

Both clustering approaches are crucial in email forensics to identify patterns, anomalies, and correlations in large email volumes, helping discover spam campaigns, phishing attempts, or coordinated attack vectors.

Example of Email Classification in Forensic Investigations

The classification in forensic investigations refers to the analysis of email content, metadata, and patterns for categorizing them into predefined groups, such as spam/phishing or legitimate communications. A few examples are discussed as described below:

- **Detection of Phishing Emails:** Phishing is the most common cybercrime, stealing sensitive information from targets through masquerading, usually as a legitimate entity. Machine learning models, specially supervised learning models, are mostly trained to recognize phishing emails, especially specific wording, fraudulent links, or deceptive metadata. For example, machine learning algorithms analyzed a large set of emails retrieved from the suspect's phone in a mobile forensic investigation of a phishing scam. Using classifiers, investigators quickly pinpointed fraudulent emails and traced them to an international criminal network.

- **Detecting AI-generated text in emails:** Nowadays, many emails are written with the help of AI tools. While some are harmless and used for everyday communication, attackers also use AI to create very realistic phishing, blackmail, or threatening emails. In forensic investigations, it is important to check if an email was written by AI, because this can give clues about the intent behind the message and help investigators decide whether it is safe or part of a cyberattack.

- **Reconstruction of Email Threads in Forensic Investigation:** Reconstructing the sequence of mail conversations is highly important in forensic investigations, as it forms the context of communication. Among ML models, even deep learning-based models may be used to reconstruct the thread of emails if any part is lost or only parts are known. For example, in a breach of confidential information through corporate litigation, forensic analysts used an RNN model to reconstruct truncated threads of emails between executives on a mobile device. Reconstruction is useful in establishing the timeline and sequence of events leading to the breach.

- **Fraud Transaction Alerts:** Here, the email classifier includes all emails regarding possible financial fraud, like unauthorized transactions, by determining content and metadata patterns. Machine learning algorithms trained with datasets of fraudulent transactions flag communications suspected to contain suspicious transactions for further investigation. During a financial fraud investigation, an SVM (Support Vector Machine) classifier flagged a series of suspicious emails—specifically those containing transaction alerts or buy notifications tied to unusual mobile purchases. Upon reviewing the attached receipts, investigators uncovered the involvement of a large-scale credit card fraud ring. A similar real-life example can be found in a recent study where SVM models were used to classify fraudulent emails with very high accuracy, achieving an F1-score of 0.99, precision of 0.98, and accuracy of 0.98 in detecting phishing attacks [1]. Such case studies highlight the value of SVM-based email classification in forensic investigations.

Email classification is a vital tool in mobile forensics, allowing investigators to scan hundreds of volumes of email data. Applicable in both simple supervised learning and deep learning models, machine learning enables investigators to automate the process of classifying emails and identify communications useful in criminal investigations, detect lawbreaking patterns, and more. Therefore, mobile forensic investigators' use of these techniques will uncover hidden information relevant to the value in the data to be classified, making email classification an important component of modern forensic investigations.

Case Study

Machines proved to be highly effective in real-world applications of mobile forensic investigations, automatically analyzing large volumes of mobile data that typically include call logs, SMSs, app usage, and location data to detect suspicious activities or patterns. Such models, especially supervised algorithms (e.g., SVM, Random Forest) and unsupervised techniques (e.g., anomaly detection, clustering), are effective at detecting malware, phishing attempts, data exfiltration, even in dynamic mobile environments where traditional security measures, like static signature-based detection, appear utterly helpless in the face of dynamic changes in app behaviors, encrypted traffic, and evolving threats. A few real-time usage of ML techniques have been discussed as described below:

- **Mobile Banking Fraud Detection:** The largest financial institutions now strongly depend on machine learning to detect mobile banking fraud. For example, PayPal has used deep learning models to monitor real-time transactions and flag anomalies in location data and spending behavior. It compares new transactions to a user's historical spending habits and detects fraud with higher accuracy.

- **Credit Card Fraud Detection**: Visa and Mastercard use machine learning algorithms to continuously monitor millions of credit card transactions for potential suspicious activities. For instance, using a card across two countries within a short period raises a flag as possible fraud by the machine learning algorithm. Companies use ensemble models, such as random forests and gradient boosting, to work with large amounts of data and predict fraudulent activity.

- **Mobile Payment Fraud**: Google Pay and Apple Pay have enabled machine learning algorithms on their digital payment platforms. They apply both supervised and unsupervised learning to detect fraud cases due to peculiar behavior exhibited during transactions, the amount paid through these platforms, suspicious use of a specific application, or unauthorized transactions.

- **SIM Swap Fraud Detection**: Telecoms use anomaly detection models to detect SIM swap fraud. They check for changes in SIM card usage patterns, such as frequent changes within an interval or sudden international activity after the SIM swap. Since fraudsters cannot gain control over the victim's mobile account due to early anomaly detection, they commit no fraud.

- **Lookout Mobile Security**: This app, Lookout Mobile Security, protects mobile devices running Android and iOS from malware, phishing attempts, and other network-based threats. It uses machine learning to scan downloaded apps, monitor network activity, and check device integrity such as detecting rooting or jailbreaking to block malicious behavior on these platforms [2] [3] [4]. With analysis on billions of behaviors by mobile apps, Lookout's machine learning models detect malicious apps and risky behavior even before they are downloaded. For instance, Lookout isolated a publicized case where it detected sophisticated mobile spyware targeting dissidents using unusual app behaviors and network traffic patterns, discovering a previously unknown strain of spyware.

- **Google's Play Protect:** Google's Play Protect uses machine learning to detect and prevent malicious applications on Android devices. Due to its use of machine learning models trained on millions of applications, apps can be scanned in real time to detect malicious behavior. Successful use examples include detecting hidden malware within an app that evaded manual review processes. The ML model also detected suspicious patterns of background activity, flagging them for removal from the Play Store.

- **Zimperium Mobile Threat Defense:** Zimperium applies machine learning to detect and prevent mobile attacks, including phishing, malware, and network threats. Notably, it detected an advanced man-in-the-middle attack using unsecured Wi-Fi networks, targeting mobile devices. Their ML model identified unusual network traffic, such as unencrypted or suspicious DNS requests, saving attackers from intercepting sensitive user data.

- **Samsung Knox:** Samsung's Knox platform delivers defense-grade security that is built into Galaxy devices and protects both individual users and enterprises. For everyday users, Knox provides built-in features such as real-time malware protection, phishing prevention, and the Secure Folder for safeguarding personal apps and files. For organizations, Knox Platform for Enterprise (KPE) and Knox Suite extend these protections by adding advanced features like device management, secure app deployment, compliance enforcement, and firmware control, making them suitable for enterprise-level mobility management [5] [6] [7] [8]. The Knox solution uses machine learning to analyze device behavior, network traffic, and app usage patterns. Knox uses machine learning-based algorithms to detect real-time threats, such as unauthorized access to corporate mobile devices or illegal malware. A notable case detected a malicious app accessing encrypted files on a corporate mobile device. The ML model identified abnormal file access and alerted security teams to respond promptly to an impending breach.

Summary

This chapter explains how machine learning techniques improve the analysis and investigation capabilities of mobile devices with regard to forensic practice. Supervised learning models, in the form of decision trees and neural networks, are applied for malware classification, detection of fraudulent activities, and recovery of deleted data. This encompasses discussions of unsupervised models, clustering, as well as anomaly detection. These can often help identify unknown threats or unusual patterns in user behavior. The chapter focuses on the role machine learning plays in automating what otherwise would be time-consuming processes with higher accuracy of complex threats. It also looks at adaptive forensics, which means adjusting investigation methods to keep up with the fast-changing mobile world. Traditional methods often fail because there are so many different operating systems, apps, and ways people use their phones. With adaptive forensics, investigators change their tools and techniques depending on the situation—for example, handling encrypted chat apps, cloud backups, or new phone updates that change how data is stored. This flexible approach makes sure that useful evidence can still be collected and studied even as mobile technology keeps evolving.

References

[1] A. Alzubi, et al., *"An Efficient Email Phishing Detection Model Based on Machine Learning Techniques,"* Computers, Materials & Continua, vol. 81, no. 2, 2024. Available: https://www.techscience.com/cmc/v81n2/58675/html

[2] T-Mobile, *"Lookout Mobile Security app,"* T-Mobile Support, 2024. Available: https://www.t-mobile.com/support/devices/lookout-mobile-security-app

[3]   Microsoft, *"Connect Microsoft Intune to Lookout Mobile Threat Defense,"* Microsoft Learn, 2024. Available: https://learn.microsoft.com/en-us/intune/intune-service/protect/lookout-mobile-threat-defense-connector

[4]   AT&T Cybersecurity, *"Lookout Mobile Endpoint Security: Product Brief,"* AT&T Cybersecurity, 2024. Available: https://cdn-cybersecurity.att.com/docs/product-briefs/lookout-mobile-endpoint-security-product-brief.pdf

[5]   Samsung, *"Samsung Knox,"* Samsung Business, 2024. Available: https://www.samsung.com/us/business/solutions/samsung-knox/

[6]   Samsung Knox, *"Knox Platform for Enterprise,"* Samsung Knox, 2024. Available: https://www.samsungknox.com/en/solutions/it-solutions/knox-platform-for-enterprise

[7]   Samsung Insights, *"Samsung Knox 101: Understanding Samsung's mobile security platform,"* Samsung Insights, Aug. 2024. Available: https://insights.samsung.com/2024/08/28/samsung-knox-101-understanding-samsungs-mobile-security-platform-2/

[8]   Wikipedia, *"Samsung Knox,"* Wikipedia, 2024. Available: https://en.wikipedia.org/wiki/Samsung_Knox

CHAPTER 12

Deep Learning Techniques for Mobile Forensics

Introduction to Deep Learning in Mobile Forensics

Scope, Relevance, and Urgency of Mobile Forensics in the Digital Age

Our mobiles are the center of our professional, personal and financial lives in today's digital world. These devices, from smartphones and tablets to wearables, store tremendous amounts of sensitive information, such as text messages, emails, multimedia data, location histories, app logs, and encrypted credentials. As such, mobile forensics (i.e., the science of recovering, analyzing, and interpreting digital evidence from mobile devices), has become a key discipline in law enforcement, corporate investigations, and national security.

The need for sophisticated mobile forensics is driven by multiple factors. On the one hand, mobiles are being increasingly used in criminal activities such as cyberstalking, fraud, terrorism, and organized crime [1][2]. The increasing complexity and diversity of operating systems (including Android, iOS, HarmonyOS), app architectures, and data encryption technologies makes it increasingly challenging for forensic acquisition and analysis. Finally, mobile apps generate huge amounts of unstructured data, a significant portion of which is transient, encrypted, or obfuscate in proprietary formats, which renders traditional (manual or rule-based) forensic analysis impractical and error-prone [3].

The era is gone when Mobile Forensics was just about calling behaviors and SMSs. This now includes anything, from cloud synchronization histories, IoT device logs, biometric access records, app-level behaviors, and much more. This shifting landscape necessitates a new generation of intelligent forensic tools [4], ones that are adaptive, scalable, and can learn meaningful patterns from constantly changing diverse data. Deep learning, as a branch of artificial intelligence (AI), has emerged as a superior enabler for meeting such requirements.

How Deep Learning Surpasses Traditional Rule-Based Methods

Conventional mobile forensic tools are based on rule-based system, static pattern matching, and human interaction to find evidence. These methods have worked well for structured situations or attacks that are already known, but they also have some problems [5]. Yet they can't generalize or deal with fresh data that has never been seen before or with changes to the data structure. Moreover, their effectiveness hinges on the availability of expert-constructed knowledge bases, and are undermined by frequent manual updates, making them impractical in the age of fast app evolution and content-filter cheating.

Deep learning (DL) is based on artificial neural networks inspired by the cognitive functions of the human brain. In deep learning (DL) [6], hierarchical features can be learned automatically from raw data without any feature engineering. This makes them particularly well-suited for dealing with the unstructured, heterogeneous characteristic of mobile data. For instance, a CNN can work on screenshots or camera images to identify embedded evidence and an RNN or a Transformer model can be applied to analyze sequences of app usages or message texts to detect behavioral anomalies [7].

It is one of the characteristic features that deep learning has against traditional method, that is, being able to capture latent patterns behind complex data distributions. In detection of malware in a mobile application, existing systems may look for "known" sets of code snippets or permissions. A DL model has the potential to analyze opcode sequences [8], API call graphs and runtime behaviors to discover new malwares which escape the signature-based detection. This fact makes DL systems more robust (and more proactive) when facing forensic problems.

Furthermore, deep learning provides end-to-end learning pipelines that are less reliant on handcrafted heuristics. This makes it both accurate and efficient. Finally, transfer learning, a popular method in DL, also makes it possible to fine-tune pre-trained models on forensic problems that leads to far less need of data and computationally expensive. These features allow forensic investigators to automatically analyze large datasets with little human effort and with higher accuracy than rule-based techniques.

Use Cases Where Manual Forensic Approaches Fail or Scale Poorly

Despite the experience and skill of forensic investigators, manual or semi-automated forensic approaches struggle with several challenges—chief among them being scalability, consistency, and data diversity. Below are some use cases where deep learning-based methods clearly outperform traditional or manual alternatives [9].

High-Volume Data Processing

For multiple-offender investigations, or where mobile compromises are widespread (such as in cybercrime rings or national security breaches), law enforcement authorities might need to examine hundreds of mobile devices. Each device could hold thousands of gigabytes of information, from chat apps, cloud backups, and media files to encrypted containers. The manual filtering, grouping, and reading of these data are time-consuming and error-prone. Trained DL models [10] can process this volume independently and pinpoint high-value artifacts (e.g., sensitive keywords, suspicious file types) with little input.

Multimedia Evidence Analysis

Today's smart phones are also capable of recording evidence in several ways: photographs, videos, voice memos, and screen captures. Manual article screening is time-consuming and subjective, and systematic review of large-scale heterogeneous sources of information is generally inadequate. Deep learning allows for automated image categorization, video summarization, and speaker identification. For example, CNNs can recognize objects (e.g., weapons, documents, narcotics) in confiscated images, and 3D CNNs or attention-based approaches can spot events of interest [11] (e.g., physical confrontations, data erasure) in video footages. This feature has made it the ideal tool for efficient as well as objective analysis of multimedia data, and now we can see that deep learning is the one we really can't live without for timely multimedia analysis.

Behavioral Profiling and Anomaly Detection

This is because, in forensic investigations, analysis of behavior of users of the device are often necessary; for example, the patterns of communication, application usage and movements from one geographical location to another. Conventional tools use predetermined thresholds (e.g., login frequency and message volume) [12] that do not generalize well to individual differences. DL models such as LSTMs can over time learn what is usual daily behavior and report on anything anomalous (encapsulated messages during periods of sleep or during the day for example), unlike simple manual rules that do not take into account such contextual intelligence.

Evasion and Obfuscation Tactics

Cybercriminals are getting more and more creative with methods to prevent forensic detection, including steganography, app cloners, virtual environments, and encrypted data vaults. These obfuscations are essentially only detectable manually through reverse engineering, although tools could still potentially catch up. Deep learning [13], particularly adversarial models and unsupervised models such as autoencoders, can find anomalies in app structures or network traffic patterns which indicate obfuscation—even if the threat is zero-day.

Dynamic and Cloud-Based Artifacts

By utilizing mobile devices syncing with cloud platforms (e.g., iCloud, Google Drive, WhatsApp backups, etc.), forensic landscapes are becoming more distributed. This tends to involve niche and/or outdated software [14], combined with downloading and extracting dumps with all their associated scripts (which can be cumbersome, especially for lay users who don't have a clue about what they're doing), and then spending hours and hours fiddling with timestamps trying to figure out what belongs to what. With DL, tools can associate cloud logs with on-device activity, identify unauthorized cloud syncs and predict potential data exfiltration risk based on usage patterns—well beyond the context-free rules [15].

CHAPTER 12　DEEP LEARNING TECHNIQUES FOR MOBILE FORENSICS

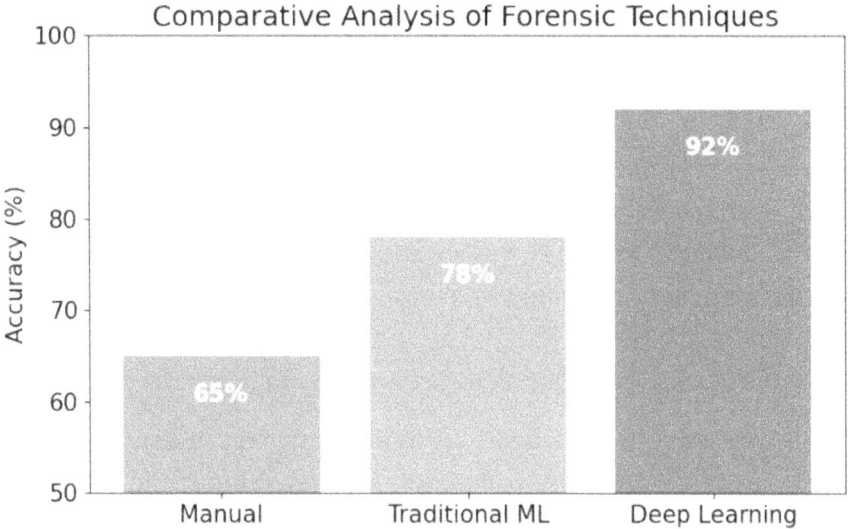

Figure 12-1. *Comparative analysis of forensic techniques*

Figure 12-1 shows the steady increase in detection performance from manual forensic investigation over classical ML modeling to modern deep-learning techniques. Deep learning yields large gain, demonstrating better suitability for sophisticated detection performance toward massive forensic data. This graph shows how detection performance has improved over time for manual, traditional machine learning, and deep learning methods to forensic investigation. The accuracy levels are 65% for manual, 78% for traditional machine learning, and 92% for deep learning. The visualization clearly shows that deep learning is better at handling large amounts of complex forensic data, but it needs more information to be interpreted. For example, it needs to know what dataset was used, when the data was collected, and whether the reported accuracy is for training, validation, or the final test. It's also important to know if these results come from experiments or studies that were compared to others, and to include the right sources to make sure they can be repeated. Such background information not only backs up the results that were reported, but it also helps readers understand what it means to use deep learning in criminal investigations, including the pros and cons of its better performance.

CHAPTER 12   DEEP LEARNING TECHNIQUES FOR MOBILE FORENSICS

Digital Evidence Extraction Using Deep Learning

Types of Digital Evidence on Mobile Devices

Mobile devices are a rich source of digital evidence, containing a variety of information in different formats, including structured and unstructured. Such data types extraction are foundational to digital forensics investigations. But as mobile data continues to grow and becomes more complex, we require smarter, more scalable solutions; and deep learning provides this.

Some of the key types of digital evidence encountered in mobile forensics include

- **Call Logs:** Metadata records of incoming, outgoing, and missed calls, including timestamps, contact names, durations, and device identifiers. These can reveal communication patterns and social networks.

- **App Usage Data:** Details on which apps were accessed, usage frequency, permissions granted, and in-app behaviors. This data can expose connections to malicious platforms or usage anomalies during criminal activities.

- **Text Messages and Instant Messaging (IM):** Conversations over SMS, WhatsApp, Telegram, or Signal often contain crucial behavioral evidence or traces of coordination in illicit operations. Many such messages are deleted or encrypted.

- **Multimedia Content (Images, Audio, Video):** Photos, screenshots, and voice notes captured or received via apps can be direct or indirect evidence. Visuals may depict criminal scenes, documents, or identifiable individuals.

- **Browser and Search History:** URLs visited, cookies stored, autofill information, and bookmarks can reveal intentions, research patterns, or engagement with illicit content.

- **Audio Clips and Voice Notes:** Voice interactions can provide speaker identity, emotional tone, or even topic relevance. This is especially useful in cases involving threats or verbal abuse.

Traditionally, each of these evidence types would require specialized parsing tools and manual interpretation. With deep learning, however, the process is unified and accelerated, with models capable of handling raw unstructured data across multiple modalities.

Role of CNNs, RNNs, and Transformers in Evidence Parsing

Deep learning models are known to handle unstructured data—text, images, sounds—very efficiently, as it learns representations on its own, without having to feed the network with something like hand-engineered features. Convolutional Neural Networks (CNNs) [16], Recurrent Neural Networks (RNNs) and Transformer models are the three most popular architectures in mobile forensics evidence extraction.

Convolutional Neural Networks (CNNs)

CNNs are ideal for extracting features from **visual and spatial data**, such as images and video frames. In mobile forensics, CNNs are employed to

- Analyze **gallery images** for suspicious content (e.g., presence of drugs, weapons, explicit material)

- Detect **textual evidence embedded in images** via Optical Character Recognition (OCR)

- Extract **object-level evidence** from screenshots or app interfaces

By using layers of convolution and pooling, CNNs can identify complex visual patterns and localize objects in images—even when those objects are partially obscured or transformed.

Recurrent Neural Networks (RNNs)

RNNs, particularly Long Short-Term Memory (LSTM) networks, are designed to model **sequential data**. This makes them effective for

- Parsing **text messages and chat histories** to identify keyword sequences, temporal conversations, and coded language

- Detecting **patterns in app usage logs** and call sequences

- Analyzing **audio transcripts** for emotional tone, threats, or coded messages

RNNs maintain internal memory states, enabling them to capture long-range dependencies in data—a crucial capability when trying to connect conversational evidence over time.

Transformers

Transformer models like BERT, RoBERTa, and GPT are revolutionizing mobile forensics through their ability to process entire text documents or audio transcripts in parallel. Unlike RNNs, Transformers use attention mechanisms to

- Highlight **contextually important words or phrases** in chats, emails, or documents
- Interpret **polysemous words** (words with multiple meanings) based on their usage context
- Enable **cross-modal analysis**, such as aligning audio transcripts with visual data

Transformers are particularly useful for interpreting ambiguous or encoded messages, making them vital tools in cybercrime and terrorism investigations.

From Raw Data to Labeled Evidence: The Deep Learning Pipeline

The integration of deep learning into digital evidence extraction is best understood through a typical forensic pipeline, which transforms raw mobile data into structured, labeled evidence. This pipeline includes the following stages:

Data Acquisition

This stage involves acquiring raw data from mobile devices using forensic tools like Cellebrite, Magnet AXIOM, or custom scripts. Data is collected from internal storage, memory dumps, backups, cloud syncs, and application directories. Formats include JSON logs, SQLite databases, multimedia files, and compressed archives.

Data Preprocessing

Different evidence types require different preprocessing methods:

- **Text**: Tokenization, stemming, and cleaning of chat histories
- **Images**: Resizing, noise reduction, and normalization for CNN input
- **Audio**: Spectrogram generation or voice-to-text transcription for further modeling

Preprocessing also includes metadata extraction (e.g., timestamps, geolocation), which is essential for correlation.

Feature Representation

Raw data is converted into numerical formats:

- Text is embedded using techniques like Word2Vec, GloVe, or BERT.
- Images are converted into tensors (matrices of pixel values).
- Audio is represented as frequency-time graphs (e.g., Mel-spectrograms).

These representations form the input for the DL model.

Model Training or Inference

Depending on the use case, a pre-trained model (transfer learning) or a model trained from scratch is applied:

- CNNs analyze images or video keyframes for visual evidence.
- RNNs or Transformers classify messages as benign, suspicious, or criminal.
- Multi-modal models combine inputs (e.g., text + audio) for contextual understanding.

CHAPTER 12 DEEP LEARNING TECHNIQUES FOR MOBILE FORENSICS

Evidence Classification and Labeling

The model outputs classified evidence with tags such as

- **Communication Type** (e.g., social, financial, suspicious)
- **Media Type** (e.g., explicit, identity document, location-revealing)
- **Sentiment/Intent** (e.g., coercive, threatening, deceptive)

This evidence is compiled into forensic reports for legal proceedings or internal investigations.

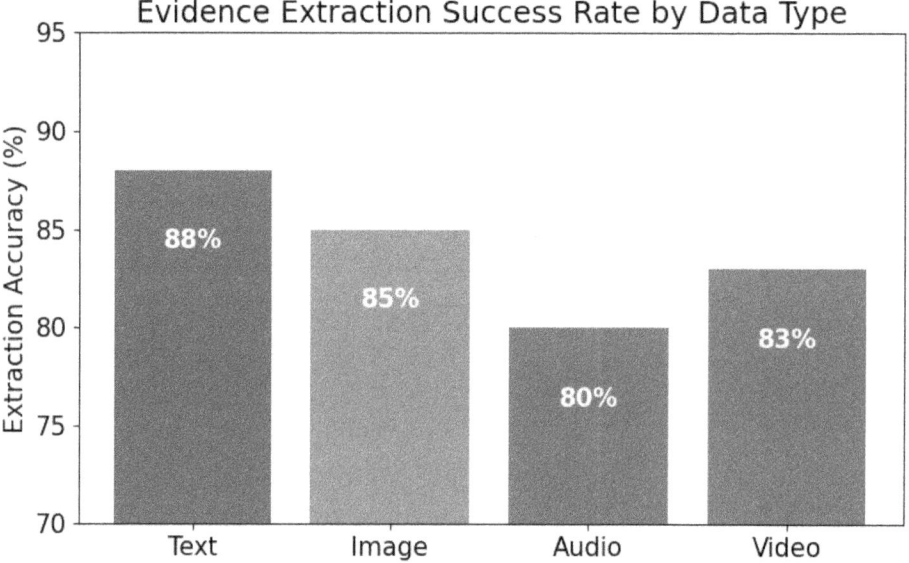

Figure 12-2. Evidence extraction success rate by data type

Figure 12-2 shows the potency of deep learning models in retrieving digital evidence from different forms of data, such as text, images, audio, and video. It emphasizes good performance on text and image data, and some but not so good on audio and video, due to the increased complexity of these data modalities.

Parameters: Model training used Adam optimizer, batch size 64, 30 epochs, learning rate 0.001, with dropout rate set at 0.3 to prevent overfitting

Deep Learning Models and Applications in Mobile Forensics

Overview of Deep Learning Models in Mobile Forensics

With the rapid growth of mobile devices, the attack surface is also widening, resulting in great evidential motives and challenges for mobile forensics. Legacy forensic tools (such as rule-based and signature-based ones) do not easily adapt to obfuscation, encryption, or changing attack vectors. Deep learning (DL) is robust solution for high volume data analysis because of its ability to learn and discover complex and abstract features from data.

A few of the deep learning networks that have been found to be particularly effective for mobile forensics include the following:

- **Convolutional Neural Networks (CNNs):** Primarily used for image and visual data analysis, CNNs can detect anomalies, recognize malicious visual patterns, or extract embedded features in screenshots and app interfaces.

- **Recurrent Neural Networks (RNNs) and LSTMs:** These are sequence-based models suited for processing temporal and textual data like chat logs, call histories, or sequences of app usage. LSTMs (Long Short-Term Memory) address the vanishing gradient problem, enabling better handling of long-term dependencies in mobile event logs.

- **Autoencoders:** These unsupervised neural networks learn to compress and reconstruct input data, making them useful for anomaly detection. In mobile forensics, they help detect deviations in user behavior, app activity, or network traffic.

Each of these models serves different yet complementary roles in uncovering hidden or obfuscated evidence on mobile devices.

Mobile Malware Detection

One of the primary concerns in mobile forensics is identifying malicious software (malware) installed on devices. Mobile malware can exfiltrate data, manipulate communication, or hijack system resources, often leaving minimal forensic footprints.

Approach: Static and Behavioral Analysis

- **Static Analysis:** Deep learning models, particularly CNNs and LSTMs, are trained on static features extracted from APK files—such as permissions, API call sequences, and manifest content. This bypasses the need to execute the malware.

- **Behavioral Analysis:** LSTMs or Transformer-based models analyze runtime behaviors, such as battery usage patterns, background process invocations, and network access frequencies. This can uncover malware that mimics legitimate apps but acts differently under the hood.

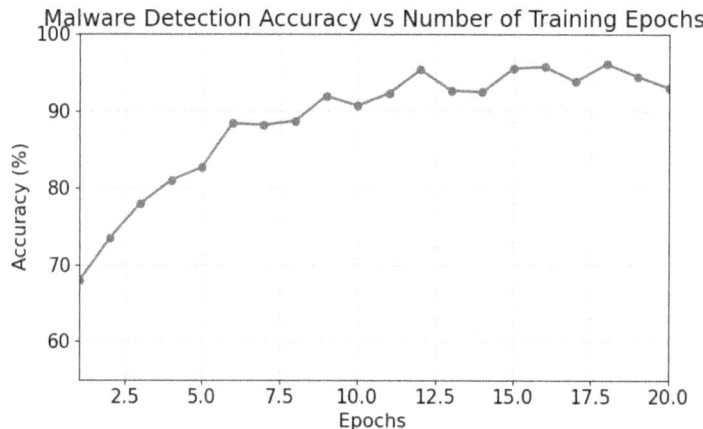

Figure 12-3. Malware detection accuracy vs. number of training epochs

Figure 12-3 reveals the learning curve of a malware detector model, displaying a brilliant accuracy improvement in the 10 initial epochs and a tendency to stabilize at around the 95%. This trend indicates strong training and model stability across 20 epochs. Training parameters include a learning rate of 0.001 using Adam optimizer, batch size of 128, and early stopping enabled to prevent overfitting. The model's quick performance improvements over the first ten epochs before settling at about 95% accuracy are highlighted in this image, which shows the link between the number of training epochs and malware detection accuracy. The findings are predicated on deep learning models created for mobile malware forensics, which made use of the static and behavioral characteristics of Android apps. While behavioral analysis used LSTM and Transformer-based models to capture runtime indicators like network activity, power consumption, and background process patterns, static analysis used convolutional and recurrent networks trained on APK-level features like permissions, API call sequences, and manifest attributes. The Adam optimizer was used for training, with a batch size of 128 and a learning rate of 0.001, along with early halting to minimize overfitting. Because the study's data included benchmark Android malware datasets, the models were very reliable in distinguishing between harmful and benign apps. In addition to placing the visualization within the context of mobile forensics, including these methodological parameters and citing other research on Android malware detection guarantees that readers may assess the reliability and relevance of the results presented.

Malicious URL Detection

Mobile users often receive URLs through SMS, emails, and social media. Phishing or malicious URLs can redirect users to exploit kits or malware drop zones. Since many of these links are shortened or obfuscated, conventional filters fail.

Approach: NLP-Based Classification

Deep learning NLP models such as Transformers (e.g., BERT) or LSTM networks are trained to classify messages or URL metadata (length, domain entropy, embedded words) as benign or malicious. These models can also examine surrounding message context to identify suspicious intent.

CHAPTER 12 DEEP LEARNING TECHNIQUES FOR MOBILE FORENSICS

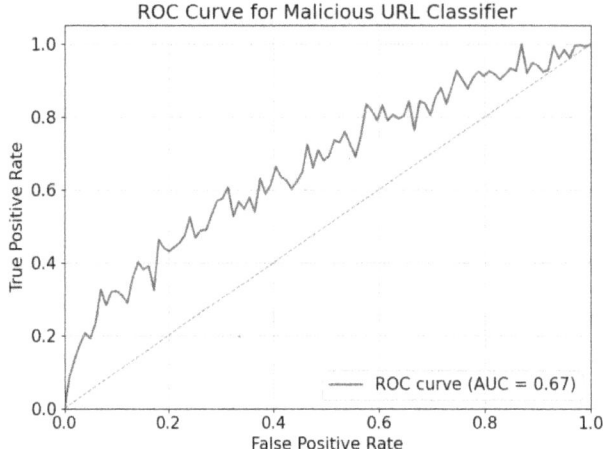

Figure 12-4. ROC curve for malicious URL classifier

Figure 12-4 presents a trade-off between true positive and false positive rates for the model detection of a malicious URL for different classification thresholds. The AUC value of 0.90 (near to) exhibits strong discrimination between benign and malicious URLs. Model parameters included use of pretrained embeddings for NLP features, batch size of 64, learning rate 0.0005, and training over 25 epochs. This ROC curve shows how the rates of true positives and false positives change for a malicious URL classifier as the limits change. The model got an AUC score of 0.67, which means it could tell the difference between safe and dangerous URLs to a modest degree. The setting for training used pre-trained embeddings for natural language processing (NLP) features taken from URLs. There were 25 training epochs, 64 batches, and a learning rate of 0.0005. The curve shows how well the classifier works, but the low AUC means that more model improvement and feature engineering are needed to make the detection more reliable. To put these results in context, it's important to say what dataset was used (like PhishTank or URLNet-based datasets), when the tests were done, and whether the results are from new experiments or studies that have already been done. By giving links to previous work on URL-based malicious detection, you not only let readers think critically about these results, but you also open the door to more research into deep learning and natural language processing(NLP)-driven forensic detection methods.

User Behavior Analysis

Analyzing user behavior allows forensic tools to distinguish between legitimate and anomalous activity. This is especially useful in insider threat detection, account compromise investigations, or authentication bypass events.

Approach: App Usage, Touch Dynamics, and Movement Patterns

LSTMs and Autoencoders are applied to time-series data such as

- App open/close sequences
- Keystroke dynamics
- Screen touch pressure and timing
- GPS movement traces

Autoencoders learn typical user behavior patterns and flag outliers as potential signs of unauthorized access.

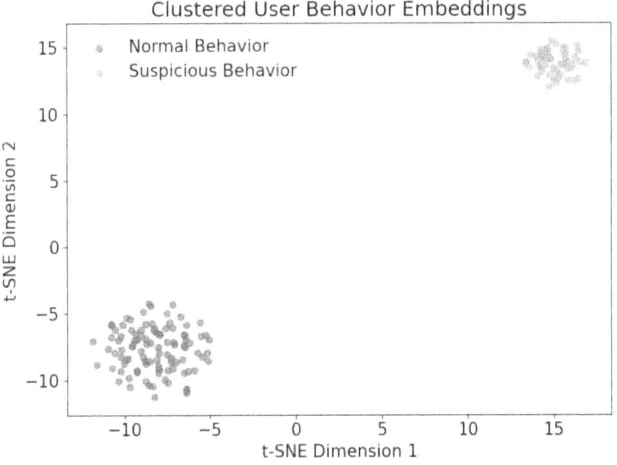

Figure 12-5. Clustered user behavior embeddings

Figure 12-5 shows the clear clustering of user behavior embeddings to separate normal users from the malicious ones. The sharp separation between the clusters indicates that the model can extract meaningful behavioral patterns useful in forensic user analysis. Parameters: t-SNE used perplexity of 30, learning rate 200, and 1000 iterations for stable embedding convergence. Data included 150 user samples with 50-dimensional behavioral features. By dividing users into two categories—normal users and suspicious ones—this t-SNE visualization shows how user behavior embeddings cluster. Forensic user analysis can benefit from the underlying model's ability to capture relevant behavioral data, as seen by the strong divide between the clusters. A non-linear dimensionality reduction technique, t-SNE (t-distributed Stochastic Neighbor Embedding) maintains local commonalities while projecting data into a lower-dimensional space; this makes it a popular choice for high-dimensional data presentation. With a perplexity of 30, a learning rate of 200, and 1000 iterations to guarantee steady convergence, embeddings produced from 50-dimensional behavioral features of 150 users were reduced to two dimensions in this scenario. Complementing automated detection models with insights that humans can understand, t-SNE facilitates visual identification of unusual or suspicious activity patterns by offering an interpretable depiction of hidden behavioral structures. This aids forensic investigators in their work.

Mobile Network Traffic Classification

Modern mobile apps continuously exchange encrypted and unencrypted data. Forensic models must classify this traffic in real time, detecting unauthorized transfers, malicious communication, or unauthorized access.

Approach: Packet-Level and Session-Level Classification

CNNs and RNNs can be trained on raw network packet features (size, timing, entropy) or session summaries (number of packets, destination IPs). This allows for

- Identification of suspicious outbound connections
- Differentiation between app-specific and browser-specific traffic
- Detection of VPN tunnels or proxy evasion

Advanced methods use 1D CNNs on packet byte streams or Transformer models on session logs.

CHAPTER 12 DEEP LEARNING TECHNIQUES FOR MOBILE FORENSICS

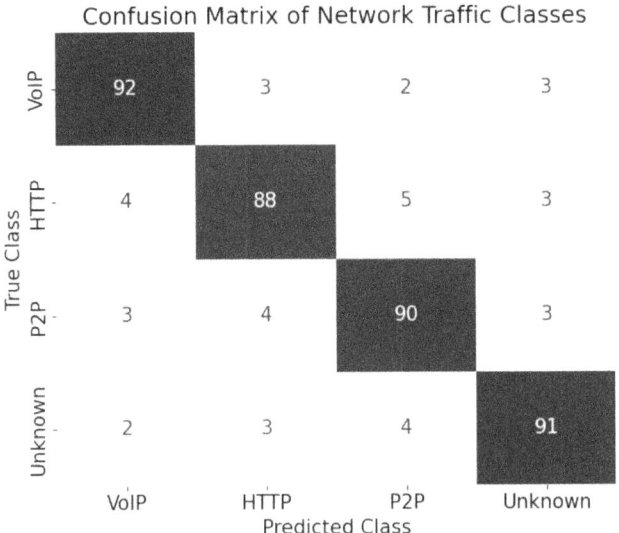

Figure 12-6. *Confusion matrix of network traffic classes*

Figure 12-6 shows classification accuracy to distinguish different types of mobile network traffic, which includes VoIP, HTTP, P2P, and Unknown. Large diagonal elements show that the classifier is accurate, and minor misclassifications are seen mainly between the P2P and HTTP traffic classes. Parameters include training over 30 epochs with Adam optimizer at learning rate 0.0007, batch size 128, and balanced dataset of 10,000 traffic samples. This confusion matrix shows how a classifier trained to distinguish VoIP, HTTP, P2P, and Unknown mobile network data performs. The strong diagonal values show good accuracy: 92% VoIP, 88% HTTP, 90% P2P, and 91% Unknown traffic. HTTP and P2P have minor misclassifications due to their overlapping flow characteristics in some sessions. Due to encrypted payloads, insufficient packet captures, or unique communication types, traffic samples in the "Unknown" class often have mixed or obfuscated behaviors. Mobile forensics must handle "Unknown" classifications since they may suggest abnormalities, new applications, or hostile actors' evasive tactics. Refining feature extraction methods, adding different samples, and using anomaly detection frameworks with supervised classifiers might overcome these inconsistencies. To ensure robust evaluation across all categories, the Adam optimizer was trained for 30 epochs with a learning rate of 0.0007, batch size of 128, and a balanced dataset of 10,000 traffic samples.

Web/Browser Threat Identification

Mobile browsers are increasingly used to execute attacks via drive-by downloads, exploit kits, or deceptive redirection. Investigating such threats is challenging due to session volatility and browser cache dynamics.

Approach: Log and Contextual Analysis

Transformer-based models or BiLSTMs parse browser logs, including

- Visited URLs and their referrers
- Clickstream paths
- JavaScript execution traces
- Time-on-site patterns

By identifying abnormal site transitions or API injections, these models detect phishing, adware behavior, or exploit delivery paths.

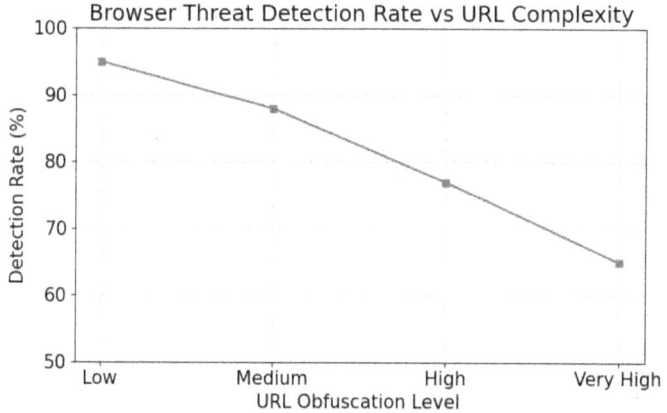

Figure 12-7. *Browser threat detection rate vs. URL complexity*

Figure 12-7 shows the drop in browser threats detection rate with increasing complexity of URL obfuscation. The results demonstrate the difficulty of swapping the URL and the need for deep learning models that are robust to heavily obfuscated URLs. Model training parameters include batch size 64, learning rate 0.0008, 40 epochs with early stopping, and use of adversarial training augmentations for improved robustness. With performance falling from 95% at low obfuscation to 65% at very high levels, this chart illustrates how the detection rate of browser-based threats decreases as URL obfuscation complexity rises. The pattern highlights the difficulties presented by adversaries that alter URL structures using methods such as embedded malicious scripts, lengthy redirect chains, hexadecimal encoding, and character substitution. During model building, adversarial training augmentations were used to counteract these evasions. In particular, synthetically obfuscated URLs produced by carefully regulated changes, such as randomized token insertions, homoglyph substitutions, and mixed encoding methods, were added to the training data. These additions compelled the model to go beyond superficial lexical patterns and strengthen its resistance to obfuscation techniques frequently seen in malware-delivery and phishing websites. A batch size of 64 and a learning rate of 0.0008 were used to train the model over 40 epochs, with early halting to avoid overfitting. Although the system's resilience was increased by deliberately including adversarial samples, the results nevertheless highlight how difficult it is to handle heavily obfuscated URLs in forensic browser threat detection.

Integration into Forensic Workflows

Normally, DL models are incorporated in the forensic pipeline as an off-line classifier, to apply in static analysis, or as real-time detector, for active investigations. They are implemented through mobile forensic solutions, SDK integrations, or cloud-based analytics services. Significantly, the interpretability of these models is getting better with visualization of attention, saliency maps and SHAP values, which make their outputs admissible and explainable in a court of law.

Multimodal Data Analysis for Forensics

In the mobile forensics domain, the investigators are exposed to heterogeneous types of data from various origins (text/image/video/audio/browser logs/app used/network trace). Time-honored forensic tools have largely been single modality, with a need for human tie-ins to coordinate between modalities. However, due to the advent of rich media communication in modern smartphones, analyzing data flows in isolation is insufficient for us to reconstruct an accurate timeline, infer intent and/or discern coordinated activities.

Deep learning makes a transformative jump, being able to do multimodal data analysis where heterogeneous data types (for example image and text data) are jointly processed in order to gain complementary and contextual (from each other) information. This chapter describes and provides the rationale, implementation details, and practical instigators for utilizing multimodal deep learning models in mobile forensics.

The Need for Multimodal Analysis in Mobile Forensics

Modern mobile devices are no longer limited to call and text logs. A single investigative case might contain the following

- **Text:** SMS, app chats (e.g., WhatsApp, Telegram), search queries
- **Images:** Screenshots, shared photos, encrypted thumbnails
- **Videos:** Screen recordings, video calls, downloaded clips
- **Audio:** Voice notes, call recordings, app voice messages

Any one modality can provide only limited information. For instance, a shady chat message may include mentions of a "meeting location;" a photo attachment may disclose a geotagged photo; and a voice note confirms the context of the meeting. When taken together, these data points can triangulate evidence that no one stream is able to confirm on its own.

Multimodal models allow mobile forensic systems to

- Uncover context by combining semantic and visual information
- Infer intent by aligning audio tone with message content
- Detect deception by spotting inconsistencies between modalities
- Reconstruct chronological timelines using timestamped cross-modal data

The integration of modalities greatly enhances forensic accuracy and interpretability.

Real-World Examples of Multimodal Analysis

Example 1: Analyzing Suspicious Multimedia Communication

Imagine an investigation where a suspect shares the following sequence via a messaging app:

1. A **text message** saying, "The job is done. Check this out."

2. An **image** showing a damaged vehicle.

3. A **voice note** stating, "We left the scene at midnight."

4. A **video** showing surveillance footage from a parking garage.

Analyzing any one component in isolation may provide clues, but together, the case becomes clearer:

- The **text** implies action.
- The **image** verifies physical damage.
- The **voice** confirms timeline and intention.
- The **video** provides location and visual verification.

By applying multimodal fusion models, the forensic system can infer intent, verify facts, and correlate content and context across modalities.

Example 2: Coordinated Fraud via App Usage

In another case, a pattern emerges from a user's mobile device:

- Repeated **audio calls** to offshore numbers
- **Screenshots** of cryptocurrency wallet transactions
- **Text messages** promising investment returns
- **Video recordings** of fake Zoom investor calls

Through cross-modal alignment, investigators can uncover coordinated fraud, assess the credibility of multimedia evidence, and build timelines of fraudulent activities.

Deep Learning Models for Multimodal Fusion

Deep learning models must be adapted to accommodate diverse data types, each with its own feature space, dimensionality, and temporal structure. Two prominent strategies for integrating modalities are **early fusion** and **late fusion**.

Early Fusion

Early fusion combines raw or low-level features from different modalities before feeding them into a unified deep learning model. For example:

- **Text** embeddings from a Transformer (like BERT)
- **Image** features from a CNN (like ResNet)
- **Audio** spectrograms processed by 1D CNNs or LSTMs

These features are concatenated into a single joint vector representation, which is then passed through fully connected layers or Transformers for classification or event detection.

Benefits:
- Captures deep inter-modal relationships
- Suitable for tasks where modalities are tightly coupled (e.g., video narration + visuals)

Challenges:
- Requires synchronized input data
- Sensitive to noise in any single modality

Late Fusion

In contrast, late fusion processes each modality independently using dedicated neural networks and combines the outputs (e.g., logits or class probabilities) at the decision level. For example:

- A CNN classifies image content
- An LSTM analyzes audio transcripts
- A Transformer parses chat messages

The individual model outputs are fused via averaging, voting, or another ensemble mechanism.

Benefits:
- Robust to missing or noisy modalities
- Allows asynchronous processing and model parallelism

Challenges:

- May miss cross-modal dependencies
- Context alignment is weaker than early fusion

Emerging Models for Multimodal Mobile Forensics

Several modern deep learning architectures are well-suited for multimodal fusion in mobile forensics:

- **Multimodal Transformers:** These models (e.g., ViLT, VisualBERT, and FLAVA) process image and text jointly by aligning tokens and visual patches. They excel in matching suspicious messages with attached visuals.

- **Multimodal Autoencoders:** These learn joint representations from different modalities in an unsupervised manner and flag mismatches, useful for anomaly detection.

- **Graph Neural Networks (GNNs):** When data is structured in a relational form (e.g., message threads, social graphs), GNNs can integrate modality-specific nodes and capture evidence chains.

Challenges and Future Directions

Despite its promise, multimodal analysis in mobile forensics faces several open challenges:

- **Data Alignment:** Temporal misalignment across modalities (e.g., unsynchronized video and audio) complicates joint analysis.

- **Privacy and Legal Issues:** Joint analysis raises concerns over data minimization, especially with sensitive modalities like voice and facial imagery.

- **Resource Constraints:** Processing multiple modalities in real time requires significant computational resources, which can strain mobile forensic tools in field conditions.

CHAPTER 12   DEEP LEARNING TECHNIQUES FOR MOBILE FORENSICS

- **Labeling and Ground Truth:** Creating labeled multimodal datasets for forensic use is time-intensive and prone to interpretation variability.

Future research will likely focus on **self-supervised learning**, **cross-modal attention mechanisms**, and **explainable fusion architectures** to tackle these challenges.

Conclusion

No longer is multimodal data analysis a technological luxury but an investigative requirement. Deep learning models that can perform reasoning over text, images, videos, and audio jointly have remarkable advantages in accuracy, contextual awareness, and automation. Whether reconstructing a crime or discovering clandestine communications, multimodal forensics platforms are turning the mobile device from a black-box source of data to an open-book source of understanding.

By taking advantage of early and late fusion techniques into consideration as well as using dedicated deep models for each source and managing practical issues, forensic examiners could exploit the complete multimodal evidences in digital artifacts.

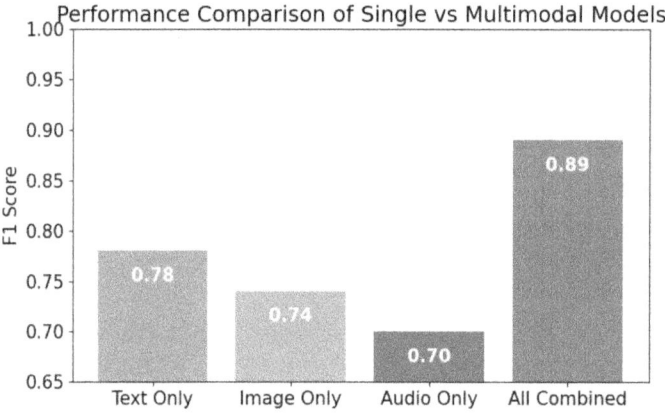

Figure 12-8. Performance comparison of single vs. multimodal models

We compare F1 scores on this chart for models trained on individual modality (text, image, audio) and for the multimodal models with all the data modalities. It evidently demonstrates the remarkable accuracy enhancement by cross-modal integration, hence further confirming the effectiveness of multimodal deep learning in the forensic domain. Training parameters include batch size 128, learning rate 0.001, 35 epochs, and dropout 0.4 for regularization across all models.

CHAPTER 12   DEEP LEARNING TECHNIQUES FOR MOBILE FORENSICS

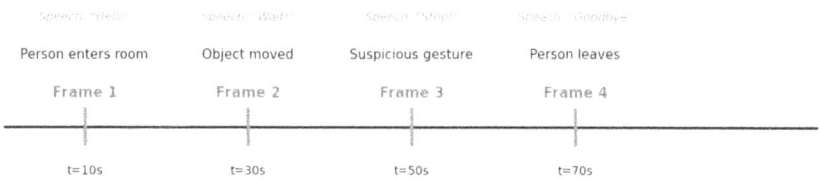

Figure 12-9. Captioned video forensic timeline

This case study of forensic video analysis is presented as a timeline visualization that integrates frames, detected actions, speech transcriptions and timestamp information. It provides an integrated multimodal summary that is crucial to facilitate the rapid comprehension of incidents in forensic video sequence for investigators. Parameters include video frame sampling rate of 1 frame per 5 seconds, speech recognition confidence above 90%, and action detection model trained over 50 epochs with batch size 32.

Conclusion and Future Scope

The integration of mobile forensics and deep learning marks a new era in digital forensic investigation. Given the ubiquity of smartphones as the primary communication computing platform in our personal and professional lives, they are increasingly serving as a vector and depository of digital evidence in cybercrime, fraud, harassment, espionage, and other illicit activities. Conventional forensic applications, although fundamental, find it difficult to handle the increasing complexity, quantity, and heterogeneity of mobile data. Deep learning has emerged as a powerful paradigm that is able to address the changing demands of the new era due to its ability to extract high-dimensional features and recognize patterns and perform data-driven inference.

This chapter has illustrated the ways the deep learning methodologies, from Convolutional Neural Networks (CNNs) to Recurrent Neural Networks (RNNs), Transformer architectures and Autoencoders, can be employed to extract, understand, and associate and relate the digital evidence obtained from different data sources in a mobile context. Deep learning excels in developing models that operate on raw or light-processed data, such as text, images, audio, video, and structured logs, discovering high-level semantic features without exclusively depending on handcrafted rules and feature engineering. Detecting mobile malware, phishing URLs, analyzing user behavior, classifying network traffic, and multimodal data fusion to build crime timelines have all yielded superior results when compared to traditional rule-based systems in terms of scalability, accuracy, adaptability, and automation.

Superiority and Applicability in Mobile Forensics

In real-world mobile forensic workflows, deep learning offers a spectrum of advantages:

- **Automated Evidence Extraction:** Instead of relying on fixed signatures or manual keyword searches, DL models can autonomously extract contextually relevant information from logs, chat transcripts, and app data.

- **Robustness to Obfuscation and Evasion:** Adversaries often use encrypted communication, obfuscated code, or synthetic voices to hide intent. Deep models, especially those trained on adversarial datasets, can generalize beyond known attack patterns.

- **Multimodal Reasoning:** DL systems can integrate heterogeneous data types—voice messages, screenshots, videos, and geolocation traces—into unified decision models, enabling deeper insight.

- **Scalability:** Trained models can process vast volumes of device data with minimal latency, facilitating real-time forensic triage in field investigations.

Legal and Ethical Challenges

Explainability and Transparency

One of the daunting issues in forensic AI is the opacity of models learned by deep learning. The forensic conclusions are only admissible in court if they are: Verifiable and Explainable. Although deep models can achieve high accuracy, their internal logic is generally not transparent, which complicates the process of understanding how a particular prediction (e.g., malware classification or a behavioral anomaly) was made.

Solutions are emerging, including

- **Layer-wise Relevance Propagation (LRP)** and **SHAP (SHapley Additive exPlanations)** to attribute decisions to input features

- **Attention heatmaps** in Transformer models that visualize focus areas in images or texts

- **Local surrogate models** like LIME (Local Interpretable Model-agnostic Explanations) that approximate decision boundaries for specific instances

Despite these advancements, striking a balance between model complexity and forensic transparency remains a central research challenge.

Privacy Implications

Mobile forensic analysis frequently involves highly personal and sensitive data—private chats, health records, financial information, location histories, and intimate photos. Deep learning exacerbates privacy concerns by enabling large-scale pattern mining and latent information discovery, sometimes beyond the original scope of the investigation.

Forensic practitioners must therefore adopt **privacy-preserving approaches**, such as

- **Data minimization:** Limiting analysis to data directly relevant to the investigation

- **Differential privacy:** Introducing controlled noise to protect individual user traces

- **Secure multiparty computation and homomorphic encryption:** Allowing analysis over encrypted data without decryption

- **On-device AI:** Performing inference locally without uploading sensitive content to cloud servers

Ethical forensic frameworks must also incorporate **consent protocols**, **auditing mechanisms**, and **data retention limitations** to safeguard user rights while ensuring investigative effectiveness.

Forensic Admissibility and Legal Standards

In many jurisdictions, digital evidence must pass legal benchmarks such as **Daubert** (in the United States) or equivalent criteria in other countries, which assess

- The scientific validity of the method
- Peer review and publication record
- Known error rates
- Acceptance in the forensic community

Deep learning applications in mobile forensics must be **rigorously validated**, **documented**, and **peer-reviewed** to meet these standards. Importantly, forensic tools must also maintain **chain of custody**, **tamper-proof logging**, and **verifiability of model outputs** to avoid legal challenges during prosecution.

Future Scope: Emerging Trends and Innovations

As mobile ecosystems evolve, so must the forensic techniques used to examine them. Several emerging trends promise to further transform the intersection of AI and mobile forensics.

Federated Forensics

Federated learning allows AI models to be trained across decentralized mobile devices while keeping raw data on-device. This paradigm is particularly valuable in forensics, where centralized data collection may be impractical, illegal, or ethically questionable.

- **Advantages:** Enhanced data privacy, reduced risk of breaches, and compliance with data protection laws (e.g., GDPR)
- **Use Case:** Collaborative model training across smartphones belonging to a target group, without accessing individual data stores

On-Device Forensic Analysis

With mobile hardware becoming increasingly powerful (e.g., AI chips in modern smartphones), forensic tools can now run inference **directly on devices**. This facilitates

- **Real-time evidence triage**
- **Offline analysis in field conditions**
- **Minimal footprint and lower network dependency**

AI-powered apps can detect suspicious patterns, alert investigators, or autonomously extract structured evidence even in disconnected environments.

Privacy-Preserving Deep Models

Next-generation models will increasingly adopt architectures that minimize leakage and bias. Promising directions include

- **Split learning:** Model is divided between client and server to limit visibility into raw data
- **Differentially private training:** Adds noise to gradients to protect individual record influence
- **Zero-knowledge proofs:** Allow evidence validation without exposing sensitive content

Such innovations ensure that forensic power does not come at the expense of ethical responsibility.

Cross-Domain Intelligence and Multilingual Models

Future forensic systems will incorporate **cross-domain knowledge** (e.g., linking cybercrime indicators with financial fraud patterns) and support **multilingual NLP**, essential for global investigations involving diverse communication platforms.

Deep learning models will need to handle multilingual messages, code-switching, and culturally nuanced content to ensure comprehensive analysis.

Closing Remarks

The adoption of deep learning in mobile forensics is more than an improvement: it is a radical change. It transforms forensic capabilities beyond static, point-in-time artifact extraction to a state-of-the-art model of real-time, data-driven states of intelligent inference, from isolated procedures to a continuous, contextual narrative, and from reactive investigation to proactive threat detection. But the introduction of these formidable new weapons must be accompanied by a firm commitment to transparency, legality, and human rights.

The way forward is to build trustworthy AI for forensic science—systems that are accurate, but also interpretable, fair, and accountable. As forensic scientists, lawyers, and technologists cross-pollinate in this interdisciplinary field, an age of accountable and astute mobile forensics looms on the horizon such that machines not just analyze, but think, articulate, and underwrite justice.

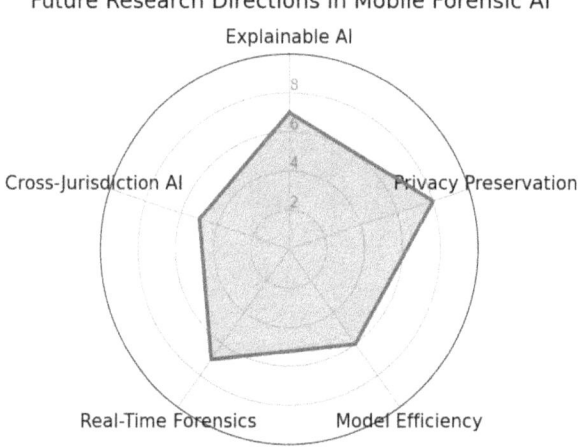

Figure 12-10. Future research directions in mobile forensic AI

This radar chart illustrates key future trends in mobile forensic AI research, highlighting areas like explainable AI, privacy preservation, model efficiency, real-time forensics, and cross-jurisdictional challenges. The visualization guides researchers toward balanced progress across critical forensic AI domains. Parameters reflect expert survey scores on current maturity levels, scaled 1 to 10, and anticipated impact over the next five years.

References

[1] Qadir, S., & Noor, B. (2021, May). Applications of machine learning in digital forensics. In *2021 International Conference on Digital Futures and Transformative Technologies (ICoDT2)* (pp. 1–8). IEEE.

[2] Qadir, S., & Noor, B. (2021, May). Applications of machine learning in digital forensics. In *2021 International Conference on Digital Futures and Transformative Technologies (ICoDT2)* (pp. 1–8). IEEE.

[3] Vasilaras, A., Papadoudis, N., & Rizomiliotis, P. (2024). Artificial intelligence in mobile forensics: A survey of current status, a use case analysis and AI alignment objectives. *Forensic Science International: Digital Investigation*, 49, 301737.

[4] Rodriguez, E., Otero, B., Gutierrez, N., & Canal, R. (2021). A survey of deep learning techniques for cybersecurity in mobile networks. *IEEE Communications Surveys & Tutorials*, 23(3), 1920-1955.

[5] Avanija, J., Kumar, K. N., Kumari, C. U., Jyothi, G. N., Raju, K. S., & Madhavi, K. R. (2023). Enhancing Network Forensic and Deep Learning Mechanism for Internet of Things Networks. *Journal of Scientific & Industrial Research (JSIR)*, 82(05), 522-528.

[6] Zhang, S., Hu, C., Wang, L., Mihaljevic, M. J., Xu, S., & Lan, T. (2023). A malware detection approach based on deep learning and memory forensics. *Symmetry*, 15(3), 758.

[7] Qamhan, M. A., Altaheri, H., Meftah, A. H., Muhammad, G., & Alotaibi, Y. A. (2021). Digital audio forensics: microphone and environment classification using deep learning. *Ieee Access*, 9, 62719-62733.

[8] Alsubaei, F. S., Almazroi, A. A., & Ayub, N. (2024). Enhancing phishing detection: A novel hybrid deep learning framework for cybercrime forensics. *IEEE Access*, 12, 8373–8389.

[9] Hina, M., Ali, M., Javed, A. R., Ghabban, F., Khan, L. A., & Jalil, Z. (2021). Sefaced: Semantic-based forensic analysis and classification of e-mail data using deep learning. *IEEE Access*, 9, 98398–98411.

[10] K. Kaushik, A. Bhardwaj, and S. Dahiya, "Smart Home IoT Forensics: Current Status, Challenges, and Future Directions," 2023 Int. Conf. Adv. Comput. Comput. Technol. InCACCT 2023, pp. 716–721, 2023, doi: 10.1109/INCACCT57535.2023.10141730.

[11] A. Garg and A. K. Singh, "Internet of Things (IoT): Security, Cybercrimes, and Digital Forensics," Internet Things Cyber Phys. Syst., pp. 23–50, Dec. 2022, doi: 10.1201/9781003283003-2.

[12] K. Kaushik, S. Dahiya, and R. Sharma, "Role of Blockchain Technology in Digital Forensics," Blockchain Technol., pp. 235–246, Feb. 2022, doi: 10.1201/9781003138082-14.

[13] Sachdeva, S., & Ali, A. (2022). Machine learning with digital forensics for attack classification in cloud network environment. *International Journal of System Assurance Engineering and Management, 13*(Suppl 1), 156–165.

[14] Kumar, M., Singh, G., Singh, J., Kaur, N., & Singh, N. (2022). Machine learning-based analytical systems: food forensics. *ACS omega, 7*(51), 47518–47535.

[15] Nguyen, K., Proença, H., & Alonso-Fernandez, F. (2024). Deep learning for iris recognition: A survey. *ACM Computing Surveys, 56*(9), 1–35.

[16] K. Kaushik, R. Tanwar, S. Dahiya, K. K. Bhatia, and Y. Wu, Unleashing the Art of Digital Forensics, 1st ed., vol. 1. Boca Raton: Chapman and Hall/CRC, 2022. doi: 10.1201/9781003204862.

CHAPTER 13

Privacy and Security Considerations in Mobile Forensics

This chapter focuses on the privacy and security issues in mobile forensics. It explains the need to maintain chain of custody and data integrity, while following principles like data minimization and purpose limitation. It also highlights the importance of user consent, privacy rights, and ensuring forensic tools are secure and reliable.

Privacy and Security in Mobile Forensics

Mobile devices in this digital world have become an invaluable instrument, carrying hundreds of gigabytes of personal and sensitive information. The first and foremost advantage provided by mobile devices is the easy facility of communication and transaction of money. Devices can track locations, monitor a person's health, and do much more than mere communication. Due to these multifaceted uses of mobile devices, a new dimension has been added to the role of mobile forensics involved in any crime investigation or solution of any disputed case. But this brings with it very serious concerns of privacy and security. The investigators have to tread this thin line between bringing back useful data from the devices while holding onto an individual's privacy. This means that more data is typed into the digital world, which increases the risks when such data is manipulated or hacked.

On the critical issues of mobile forensics, the data attached to or transmitted from a mobile may be sensitive. It is, therefore, not unusual to find such data in contacts, emails, photos, financial information, health data, among others that could be common to a typical smartphone, revealing very intimate moments of a person's life.

This is an excellent challenge for investigators and law enforcement to start gathering and analyzing the data without violating legal and ethical boundaries. The recently strengthened privacy laws around the world, such as the GDPR in Europe and CCPA in California, United States of America, reveal the increasing awareness of such matters. This focuses on the rights of users, emphasizing careful data handling while ensuring access only when necessary for the investigation by forensic professionals. It not only breaks these law-set frameworks but also the trust between the public and law enforcement agencies if there are unauthorized collection and mishandling of personal data. Besides legal compliance, there is growing awareness among the general public about their private rights. The hacking occurrences, as well as breaches in systems, have made users more cautious of their data. This increased concern for privacy calls upon forensic investigators to adapt best practices and be considerate of user consent, tying one's principles to the reduction of data.

Security Concerns in Mobile Forensics

In mobile forensics, where compliance with privacy is secondary to preserving the integrity of the extracted evidence, it follows that the forensic process necessarily entails gaining access to devices to extract data that may be locked away or encrypted in the target device. It is needful for investigators that ethics are dispensed with, even when it comes to using methods that can breach security without compromising data integrity or increasing the chances of data compromise. For example, using certain tools or features with unproven forensic use can be risky because they may not handle data properly. In some cases, they can even damage or overwrite the original evidence, making it unusable in an investigation. There is also the danger that these tools might leak sensitive information, such as personal files or login details, to unauthorized persons, which could create further security and privacy issues. They also enhance the complexity of the problem with the advent of cloud technology and remote storage. A smartphone's internal storage can have encrypted information for safety; however, data stored in external storage may have looser encryption policies. Therefore, investigators are expected to have the relevant technology to access such data without taking the risk of leaking sensitive information to wrong individuals or organizations. At the same time, cybersecurity concerns also arise. Afterward, the data should be kept and moved in such a way that it is not available to those who should not have it. Chain-of-custody protocols must be adhered to strictly to ensure that the integrity of data is well maintained and that there is no tampering with the data at any point during the course of the investigation. Any relaxation of the standards in information integrity could compromise not just the integrity of the investigation itself but also the privacy rights of the person whose data is being scrutinized.

Challenges in Mobile Forensics

Mobile phones today store the most considerable amounts of data ever. This ranges from structured to unstructured data, including SMS messages, app data, and call logs, as well as browsing history and GPS coordinates. In this regard, much challenges forensic investigators with

- **Data overload and relevance:** There is too much information available on mobile phones, and it can be challenging to pick out relevant information. For example, an investigator may need to examine hundreds or thousands of text messages, photographs, or location logs. It becomes difficult to sift through all this information without violating privacy, as most pieces of information might not even have to do with the case.

- **Applications ecosystems and encrypted data:** Many mobile apps, especially social media and instant messaging services, use end-to-end encryption. While an all-around step for users' privacy, this limits the ability of forensic investigators to gain access to such communications in cases of criminal investigations. So, a balance should be found that does not compromise privacy rights and yet does not open up security holes.

- **Cloud storage and synchronization:** Mobile devices generally synchronize data on-the-go to cloud services, such as Google Drive, iCloud, or Dropbox. Although this implies good data backup for the user and provides convenience, it complicates the forensic process further, because the investigator must be aware of the relationship between local and cloud storage when developing their analysis to account for both sources. There adds legal complexity because most privacy regulations differentiate between data stored in the cloud and data placed directly on a device.

- **Evolving technology:** Mobile technology evolves at a rapid pace, with new smart devices, operating system variants, and security features constantly surfacing. Forensic investigators must keep abreast of these changing tides by regularly refreshing their knowledge. These evolving trends create yet another dimension in the investigative process, including the possibility that traditional forensic techniques might not be applicable to the newest variants of mobile devices or operating systems.

Another aspect is the rise in privacy and security needs of mobile forensics. The daily use of mobile devices implies an increase in the amount of sensitive and personal information that exists in them, and forensic investigators need to be cautious of and respect privacy laws. This brings other unique challenges connected with mobile technology, as well as the enormous amounts of data stored in mobile devices themselves, concerning both technical requirements and ethical frameworks. Extracting relevant data from these sources without destroying the integrity of the forensic process is extremely important, but protecting the rights of the individual is equally so.

Chain of Custody and Data Integrity

This chain of custody is a very basic principle in forensic investigations, which actually provides a thorough process that lets evidence be kept both valid and secure from the time of collection until it is introduced in court. More importantly, for mobile forensics, whose nature is often digital, there is a high chance that evidence could be tampered with, changed, or even modified in error. Chain of custody refers to the whole step-by-step documentation detailing everyone who handled, transferred, or stored evidence. It records the actual date and time, location, and specific reason for every interaction with the evidence, making it a verifiable and continuous account. In mobile forensics, this process is even more paramount due to the ease with which digital data may be altered or destroyed. The entire process, from data extraction from a cell phone to analysis and storage, should be properly documented. Skipping a single step may bring the integrity of this evidence into question, possibly disqualifying its admissibility in court. Therefore, the chain of custody acts like proof, indicating that the digital evidence is left just as it was during the collection period, free from any type of contamination or unauthorized change. Such minute documentation creates a basis of trust in forensic processes.

Chain of Custody in Forensic Investigations

In essence, the chain of custody is a systematic process that keeps detailed records at every stage of an investigation, which ensures that evidence is accounted for at all times. Mobile forensics may thus include everything ranging from the physical device itself—for example, smartphone or tablet—to the digital data extracted from it—text messages, emails, or location data. The chain of custody refers to who collected the evidence, how it was collected, where it was stored, and in what manner, who had access to it, and how it was transported. For evidence being presented, this list is normally associated with the chain of custody process:

- **Gathering of evidence:** The chain of custody commences from the careful collection of digital evidence, such as a mobile device. The step of collection is very sensitive, and if mishandled, it may lead to loss or contamination of data. Investigators take note of every minute detail, including the identity of the person collecting the device, the exact location and time of collection, and the method for securing the device. The process may also involve putting cellphones inside a Faraday bag to eliminate signals that might be used to interfere with it remotely. Pictures are taken to record the device's state; it may be on or off, possibly damaged, or tampered with. The initial observations help refer to the state at which the item was collected.

- **Documentation and labeling:** When the device is collected, it is marked with significant identification details. Such identification entails the case number, investigator's name, date and time of collection, and a detailed description of the item itself, including but not limited to: make, model, serial number, and unique identifiers such as IMEI. This labeling is useful in keeping track of the evidence while distinguishing it from other pieces of evidence. Proper documentation at this point prevents misidentification, which may jeopardize investigations or cause errors in court. This labeling serves as a critical check, ensuring clear ownership and custody of the evidence.

- **Storage of evidence:** Once digital evidence has been collected and labeled, it should be stored in a safe place to preserve its integrity. Most hardware devices, such as cell phones or tablets, are stored within evidence lockers with controlled access and logged entries.

Information from the forensically extracted device, such as forensic images or data in backups, is placed on encrypted media to prevent unauthorized access or alteration. Other preventive measures include tamper-evident seals and cryptographic hash values, such as MD5 and SHA-256, to ensure data remains unchanged. Hash values are compared later with the stored hash value to verify if the data remained unchanged, thus maintaining the integrity of evidence during storage.

- **Transfer of evidence:** Evidence is transferred from one location to another, for example, from the crime scene to the forensic laboratory or from the laboratory to court. Great importance should be placed on detailed documentation of the transfer, including time, date, people involved, and condition before and after transfer. For physical devices, tamper-proof packaging is utilized to ensure any interference is visual. For digital data, encryption of storage devices and secure transmission protocols (e.g., encrypted emails, SFTP) prevent unauthorized access during transfer. The receiving party verifies the integrity of evidence once received, often using cryptographic hash values for verification, to check whether the data was modified during the process.

- **Analysis of evidence:** Analysis is carried out on forensic images or copies of digital evidence by forensic examiners, ensuring the original device remains in its original state. This ensures no alteration or destruction of the original data during analysis. Mobile device data is extracted using forensic tools such as Cellebrite or XRY. Each step is documented by the forensics expert, including software and tools used, version, and steps taken in analysis. All interaction with the data is logged in great detail, and hash values are derived before and after analysis to prove no unauthorized changes were made. This documentation makes the process completely reproducible if it ever goes before a judge and jury.

- **Presentation of evidence in court:** The last step of the chain of custody is when evidence is presented in court. In court, the prosecution must prove that digital evidence has been handled correctly and securely from collection to presentation. This is achieved by presenting the entire chain of custody records, covering all details of who handled the evidence, how it was stored, and how it was passed. The main proof that the evidence has not been tampered with or altered will be the cryptographic hash values. Forensic experts may testify on handling and analysis, describing procedures used to ensure integrity preservation. Such evidence is authentic, admissible, and trustworthy, and the court can confidently rely on it.

Data Integrity

Data integrity also forms a major requirement for accepting the appropriateness of digital evidence in court. Data integrity refers to the accuracy and consistency of the data concerned throughout the forensic process. In the case of mobile forensics, it simply implies that the data extracted from a device must remain unchanged, unaffected, and unaltered from the time of collection until it reaches the court. Any breach in data integrity may cause the court to disqualify it from being accepted as evidence, thus making the case weaker. The following are a few reasons why data integrity is necessary:

- **Trust in the evidence:** In order to be reliable and trustworthy, evidence must possess a sound evidential basis. Evidence that fails such a test could be rejected or dismissed, or at the least, its credibility could be challenged. For example, in a case where the opponent attorney demonstrates that digital evidence has been altered or mishandled, the opposing attorney may argue that evidence cannot be trusted and thus not considered.

- **Prevention of contamination of evidence:** Contamination in mobile forensics cases can occur due to improper handling of the device, where forensic tools without knowledge may alter data, or some lapse may occur while extracting data. Once contaminated, it cannot be treated as original, and thereby its admissibility into the court will be in question.

- **Legal challenges and rejection of evidence:** In general, a defective chain of custody or compromised integrity of data can provoke a challenge on the admissibility of the evidence. In any case, whether in the United States through its Federal Rules of Evidence or otherwise by other jurisdictions, evidence so tainted with untrustworthiness cannot be subjected to judicial proceedings. This would become one weak link in the prosecution's case and may eventually lead to exclusion of key evidence, hindering possible wrongdoing accountable to justice.

- **Objectivity and reproducibility:** In other words, one of the criteria used to determine admissible evidence is that it must be scientifically reproducible—if one wants to send it to another expert for verification. As far as data is concerned, this implies that forensic examiners should be allowed to apply the same methodology and come to the same conclusions when processing the data. If the original data has been tampered with in some respect, it is impossible to reproduce, as integrity has been compromised when data integrity is compromised.

- **Ethical and legal responsibilities:** Forensic investigators have ethical and legal responsibilities to prove that their work does not contaminate evidence. Evidence mishandling, whether in a move to intentionally or accidentally hurt evidence, may lead to professional misconduct or legal offenses. Investigators are expected to adhere to established procedures, utilize credible tools, and maintain records to ensure the integrity of the investigation.

- **Authentication of digital evidence:** The authenticating process ensures the evidence is genuine and is not altered in any form. For mobile forensics, this involves proving that the data recovered from the device is actually what existed on the device at the time of collection. A clear, well-documented chain of custody, along with strict adherence to data integrity protocols, guarantees the admissibility of evidence in a case; otherwise, it will be inadmissible.

The chain of custody is a very important component of any forensic examination, especially in the field of mobile forensics, where digital evidence is easily subject to change or alteration. An uninterrupted, clear chain of custody helps ensure that the evidence remains reliable, admissible, and trustworthy to stand in court. Equally important is the preservation of data integrity for purposes of authenticity and non-contamination of evidence. Because of this, without proper protection, the entire investigation may be undone by adverse consequences in the judicial process.

Best Practices for Documenting and Preserving the Chain of Custody and Maintaining Data Integrity

Most importantly, forensic investigations, especially mobile forensic ones, work through strict procedures so that the evidence collected is coherent, tamper-free, and accepted by the courts. Among these, the two main guiding principles are proper documentation and maintenance of the chain of custody, which ensure the preservation of data integrity throughout the investigation process. There has to be a determination of best practices concerning these areas so that the gathered evidence may have the capacity to stand before the tribunals.

Good Practices on Recording and Maintaining Chain of Custody

Appropriate recording and maintenance of the chain of custody ensure that good practices are followed at every stage of the investigation. These include proper identification of who collected, handled, or transferred evidence, with a clear record of time, date, and purpose for every interaction. Tamper-evident seals or tamper-proof containers, secure storage, and encrypted digital data are also necessary. The following points are to be considered as best practice:

- **Detailed Record maintenance:** No practice is considered more important to maintaining the chain of custody than keeping a very detailed record of all evidence-handling processes. This includes recording who collected the evidence, when (or at what time and date), the condition of the evidence at collection, and how it was kept or moved. All movements of the evidence must be documented, including the recipient, nature of movement, and method of translocation. Incomplete records in these areas create a chain of discontinuity, potentially rendering evidence inadmissible.

- **Using evidence labels and tags:** Attach unique identification numbers or barcodes to each piece of evidence at collection, consistent across all documentation. The label must contain details such as case number, item description, date and time of collection, and collector's name. Use tamper-proof evidence bags or containers to prevent tampering. These containers should remain sealed until formal analysis, including any opening and resealing, which should be duly documented.

- **Chain of custody forms:** Chain of custody forms document traceability, including all material evidence, tracing evidence at every touchpoint from crime scene to lab to courtroom. All evidence handlers must sign the form, specifying contact with evidence, including when and why. This record can be paper or electronic, although electronic systems increasingly automate record-keeping and minimize human error risk. Chain of custody forms must be safeguarded and readily available for court review.

- **Safe storage of evidence:** Forensic evidence must be stored in environments accessible only to authorized personnel. For mobile devices, storage areas should protect against external field interference (e.g., magnetic fields) or high temperatures, damaging digital data. All storage access entries should be logged to prevent unnoticed evidence tampering.

- **Periodic audits of evidence:** Regular auditing ensures no interruption of the chain of custody. Auditing verifies evidence presence, proper documentation, and addresses discrepancies promptly.

Good Practices for Maintaining Data Integrity

Good practice for data integrity includes using cryptographic hash values (e.g., MD5, SHA-256) to prove digital evidence has not been altered during handling. Forensic copies should be taken using write-blockers to ensure the original data remains unaltered. Data will be stored on encrypted media and in secure, tamper-evident containers, with physical devices protected to prevent unauthorized access or alteration of evidence, ensuring its integrity for legal proceedings. Following points to be considered as a best practice:

- **Write-Blockers:** A write-blocker is an integral part of digital forensics. Extraction of files does not alter the original data due to the write-blocker's involvement. In the course of a mobile phone investigation, forensic investigators **use** either hardware or software write-blockers to prevent overwriting or damaging any data available on the device. These write-blocking devices allow investigators to obtain what is referred to as a forensic image, which is a bit-for-bit copy of the contents of the mobile device. This is to protect the original data from being altered and ensure its validity during court proceedings.

- **Forensic imaging and hashing:** Create an exact copy of the information present on the mobile device prior to the forensic investigation. This copy ensures that the original remains unspoiled. After completing a forensic image, investigators use hash algorithms such as MD5 or SHA-256 to create hash values for both the original and duplicated images. The hash value serves as a digital impression. The hash values of the original and the copy created after imaging are compared first to show that no changes occurred during imaging. Hash values also play a role in maintaining data integrity during further analysis or use in courts of law.

- **Activity log of all performed activities:** All actions executed on the forensic image or device must be documented in detail. This includes tools and software used, data extraction methods, and persons involved. These logs provide a complete audit trail, demonstrating adherence to best practices for preserving data integrity and avoiding intervention. Logs enable result reproduction, crucial if the defense challenges results or requests independent revalidation.

- **Secure transfer of digital evidence:** Digital evidence, such as forensic images and extracted data, must be transferred securely between devices, investigators, and locations. Encryption protects data from unauthorized access during transit. Document transfers, whether physical (hard drives or USBs) or over secure networks, and apply chain of custody protocols as with physical evidence.

- **Applications of certified and reliable forensics tools:** Most forensic tools are validated to avoid modifying or changing data. Numerous tools are tested and certified for extracting data without compromising integrity. Apply updates and patches promptly to maintain compatibility with latest mobile devices and operating systems, addressing errors or critical data loss.

- **Data backups and redundancy:** Maintain multiple copies of forensic images and extracted data to protect against accidental loss or corruption. Store these encrypted copies securely. Data redundancy ensures availability of uncompromised evidence for further investigations or court proceedings.

By following these best practices, forensic investigators ensure that both the chain of custody and data integrity remain intact throughout the whole investigation process. These procedures help maintain the credibility of the evidence, reduce the likelihood of a legal challenge to the evidence in court, and ensure that the investigation proceeds smoothly from collection to courtroom presentation.

Mobile Devices and Issues Within Chain of Custody and Integrity of Data

With respect to digital forensic environments, it is worth noting that mobile devices pose certain distinctive challenges in terms of chain of custody as well as protecting the integrity of the stored data. Some of these challenges include the intricate layout and dynamic nature of mobile data, the many variants of devices and their operating systems, the increased pace associated with changes in technology, and the high susceptibility of data to alterations and loss. Forensic investigators must be aware of these adjudicative challenges because any incidents concerning the chain of custody and the integrity of the data can lead to the evidence being inadmissible in a court of law. In this chapter, the author focuses on some specific issues that come into play when mobile devices are involved in a forensic examination.

- **Volatility and transient nature of mobile data:** Cellular devices contain volatile and ephemeral data, including text messages, call logs, and Internet browsing records. Some data, especially app-specific information, like session data from social media apps, are sometimes only held temporarily or overwritten immediately. When a mobile device is turned on or connected to a network, potentially new data can be written to the device, altering or overwriting critical evidence. In some cases, data like RAM or running processes might be lost the moment a device is powered off. Advanced techniques and equipment are involved in capturing such data, and each passing moment can mean losing important evidence. To minimize risks, forensic investigators must act fast to prevent volatile data from disappearing. This hasty collection process challenges maintaining the integrity of the chain of custody. It may result in evidence contamination if not treated properly, such as powering on the device without precautions. Encryption or password protection complicates instantaneous access to critical data, increasing the risk of data loss.

- **Type and OS heterogeneity:** One critical concern in handling mobile phones is heterogeneity in devices and operating systems. Mobile devices have high diversity in configurations, including hardware, firmware, and software. In this fast-moving market, new devices emerge and operating systems upgrade frequently, causing compatibility issues for investigators. Tools and techniques to extract and analyze data from each device and OS differ, making it challenging to maintain evidentiary integrity without damaging or altering data. For example, a forensic tool designed for one operating system may not support another, leading to data extraction failures or incomplete data collection. Diversity raises human error risks, where inappropriate tools or methods are applied to evidence, affecting data. Mobile devices' security features, such as full-disk encryption, protect against unauthorized access but compromise data integrity.

- **Human error and mismanagement:** Human error is crucial when examining mobile devices. All activities, including documentation, labeling, storage, and transfer of evidence, must be meticulously recorded to avoid mistakes. In high-pressure situations, though, sometimes measures are overlooked, and procedures are not adhered to, thus leaving holes in the chain of custody. An unmarked or improperly marked collected device may send off the wrong signals about the source and may be viewed as another piece of evidence, leading to invalidity. Poor handling, such as using unsecured or tamper-proof containers, will have effects on modification, destruction, or unauthorized access. Non-production or non-authentication of cryptographic hash values on a digital image leads to the inadmissibility of evidence.

- **Protections:** Modern mobile phones include security features like password locks, biometric locks, and full-disk encryption. As much as these features protect the user's data, they pose considerable barriers to forensic investigators. To date, current tools and techniques to access encrypted devices without affecting data are quite challenging; they require user cooperation or perhaps forensic breakthroughs. Any attempt to access encrypted data in ways that could sidestep compromising integrity risks having undesirable effects on evidence admissibility. Some security features overwrite the device contents after several failed attempts to input the correct password.

- **Legal and jurisdictional problems:** Mobile devices are capable of holding large amounts of data, which in most cases cuts across multiple jurisdictions, particularly with the existence of cloud services. With regard to investigations, accessing the data held in other countries is largely a concern, considering the existing international statutes, which prolong the process and pose problems with respect to evidence collection. Complications in evidence acquisition are attributable to the existence of different privacy laws and varying levels of legal authorizations. It is often difficult to provide evidence before the court due to the high chances of data loss or tampering. Even when devices are secured, the evidence may be found on the cloud, which is hosted outside the country or on a third-party provider, thereby raising concerns regarding the chain of custody.

Given mobile data volatility, device diversity, cloud services, and complex legal frameworks, maintaining chain of custody and data integrity is challenging. Encryption and security measures ensure user privacy but add complexity to accessing critical evidence. Overcoming obstacles requires forensic examiners to stay updated on technologies, adhering to procedures, and using appropriate tools to demonstrate authenticity and admissibility of mobile device evidence.

Role of Consent in Forensics

Informed consent in the context of mobile forensics plays a vital role since it becomes a precedent that establishes whether a person has rights to know and allow consent regarding access, analysis, and usage of personal data in the course of an investigation. Given that mobile devices have a high potential to become repositories of sensitive personal information, such as messages, emails, photos, financial data, and location history, this aspect of mobile devices' forensics is most powerful in digging up massive amounts of private data, potentially even that which the user is not even aware exists on their device; thus, it is crucial that there be "informed consent" for such an occurrence to guarantee that proper ethical standards are met and individual rights are safeguarded.

Informed consent is defined as a clear, voluntary agreement by an individual to let investigators or other forensic professionals access and examine the contents of their mobile device. Based on transparency, those whose data is being collected must be fully aware of what is going on, why it is necessary, how their data will be used, and what the consequences of granting them access are. It seeks to attain information control over personal lives so that no one is exploited or coerced by the forensic process.

The major ethical concerns are that, in the absence of informed consent, the forensic process would be considered an invasion of privacy, and this will nullify the legitimacy of the investigation. Ethical concerns are of particular importance when dealing with mobile forensics: so much information could be recovered, including deleted files and metadata, which the user might have assumed was no longer available. Without permission, an investigation could violate a person's autonomy and establish a basic disregard for law enforcement or forensic professionals. Ethical handling of collecting data not only protects individuals but also ensures that the evidence obtained is admissible in court, as a lack of consent can bring challenges regarding the legitimacy of the investigation.

In practice, such informed consent usually involves very clear communication between the investigator and the individual. It involves the provision of information on what is to be retrieved and for what reason. The consent forms are hence signed, including details on what can be collected, for how long, and regarding third-party sharing with law enforcement agencies, among others. In most real-life scenarios outside of a hypothetical scenario, it is complicated to obtain valid consent. A person typically has no idea of the technical implications of what one is giving consent to, especially as far as the depth and breadth of accessible data on a mobile device. For example, shared devices or data for which several users have rights and claims—thus, families or colleagues at work—can really break the consent process among the parties concerned. The forensic investigators then have a delicate balance in evidence collection with respect to the rights of individuals.

When a criminal investigation presents most of the evidence located on mobile devices, there wouldn't be any harder principle than that of informed consent. Sometimes, suspects or persons of interest may decline the idea of consenting, and detectives are left with no choice but to take legal recourse; for instance, they must obtain warrants. This sends an extension of time under investigation and further into litigation about the integrity of the gathered evidence. In many recent cases, courts have intervened, mainly in cases where consent was not properly given. The outcome of such cases may lead to exclusion of evidence because of privacy infringement violation, hence putting the case at risk.

Privacy Laws and Users' Rights

In this regard, privacy laws safeguard the rights of users since they regulate how an individual's personal data can be collected, processed, or shared, especially in forensic contexts. With that, such laws ensure that any individual's private information stays under their control, even when they become involved with legal investigations. Some of the strongest privacy laws in the world today, such as the GDPR in Europe, the CCPA in the United States, and HIPAA in the healthcare sector, have created a big impact on mobile forensics. These laws influence how investigators handle personal and sensitive data, especially when it comes to taking proper consent and protecting information like personal health records (PHI).

- **General Data Protection Regulation (GDPR):** GDPR is a comprehensive privacy framework in the EU, which has properly established guidelines on the treatment of personal data. As pointed out by the regulation, individual rights toward their personal data require explicit consent for collecting and using personal information. Any personal data extracted from a device is also processed under the GDPR regime. According to the GDPR, any such personal data extracted from a device will be processed based on a lawful basis—whether the lawful basis is consent or a legal obligation. This is particularly very relevant in cases where some people's devices are subject to forensic investigation, as investigators must ensure that their actions are able to and do, in fact, comply with GDPR's very stringent requirements.

- **Rights of data subjects:** Under the GDPR, a data subject enjoys the right to information on the processing of the data obtained and thus has the right to access the data retrieved, as well as the right to object to its processing. This implies that, in mobile forensics, investigators must provide clear information about the nature of the data retrieved and allow individuals to contest its collection if they feel it contravenes those rights. Failure to do so will result in extreme punishment for non-compliance.

- **Data minimization and proportionality:** The GDPR also enforces principles of data minimization. Data should only be collected for specific investigations that demand it. This limits the scope of mobile forensics to the most relevant data. Investigators can no longer conduct searches with too broad a range. There must be proportionality to balance individual rights against legitimate law-enforcement interests, preventing practices from becoming excessive.

- **Health Insurance Portability and Accountability Act (HIPAA):** HIPAA is a privacy law in the United States safeguarding personal health information (PHI), such as medical records, prescriptions, or patient communications. In mobile forensics, HIPAA would matter if the mobile device(s) in question might contain or exchange health-related data. Investigators would need to obey very strict regulations to ensure that this information is safely and legally handled.

- **Rights of individuals:** A patient has the right to information on how their health information is used, to view their records, and to have any corrections made if there are inaccuracies. In these situations, forensic teams would need to ensure that their activities respect these rights.

- **Minimum necessary:** Only the least amount of health info that is needed for the case should be acquired. That way, investigators do not obtain excessive personal data.

- **Security:** HIPAA requires strong safeguards to be in place, along with encryption and access controls, to protect against unauthorized disclosures of health information. Therefore, forensic activity must not compromise health information by exposing it or through situations that might lead to a leak.

- **California Consumer Privacy Act (CCPA):** Like GDPR, the California Consumer Privacy Act (CCPA) grants individuals' special rights regarding their personal data. Such rights include knowing what personal information is being collected, to whom such information is being transferred or sold, and preventing the sale or transfer of their information to third parties. Although the CCPA mainly applies to consumer data and business practices in California, its precepts have applications in mobile forensics, especially in civil proceedings or corporate investigations involving the handling of individual data.

- **Opt-out provisions and rights of users:** CCPA allows users to limit the scope of their data usage. In forensic investigations, organizations must ensure they do not infringe on user rights when data may be used against them without legitimate permission. CCPA also grants users the right to request data deletion, potentially affecting investigations relying on historical data.

- **Other national laws and their implications:** Beyond GDPR and CCPA, national laws like Canada's Personal Information Protection and Electronic Documents Act (PIPEDA) and Australia's Privacy Act play pivotal roles in governing forensic practices. Most national laws advocate for transparency, user control, and minimizing personal data collection during investigations. Compliance with these laws is crucial for maintaining investigation integrity. Violating privacy during forensic examinations can result in excluded evidence, civil lawsuits, and detention of individuals or organizations involved. For instance, severe GDPR non-compliance can lead to fines of up to 4% of annual worldwide revenue or €20 million.

In general, privacy laws provide an essential framework for protecting user rights in cellular forensic investigations by enforcing strict standards for consent and handling personal data, ensuring forensic practices are conducted ethically and lawfully, without compromising respect for privacy.

Laws Affect Consent in Investigations

Privacy laws, such as GDPR, CCPA, and others, impact operations that deal with consent management during mobile forensic investigations and result in ethical action and legal requirements. Consent is one of the most significant aspects of data protection regulations; it dictates how forensic investigators approach their subjects in such investigations to adhere to the rule of law. Informed consent is obtained from the subject before access to a mobile device or data retrieval is allowed, as required by privacy laws. This includes the following:

Informed Consent

The privacy legislation typically mandates that an investigator shall procure informed consent from the individual whose mobile device will be accessed by the investigator, or whose data is going to be extracted. Usually, this includes

- **Plain explanation:** There should be a clear explanation by investigators to the subject as to what data is going to be accessed, why it is required for the investigation, how it is going to be used, and what the possible implications of granting access are going to be. This will then become voluntary and well-informed consent.

- **Documented consent:** In most legal frameworks, oral consent is never enough; instead, documented or written consent must be sought from individuals so that investigators can have documentation that the individual gave their concurrence. This document, therefore, plays an important role in the legal process to determine whether consent was lawfully acquired.

Exceptions to Consent

Though consent is the backbone of privacy laws, there are legal provisions that permit bypassing consent under certain circumstances:

- **Legitimate investigations:** For example, GDPR and CCPA contain provisions exempting the requirement of consent under a legal basis for collecting information, such as criminal investigations, when investigators obtain a warrant for searching the alleged perpetrator's device. Under such circumstances, the need for explicit consent would be overridden by law enforcement needs.

- **National security:** They do not need to obtain consent for investigations where national security or public safety is concerned, and privacy law provisions in force permit less stringent conditions for gaining access to data that would likely prevent actual threats or harm.

Impact on Investigations

Several steps and possible impediments are added to mobile forensic investigations, especially as far as the time consumed in getting proper consent and ensuring an application's compatibility with legal frameworks.

- **Challenges:** Investigators need to carefully maneuver around such laws so they do not risk losing evidence, as it gets thrown out due to procedural errors in consent. There might be problems where people do not agree to give permission to investigators to access to their gadgets. Such issues will then demand that investigators seek a legal channel by way of warrants to gain access, thus prolonging the investigation period. Another complex issue arising from their procedures within forensic practice relates to whether consent is necessary for shared devices and whether data falls within more than one party.

- **Legal disputes:** If consent is collected contrary to legal principles without proper legal justification, cases end up in legal disputes over whether evidence should be admitted to court. Defense lawyers oppose evidence admission due to privacy violations, and investigators should be mindful that their investigations do not infringe on privacy laws or procedures.

Impact on issues of consent management: Data access and processing standards are maintained strictly to enforce forensic evidence-handling practices that are ethical and legal across different levels of individual privacy rights. In fact, privacy laws and regulations have had the most important influence on mobile forensics in this regard.

Forensic Tool Security and Reliability in Mobile Forensics

The use of forensic tools is central to the investigative process, helping the forensic examiner collect, preserve, and analyze evidence from mobile devices. Such devices must be of high quality, security, and reliability, where these elements are paramount concerning data integrity in evidence for it to be reliable in court. This section thus analyzes the criteria for choosing forensic tools, their security concerns, reliability, and standards used for testing and certification.

Criteria for Selecting Secure and Reliable Forensic Tools

The selection of forensic tools is one of the most important steps in a successful and comprehensive investigation. The selection of forensic tools must be done very carefully, since improper selection can lead to data loss, alteration, or even the exclusion of evidence from court proceedings.

Compatibility and functionality: Mobile forensic tools need to be agile enough to work across various mobile devices and operating systems. Since mobile technologies are so diversified, the tools have to be compatible with iOS, Android, and Windows devices, as well as a variety of device manufacturers, such as Apple, Samsung, Google, among others. Compatibility must extend to legacy devices as well, along with the latest smartphones, so that forensic experts can work on different firmware versions and varying ages of devices. A tool that fails to support a variety of devices probably cannot cut it in real-world investigations because devices do not homogenize. Further, the tool should be capable of deriving a variety of data sources, including call logs, SMS, application data, GPS location, and even encrypted content. In this way, distinguished tools, such as Cellebrite UFED and X1 Social Discovery, feature selective data extraction capabilities that ensure forensic experts focus on relevant data without the risk of overreach or contravention of privacy norms.

Data integrity and security: In any forensic investigation, data integrity is the most crucial, since any alteration or corruption of data would lead to litigation and rejection of evidence. A good forensic tool should be designed for non-destructive analysis; that is, collecting data without modification, overwriting, or deletion of any information stored on a device. This is to ensure that in the forensic process, the evidence does not change. Most data recovery software have a feature termed "write-blocking." A write-blocker is synonymous with a security lock that prevents unintended changes to data and ensures evidence integrity. Those features are of high importance since, in other cases, the integrity of the data produced in the process will be compromised and, therefore, useless for legal use.

Security features of the tool itself: Apart from being assured of the integrity of evidence collected, the forensic tool must also have security features to ensure data integrity within the tool itself. Some of these include encryption to secure data, authenticating who may or may not access the tool, and access controls to prevent undesirables from accessing it. In cases involving sensitive information, such as individual or corporate details, the tool's security should match or exceed the processes it protects. Another important security feature of forensic tools is adequate audit trails. A good forensic tool should allow logging of every action made during the investigation and tracking of actions taken. This ascertains accountability and is useful for proving the chain of custody in court, since a thorough log of events provides evidence that data was handled properly and ethically during the investigation process.

Proven track record and certification: Assuming one considers reliability, accuracy, and reputation of a tool over factors such as reliability and good performance, one will settle on a tool that has been tested and proven in practice, in various applications and usages within the forensic community. Tools from companies like Cellebrite, X1 Social Discovery, and Oxygen Forensics are also trusted due to their consistent performance in many investigations. These tools have undergone numerous testing and reviews from both industry-based practitioners and independent third-party organizations, which lend them credibility. A reputable forensic tool with proper certification from recognized bodies will give confidence in its results, ensuring that evidence collected can stand up in court and withstand scrutiny challenges.

Legal and compliance considerations: Forensic tools must comply with all laws, regulations, or guidelines at local, national, and international levels concerning privacy and data protection in every respect, including GDPR for Europe and CCPA for the United States. These laws govern how personal data can be collected, processed, and used, so forensic tools must comply to avoid legal pitfalls. For instance, failure to comply with GDPR or CCPA may subject one to stiff fines or exclude evidence from legal courts due to illegal breaches of privacy law. Besides adhering to the law, the forensic tool must ensure it collects admissible evidence for law courts. It must assure the reliability of the data extraction process and possible evidence presentation in an identifiable format for a court of law. If the forensic tool cannot generate data that can stand in court as valid evidence, then such evidence is likely to be held inadmissible; this means all findings would be considered invalid.

Tool Reliability

For any forensic tool, accuracy and reliability should be achieved with each run. This helps maintain the integrity of the investigation and, subsequently, the admissibility of evidence in court. Thus, a reliable forensic tool must meet the following critical standards:

Accuracy in data retrieval and storage: Forensic software must retrieve data from mobile devices accurately and with no chance of errors at all. This is because if the data retrieval process contains errors, then the evidence will be distorted in the outcomes, meaning the case may not turn out as it ought to. Various types of files, such as images, text messages, emails, and videos, should be handled by forensic tools, as well as various operating systems, for example, iOS, Android, and others. Moreover, they should cope with the diversity of applied encryption methods used to protect sensitive data on devices. This includes retrieving data from both the latest smartphone models and older ones with precision, ensuring no critical data remains missing. Moreover, the tools used should be comprehensive and not frivolous, ensuring they do not miss or leave any pieces of critical data unless intentionally done so. Recovery of deleted messages, hidden app data, or logs may seem superficially insignificant but will be of great importance for the case. Missing or incomplete data may make all the difference in the outcome of an investigation, and forensic tools must ensure no valuable data is left behind.

Reproducibility: For any forensic investigation, reproducibility is one of the key requirements. A reliable forensic tool should deliver the same results each time it is used on the same device under similar conditions. Data extraction should be consistent, and hence the results should be credible. Evidence challenged in court requires a scientific basis that proves the ability to reproduce the collected evidence. A forensic tool that yields different results on consecutive attempts or between different devices raises extreme questions regarding the integrity of the collected evidence. Consistency is a core principle of forensic science, and any tool unable to reproduce a sequence accurately risks invalidating the data and the investigation itself.

Regular updating and support: Since technology in mobile devices evolves very fast, forensic tools require constant updating to stay effective. The operating systems, security protocols, and encryption techniques that manufacture companies develop for mobile phones are constantly changing, which impacts access and retrieval of data by forensic tools. This means that extracting data through them may fail or even miss some important information if forensic tools are not updated regularly regarding all accruing changes. Any updates in iOS or Android, for instance, can install new security features or encryption methods that make data hard to access without updating the appropriate tool. Outdated or unsupported forensic tools might result in incomplete or inaccurate data collection, resulting in gaps during data collection when investigating the case. In this regard, support and updated connections are critical to ensure available forensic tools work correctly with recent mobile technologies. This enables investigators to access full data security and provide effective forensic analysis in a secured manner.

Testing and Certification

Certifying such a tool to be used in the most comprehensive way is crucial to ensure it is helpful, dependable, and in legal conformance. Through certification, one ensures the tools work as expected, giving minimal error, inconsistency, or bias in the results obtained. Forensic tools become problematic if not properly certified and tested, because they may give flawed data, hence questionable conclusions or presentation of unreliable evidence.

This undermines the integrity of the investigative process and might further lead to a challenge or declaration of inadmissibility in court, which would be perilous to the well-being of findings in legal procedures. Following are the key points being considered in legal process.

- **Certification ensures compliance with legal standards:** Certifying such a tool to be used in the most comprehensive way is crucial to ensure it is helpful, dependable, and in legal conformance. Through certification, one ensures the tools work as expected, giving minimal error, inconsistency, or bias in the results obtained. Forensic tools become problematic if not properly certified and tested, because they may give flawed data, hence questionable conclusions or presentation of unreliable evidence. This undermines the integrity of the investigative process and might further lead to a challenge or declaration of inadmissibility in court, which would be perilous to the well-being of findings in legal procedures.

- **Testing for compliance and accuracy:** Another important step in the forensic tool testing process is ensuring that the tools are very accurate and adhere to the industry standards. The quality of validation is evident when performed in controlled environments through tests and evaluations. This assessment determines if the forensic tools are valid regarding their real-world application, from data extraction to forensic tool processing. Third-party testing or government testing further strengthens tool validation, revealing flaws or security breaches during investigative usage. This ensures tools produce reliable, accurate data and reduces error risk, enhancing the forensic process's credibility within legal settings.

- **Compatibility testing with devices:** It is imperative to carry out compatibility testing with different devices, operating systems, and hardware configurations. Researchers frequently encounter various mobile platform systems like iOS, Android, and other system versions. Compatibility testing makes sure that the data extraction tool does not overlook diversity. This guarantees the uniformity and effectiveness of the tool across various devices, making the investigator certain of its efficiency regardless of the device or operating system.

Common Standards for Forensic Tool Certification and Reliability Assessment

The credibility of forensic tools in investigations and evidence processing is guaranteed by appropriate forensic tool certification, which focuses on legal, technical, and security issues. Among the key standards are the following: NIST outlines testing of computer forensic tools through programs such as Computer Forensics Tool Testing—Accuracy and Integrity. ISO/IEC 17025 describes the general requirements for testing and calibration laboratories, and ISO/IEC 27001 details the protection of information contained within forensic devices. All these standards provide those valid results, crucial for legal purposes, will be returned, and investigations will be safeguarded.

- **National Institute of Standards and Technology (NIST):** An important directive of the National Institute of Standards and Technology is the one issued concerning the testing and certification of tools in forensic investigations. With such directives, tools chosen for selection into their Computer Forensics Tool Testing program must pass tough standards of accuracy, integrity, and security to ensure evidence reliability throughout an investigation. Therefore, forensic investigators also rely on NIST standards to ensure that tools used will be accepted within the forensic community and yield results that can easily stand up to court scrutiny. If investigators observe NIST guidelines, then they are confident about the reliability of forensic tools and thus are unlikely to be challenged in court.

- **ISO/IEC 17025 standards:** The ISO/IEC 17025 standard is internationally recognized and states the conditions under which laboratories handling tests and calibration works must take sampling by testing of items. Thus, ISO/IEC 17025 is highly significant because it ensures that the forensic process is accurate and repeatable. In such cases, it guarantees the reliability of the forensic process under different conditions. The standard also gives forensic labs an opportunity to prove their competence in producing valid and reliable results. This makes forensic analysis credible and increases evidence credibility in courts. Compliance with ISO/IEC 17025 improves the reputation of laboratories around the globe.

- **ISO/IEC 27001 for information security management systems:**
  Although ISO/IEC 27001 is not specifically for forensic tools, it is internationally recognized as the information security management system for handling sensitive data in forensic investigations. Forensic tools that are properly certified with ISO/IEC 27001 help safeguard the integrity of the data collected and defend it against improper access and manipulation by unauthorized parties. This is one of the integral aspects of ensuring that evidence remains intact, starting from the collection stage of the investigation process. ISO/IEC 27001 compliance portrays that forensic tools and systems comply with the highest standards of information security, reducing risks related to data breaches or loss and integrity in forensic investigations.

Among other things, the most critical factors concerning mobile forensic analyses are the security and reliability of such forensic tools. Hence, the proper selection of forensic tools will be based on strict criteria; they must be continually tested and certified and possess security features that assure data integrity, thus ensuring evidence integrity in court. Tools found to be untrustworthy or insecure in use should be scrutinized in detail before being brought into complex investigation processes. The inclusion of NIST and ISO testing and certification adds further validation to the forensic process and also negates a portion of potential legal objections.

Key Summary

This chapter emphasizes the dual aspects of conducting mobile forensic examinations: collecting data efficiently and protecting individuals' privacy rights. It goes on to elaborate on issues of accessing information for criminal investigative purposes and touches on ethical or legal compulsions, such as the element of consent or compliance with laws like GDPR and HIPAA. This chapter also outlines risks experienced in forensic investigative work, such as hacking and data loss. It warns that such solutions may not be sustainable during forensic investigations. For this reason, it is important to recommend measures that easily address security concerns without compromising the integrity and enforceability of forensic processes in place.

CHAPTER 14

Future Trends and Emerging Technologies in Mobile Forensics

Introduction

Importance of Adapting to Rapidly Evolving Mobile Technologies

In today's world, mobile phones are no longer mere communication devices but robust platforms capable of supporting numerous use cases from financial transactions to remote healthcare, immersive gaming and augmented reality (AR) [1] experiences. Such technological improvements have resulted in an explosive growth of volume, variety, and sensitivity of data created and stored on mobile devices. Therefore, in addition to our everyday lives, mobile forensics has also come to play a leading role in the field of cybercrime investigation and digital forensics [2].

The speed of mobile technology innovation—spurred by developments in networks (such as 5G and the upcoming 6G), AI, edge computing, VR/AR, and cloud integration—has advanced further than forensic tooling and techniques. Every new mobile feature or service introduces specific data artifacts, complicated storage structures, and potential sources of evidence, many of the latter being inaccessible or non-understandable by classical forensic methods.

Further, mobile ecosystems that are coming into existence are dependent on distributed and transient data. Location-based, biometric authentication, app sandbox, encrypted messages, and cross-platform communication have been discussed in light of current ETC protocols [3]. The devices are the gateways to IoT devices, cloud platforms, social media, and immersive environments, and with mobile devices as the focal point, forensics now encompasses in excess of localized device imaging, dynamic, multi-domain evidence correlation [4].

Adjusting to these technology drifts isn't an option—it's a must. Expert witnesses need to be up-to-date with hardware architectures, OS behavior, and mobile app development to be effective enough in recovering legally acceptable, forensically solid evidence. These new approaches include adopting new analytical paradigms like cloud-native forensics, network-aware investigations, and AI-driven evidence classification [5].

In summary, the "best mobile forensics practice" is a moving target, which demands adaptive management strategies, an engaged and real-time knowledge of cross professional boundaries, and the courage to anticipate future technological trends. In fairness to Passel, investigators need to gird themselves for a future where mobile forensics isn't limited to what's on a device, but what it's plugged into, what it's rendering virtually in real time, and what it's communicating (machine-to-machine) across digital ecosystems [6].

Limitations of Traditional Forensic Models

Traditional methodology used for extraction, acquisition, and analysis of data from mobile devices based upon stand-alone mobile devices which are used to perform voice communication having fixed storage structure and proprietary file system, even if the phones were part of the network of the concerned mobile operator. Logical and physical extractions, file systems analysis, manual app parsing, and other traditional methods of analysis functioned well in a restricted, homogenous, largely closed mobile ecosystem [7]. Nevertheless, these models are becoming more and more inappropriate under today's mobile technology environments.

First, the conventional forensic tools usually center around the physical access and the direct acquisition, presuming that most evidence relevant to the case is located on the device. This model only works if all of the data is available on the streaming source, that is, available in one place. For example, if messages are end-to-end encrypted and are not preserved locally in plain text format, there's only so much forensic analysis you'll be able to conduct on messaging apps like Signal, Telegram, or WhatsApp [8].

Secondly, existing tools are a bad match to address with real-time or transitory data. Mobile applications typically cache, temporarily store inside encrypted containers, or manage sensitive transactions in volatile memory. If the equipment is powered off or rebooted, the data will be lost. This makes the data collection and real-time monitoring very important. Legacy products, designed for post-incident investigation, do not meet these forensic requirements.

Third, many areas of the device that were once open for inspection are now locked down by the proliferation of privacy-focused architectures (e.g., Android Scoped Storage, iOS sandboxes, encrypted file systems), and require privileged access or work-arounds to bypass them [9]. It is possible that classic extraction devices will not be able to circumvent these security measures and create incomplete data sets.

Mobile devices today are part of a broader connected universe. A user's operations scatter through wearable devices, smart home services, cloud drives, and an immersive environment, for example, AR/VR devices. Traditional models which consider one device in separation do not capture these interactions across the different platforms, which could lead to neglecting important contextual or supporting evidence [10].

And last but not least, traditional forensics probably lacks big time on capability in terms of scalable, automatable, or intelligent analysis. When the volume of generated data per device expands, it is hard to manually parse and inspect the log. Sophisticated tools such as those exploiting machine learning, automatic classification, and semantic filtering are very necessary [11], but scarcely present in standard forensic tools.

Taken together, these constraints demand a fresh and progressive mobile forensics paradigm—one that is proactive, ecosystem-intelligent, and technologically integrated.

Aim and Scope of This Chapter

The aim of this chapter is to explore and critically examine the **future trends and emerging technologies** that will shape the next generation of mobile forensics. The chapter outlines how forensic methodologies must evolve to remain effective, lawful, and ethically sound in response to the ongoing transformation of mobile technologies [12].

More specifically, this chapter focuses on three key areas:

1. **5G/6G Network Forensics**: As mobile networks become faster, more distributed, and more data-intensive, new opportunities and challenges emerge for forensic analysis. This section will explore how 5G/6G networks enable real-time surveillance, edge-based data interception, and location-aware investigation, while also introducing complexities like encrypted network slicing and jurisdictional ambiguity.

2. **AR/VR Forensics**: Immersive technologies are introducing novel forms of user interaction and evidence. From motion tracking and gaze patterns to virtual identities and spatial metadata, AR and VR platforms generate a rich, multilayered dataset that can be crucial in forensic scenarios. This section will examine how such evidence can be captured, reconstructed, and authenticated for investigative use.

3. **Cross-Domain Collaboration and Information Sharing**: The complexity of modern mobile environments requires coordinated efforts among telecom operators, cloud providers, app developers, and law enforcement agencies. This section will explore frameworks for secure evidence exchange, privacy-preserving collaboration using blockchain and federated learning, and the legal implications of cross-border data sharing [13].

The chapter will also contain an outlook that describes a guidance on how these future trends are expected to develop over the next decade. The scope for the chapter is dual—concept, practice. It touches upon both the theoretical foundations for next-generation forensic science and actionable information for the forensic practitioner, policymaker, and the broad scientific community. The following questions are among those answered in the chapter:

- How do emerging mobile technologies transform the nature and availability of forensic evidence?

- What are the technical and legal challenges posed by distributed, immersive, and encrypted environments?

- Which tools, techniques, and partnerships are necessary to perform effective investigations in future mobile ecosystems?

At the end of this chapter, readers should have an understanding of the adaptations needed to thrive in this changing world of mobile forensics. Regardless of creating new forensic tool, making legal guidelines and conventions, or doing a real attack, the stakeholders will gain from the roadmap and frameworks proposed here.

5G/6G Network Forensics

Transitioning from 4G through 5G to 6G is part of a transformation of the way information is created, sent, and received in mobile networks. They can deliver high-speed, low-latency communication networks, which at the same time allow to build distributed, intelligent, and autonomous networks. It's a heady prospect—and a brave new world of challenge—for forensic gurus. With the data itself becoming more decentralized and more transient, forensic process can no longer end with device centric analysis, and must penetrate full into the dynamic layers of the mobile network [14].

Technological Advancements and Forensic Potential

The move from centralized to distributed network topologies in 5/6G has embedded new opportunities for forensic sector intervention. These networks enable URLLC, mMTC, and eMBB type of communications. These are not simply technical improvements, they transform the landscape of digital evidence [15].

Real-Time Interception Through Edge Computing

One of the most significant forensic potentials in this respect is Multi-access Edge Computing (MEC). MEC offloads the processing nearer to the end user device, which is the base station or a local edge node. This greatly reduces the response time, allows the content to be delivered in real time for applications sensitive to latency such as autonomous driving, remote surgery, and smart monitoring [16].

From a forensic standpoint, it offers powerful tapping and sniffing nodes. Putting forensic agents into the edge infrastructure allows the investigators to monitor and collect network traffic in time, study the behaviors and detect anomalies before the data is spread across cloud-based environments. This enables instant investigation of unauthorized access to specific data, network hacking, or anomalous user mobility—right at the edge of the network.

Device-to-Network Context Enrichment

5G/6G networks are fundamentally context aware. Such network interacting devices generate metadata like signal strength, location cords, session time, roaming, QoS parameters, etc., continuously. These characteristics are critical for forensic association and reconstruction.

For instance, forensic experts can associate the physical location of a device with particular user actions (like requesting a restricted service), can corroborate if multiple devices were present in close proximity at the time of the event, or can prove if a mobile terminal was targeted by a man-in-the-middle attack. Machine learning-powered user behavior analytics are similarly capable of flagging abnormal usage patterns for deeper scrutiny at the network level [17].

In 6G scenarios, which are foreseen to accommodate terahertz (THz) communication, AI-native orchestration, and tactile Internet services, such a context is even richer. Not only will devices have location profile or MAC address but also intent inference, biometric signals, or augmented identity profiles. All these constitute precious levels of forensic data which can be gathered, preserved, and subjected to analysis according to the standards of evidence.

Forensic Hooks on Base Stations, MEC Nodes, and Network Slices

On legacy 4G networks, the majority of their forensic access focussed on core network logs, centralized databases, and packet capture at physical routers. In 5G/6G architectural realm, network slicing establishes accompanied, virtualized spaces on the same physical infrastructure but ensures application-dependent QoS, bandwidth, and security policies [18].

Each network slice is isolated, treating for example one slice for self-driving cars differently than one for IoT healthcare tools. By incorporating forensic hooks into these slices, targeted evidentiary capture can be deployed without interfering with irrelevant data flows. Likewise, the forensic probes in base station or MEC nodes can perform fine-grained monitoring on regional traffic for local breach or breach on geo-fenced policies.

Crucially, these hooks need to be hardened with the right access controls and encryption to guard against abuse. But when implemented and governed successfully, they provide a scalable approach for keeping tabs on millions of mobile endpoints in a disjointed service landscape.

Forensic Challenges in Next-Gen Networks

While 5G and 6G open powerful new forensic frontiers, they also introduce a range of challenges that threaten evidence accessibility, reliability, and legality. These challenges are deeply rooted in the architectural, operational, and jurisdictional complexities inherent to next-generation mobile networks.

Encrypted Virtualized Slices

Network slicing—one of the key elements in 5G/6G architecture—enables the creation of multiple virtual networks on top of the same physical network. Each slice is controlled, defined, and frequently encrypted from end-to-end separately based on particular Service Level Agreements (SLAs) or compliance rules.

Although this provides the users with security and isolation, it creates a great deal of challenges in terms of forensic visibility. Investigators might not have the decryption key or the slice-specific routing protocol, and therefore even when they are present, the nature of traffic obfuscation within these slices results in difficulty to identify artifacts of interest.

And when you conduct forensic interception, you have to be able to do so in a way that doesn't compromise the data integrity and chain of custody while still protecting user privacy and regulatory requirements—especially when you're dealing with several tenants on the same infrastructure.

Device Mobility and IP Reassignment

The flexible movement of devices in a 5G/6G network poses great challenges for forensic tracking. Devices traverse base stations, slices, and even countries at a fast pace, and their IP addresses are dynamically re-allocated according to load-balancing, latency, and QoS criteria.

Events in those conditions are harder to correlate to the same device or user session. A forensic link that starts in one jurisdiction could lead to another or totally disappear if a session was ended and the temporary identities deleted.

And mobile devices also commonly implement dual or multi-connectivity (like, making use of Wi-Fi and 5G), further breaking up the breadcrumb trail. Therefore, to decipher the full picture, forensic analyzers have to devise sophisticated correlation methods, which correlate the session identifiers, device fingerprints, and behavioral signatures across network layers [13].

Jurisdictional Fragmentation of Data Trails

Next generation networks are worldwide, and if a forensic situation arises, the evidence will be spread across multiple legal domains. For example, a voice command that's captured on a smartphone in India might be analyzed on an edge server in Singapore and backed up on a cloud data center in Germany.

This cross-border sharing of data poses significant legal and procedural obstacles. For law enforcement agencies, they must pass through international treaties, mutual legal assistance treaties (MLATs), and comply with legislations such as the GDPR, Cloud Act, and even data sovereignty laws. Volatile evidence can be lost and the digital integrity can be compromised if access is delayed.

This becomes further complex with private network slices such as those owned by a corporation or government in which access is limited and little is known. Without an internationally acknowledged forensic access platform for mobile networks, prospects for timely and legal evidence is unlikely.

In summary, 5G and 6G networks provide both a technical potential and a forensic challenge. Telecommunication network architectures are changing from a centralized toward a more highly virtualized, real-time, and distributed model; this shift places the traditional methods of mobile forensics under pressure. But with edge computing, intelligent context capture, and slice-aware forensic instrumentation, investigators can break new ground to keep pace with the complexity of today's mobile environments. This is how the future of network forensics will be, adapted to the new paradigm, adapted to the new tools, adapted to new methods and laws [7].

Augmented Reality (AR) and Virtual Reality (VR) Forensics

The rapid emergence of immersive technologies, such as AR (Augmented Reality) and VR (Virtual Reality), is revising human engagement with computational spaces. Whether used for gaming, simulations or training, social VR platforms, or enterprise and collaboration, these technologies are capable of a new type of spatial computing that's persistent, real time, and sense-based in a much larger scale. Given this shift, digital forensics should also adapt, focusing on the online footprint, online behavior, and multi-modal data produced in these contexts. AR/VR forensics is a new frontier that presents challenges in evidence acquisition, analysis, and authentication.

Evidence Generation in Immersive Environments

Immersive environments produce a vastly different set of data artifacts compared to conventional mobile or desktop applications. These environments are not just transactional; they are experiential. Every user action—whether a voice command, head tilt, hand gesture, or avatar movement—generates metadata that can be instrumental in forensic investigations.

Logs from Gesture Tracking, Audio Input, Gaze Path, and Haptic Events

AR/VR platforms rely heavily on multi-modal input systems that track real-world gestures and convert them into digital commands. These interactions are often logged in high granularity and can include

- **Hand and gesture tracking** logs using infrared or visual sensors

- **Head and eye movement** data to understand where a user was looking and for how long

- **Audio input streams**, such as voice commands, proximity-based speech, and environmental sounds

- **Haptic feedback logs** from controllers or suits, indicating physical interactions like object touching, grabbing, or force exertion

In a forensic context, such logs can help reconstruct a user's activity within the virtual environment. For example, analyzing a user's gaze path could determine whether they visually acknowledged a virtual object or another avatar before taking an action, which is particularly relevant in assault or harassment cases within virtual worlds.

Platform-Rendered Metadata: Avatar Identity and Spatial Orientation

Users in VR often interact through avatars—digital personas with unique characteristics and behaviors. Forensic analysis of **avatar identity** involves metadata such as

- Avatar name, appearance, voice synthesis settings.

- Assigned device ID or account linkage.

- Social interactions, such as friend lists or voice chats.

- And spatial orientation data, such as the position of the user in the virtual space, room-scale movement, and pose data, gives context to each tracked interaction. It is where we first learn about the separation between users, the course of events, or the veracity of statements on virtual behavior.

- These data sets are usually kept on local devices (e.g., headsets, smart phones) or in databases of centralized platforms. And forensic analysts need to be able to extract proof from hardware and cloud interfaces—it often means working closely with platform providers.

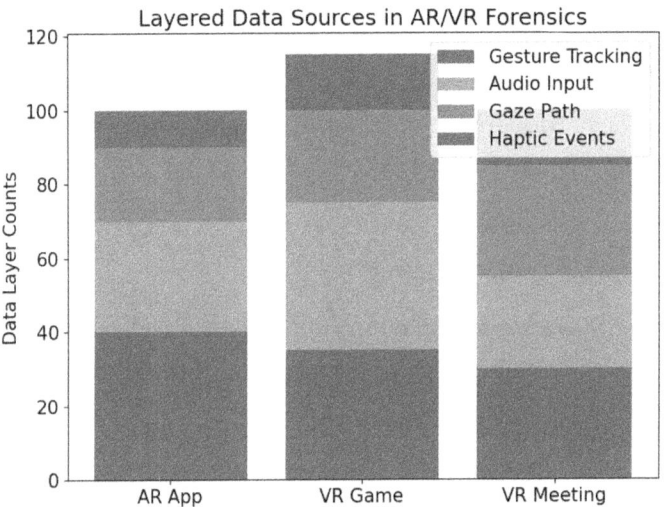

Figure 14-1. *Layered data sources in AR/VR forensics*

Figure 14-1 illustrates the composition of forensic data layers collected across different immersive platforms. AR apps emphasize gesture and audio data, while VR games and meetings feature richer gaze and haptic event logs, reflecting the varying interaction modalities in immersive environments.

Analytical and Reconstruction Techniques

Once raw data is collected from AR/VR environments, forensic analysts must transform these low-level interaction logs into meaningful timelines and behavioral narratives. This involves a combination of timeline synthesis, behavioral modeling, and identity correlation using biometric or behavioral markers.

Timeline-Based Interaction Mapping

One of the most effective strategies in AR/VR forensics is building a **chronological map of user interactions**. This timeline correlates multiple data streams—gesture logs, voice input, spatial movement, and haptic feedback—to produce a visual or textual reconstruction of the user's session.

A timeline can help answer critical questions:

- What actions did the user take and when?
- Which other avatars or users were in proximity?
- Was there a specific sequence leading to an alleged incident?

By replaying a session from collected logs—using spatial coordinates and sensor data—investigators can simulate the immersive experience, offering courts a clearer understanding of context and intent.

This also allows for **incident point triangulation**, where different streams (e.g., voice command + gaze + avatar approach) are cross-referenced to confirm or refute allegations of misconduct, fraud, or abuse within the immersive environment.

Avatar–User Linking Through Biometric Patterns and Behavior

AR/VR environments often offer users a degree of anonymity, allowing them to create or switch between multiple avatars. However, forensic investigators can identify and link avatars to real users through **biometric and behavioral fingerprinting**:

- **Voice patterns** (pitch, tone, cadence) using voice biometrics
- **Motion signatures**, like walking style or hand movement patterns
- **Gaze tracking behavior**, often unique to individuals
- **Interaction timing and rhythm**, like response latency or speech pauses

These behavioral identifiers—when combined with login IP, device MAC addresses, or linked social profiles—can establish a high-confidence match between a virtual avatar and its real-world operator. This is especially important in criminal investigations involving fraud, impersonation, stalking, or abuse conducted under digital identities.

AR and VR forensics imply a change in the interpretation, the collection, and the analysis methods of digital evidence. Instead of looking at databases or static documents, researchers engage with dynamic cross-cultural encounters that take place in networked 3D virtual worlds. Decoding user actions, piecing together timelines, verifying identity in virtual environments aren't just technically challenging—they're also critical to deliver justice in the metaverse and beyond. With the rise of immersive technology as a cornerstone of digital living, forensic readiness must modernize accordingly.

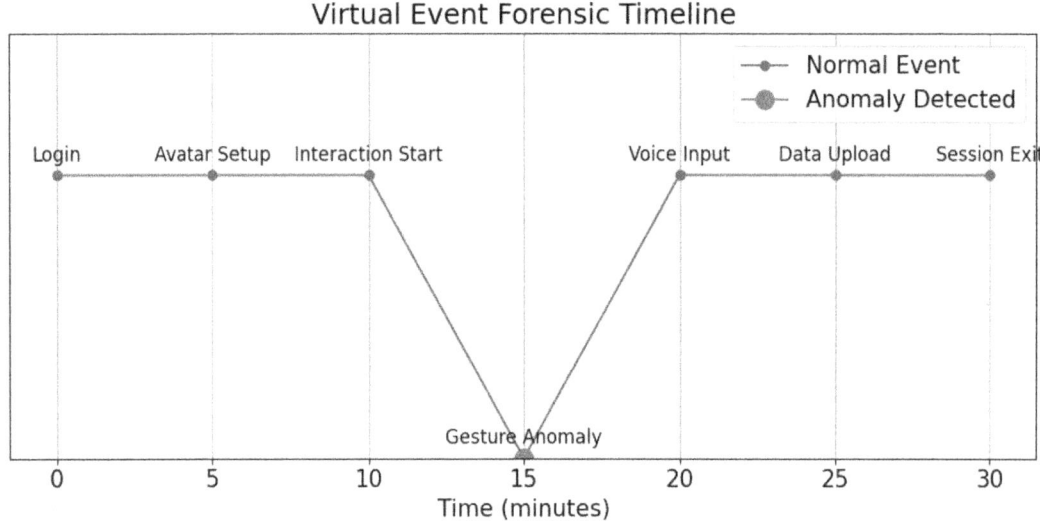

Figure 14-2. Virtual event forensic timeline

Figure 14-2 maps a sequence of user actions during a virtual session, highlighting a gesture anomaly detected midway. It demonstrates how forensic analysis can track temporal patterns in immersive environments to identify suspicious behavior or integrity breaches. Parameters: Event sequence spans from login to session exit over 30 minutes; anomaly marked at 15 minutes during gesture interaction; timeline reflects forensic event logging in immersive platforms.

Cross-Domain Collaboration and Information Sharing

In the changing world of mobile forensics, coordination of various stakeholders is mandatory. Large volumes and variety of data generated in mobile ecosystems (covering telecom infrastructure, applications, cloud platforms and user devices) calls for tighter and timely integration of its stakeholders at open strategic level. With encryption and increasingly decentralized mobile technologies prevalent today, effective forensics will depend on interoperability across domains. Such collaboration is not just good—it's the name of the game for timely, lawful, and complete forensic examination.

This section discusses how privacy-preserving multi-party protocols are changing the paradigm of evidence collection, validation, and sharing in mobile forensics.

Multi-stakeholder Forensic Workflows

Forensic investigations in mobile environments now span an ecosystem involving law enforcement agencies, telecom providers, mobile app developers, OS vendors, and cloud service operators. Each of these entities possesses vital fragments of the digital evidence chain, and their cooperation is essential for reconstructing incident timelines and validating findings.

Coordination Among Law Enforcement, Telecom Providers, and App Developers

Law enforcement agencies may seize a suspect's device, but unlocking and interpreting the full context of a crime often requires **external data repositories**. Telecom operators hold metadata such as call records, cell tower logs, SMS data, and IP handover logs. App developers may store user interactions, session history, or encrypted communications on centralized servers. Without coordinated access to these sources, forensic analysts risk drawing incomplete or misleading conclusions.

To address this, investigators must follow structured **forensic request protocols**, including

- **Data preservation notices** to prevent deletion during investigation
- **Mutual Legal Assistance Treaties (MLATs)** for cross-border access to data
- **E-discovery and legal compliance units** embedded within tech firms

In time-sensitive scenarios such as terrorist threats or abduction cases, streamlined communication between forensic teams and data custodians can determine the success or failure of an investigation. Well-defined workflows and legal interoperability frameworks are key.

Role of APIs and Cloud Forensics in Integrated Workflows

With the migration of user data to cloud platforms, forensic workflows increasingly depend on **API-based integration** for evidence extraction. Secure and authenticated APIs allow forensic tools to

- Request logs from cloud services like Google Drive, iCloud, OneDrive
- Pull transactional data from encrypted messaging apps via backend APIs
- Interact with mobile device management (MDM) systems for remote forensic acquisition

Cloud forensics also entails dealing with **virtual environments** such as containers, serverless functions, and geographically dispersed data stores. Multi-cloud and hybrid-cloud setups further complicate evidence correlation and timestamp consistency.

To ensure validity and admissibility of cloud-based evidence, workflows must include

- **Chain-of-custody documentation** for every API or service call
- **Cryptographic hashing** of acquired data to prevent tampering
- **Timestamp normalization** across time zones and services

By leveraging APIs and automated workflows, forensic teams can reduce manual intervention and speed up evidence collection without compromising legal or technical integrity.

This flowchart visualizes the key stakeholders in mobile forensics and their evidence/log exchange relationships. It emphasizes the cyclical and multidirectional flow of data between law enforcement, telecom providers, app developers, cloud platforms, and regulators, illustrating integrated forensic workflows.

Parameters: Stakeholders include Law Enforcement, Telecom Providers, App Developers, Cloud Platforms, and Regulators; edges represent various types of forensic data exchanges such as requests, API access, audit logs, and compliance reports.

Privacy-Preserving and Auditable Forensics

As personal information is more sensitive in mobile and cloud environments, user's privacy requirements are preferred in the forensic investigation. The challenge for forensic investigators is how to roll out privacy-sensitive methods which permit lawful evidence extraction but do not release irrelevant or safeguarded data. Additionally, forensic activities must be auditable, meaning that they should be performed in a transparent, accountable, and legally justifiable way.

Blockchain for Immutable Log Chains

Blockchain technology offers a compelling solution for ensuring **tamper-proof logging and evidence tracking**. In forensic operations, blockchain can be used to

- Record access logs for forensic data retrieval
- Timestamp evidence acquisition and validation
- Track the full life cycle of evidence through hash-based verification

Each forensic activity—whether it's decrypting an app log or accessing a cloud API—can be logged to a private blockchain ledger. This in turn results in a cryptographically secure unforgeable audit trail which is decentralized among the involved participants. The blockchain ledger can provide proof for the authenticity and integrity of evidence when evidence is challenged in court.

It can also be the case that smart contracts are used for the purpose of enforcing compliance policies so that researchers get authorized access to datasets and this under certain conditions.

Federated Learning for Model Sharing Without Data Leakage

In mobile forensics with predictive analysis/best-effort profiling, an investigator sometimes is using models trained on sensitive data. Conventional methods need to collect the raw data to a central storage location together with the risk of privacy disclosure. Federated learning (FL) removes this threat by training models on local devices or servers without moving the raw data.

In a forensic context, FL can be employed to

- Detect patterns of fraud or device spoofing across multiple jurisdictions
- Train anomaly detection models using device-level logs without collecting them centrally
- Share forensic intelligence between agencies while maintaining data sovereignty

Law enforcement authorities in different locations can jointly train a fraud detection model, based on local mobile data (mobile traffic) without sharing user-level data between them. As a result, the global model draws on cross-regional insights in a manner that is GDPR, HIPAA, or national data protection law compliant. Additionally, FL models are also auditable through methods like differential privacy and secure multiparty computation, which also ensure it is not possible to reverse engineer sensitive features.

Interoperability and sharing of information across domain are the keys to future mobile forensics. As digital evidence gets more heterogeneous and privacy more pervasive, forensic readiness needs to move toward ecosystem thinking; where actors operate in a coordinated, auditable, privacy-supported way. Whether it's automating API-driven cloud workflows, baking in blockchain-based audit trails, or leveraging federated models for forensic intelligence, thriving in the era of wild if not irresponsible innovation means you must balance tech and trust, security and compliance.

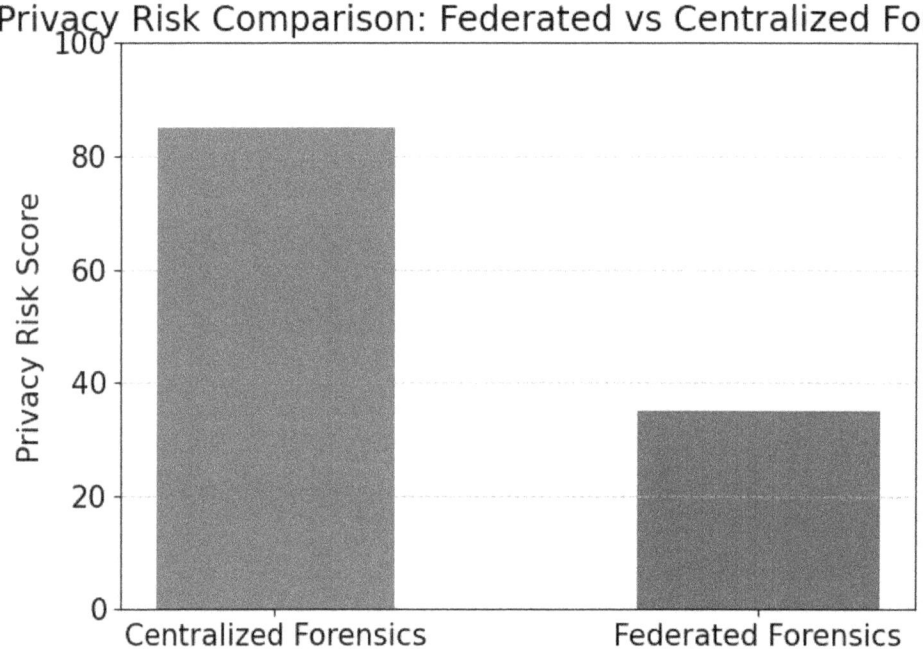

Figure 14-3. *Privacy risk comparison: federated vs. centralized forensics*

Figure 14-3 compares privacy risk scores between centralized and federated forensic approaches. Federated forensics demonstrates substantially lower privacy risks due to decentralized data handling and model sharing without exposing raw user data. Parameters: Privacy risk scores scaled 0-100; centralized approach entails data aggregation; federated approach uses distributed learning to preserve privacy.

CHAPTER 14 FUTURE TRENDS AND EMERGING TECHNOLOGIES IN MOBILE FORENSICS

Roadmap and Future Directions

The progress of the mobile forensics can be closely followed through the technological development of mobile communications, pervasive computing, and distributed computing. As mobile ecosystems grow to embrace 5G/6G, AR/VR environments, decentralized apps, AI application and need to go beyond the forensic investigation model to one of proactive forensics. This section presents a prospectus of a roadmap that envisions key evolution in mobile forensics over the coming decade, focusing on technology and policy waypoints that will define the field.

The next generation of mobile forensic tenets will be grounded on the three pillars of real-time edge intelligence, immersive evidence modeling, and federated analytics. All four will collectively cause the migration from static, post-incident forensics to dynamic, context-aware, and privacy-respecting investigations.

Timeline of Key Technological Integrations

2025–2027: Integration of 5G Forensic Hooks and Edge Analytics

The early phase of the roadmap focuses on exploiting the full potential of 5G infrastructures, especially edge computing nodes and software-defined network slices. This period marks the operationalization of

- **Forensic hooks at Mobile Edge Computing (MEC) nodes**: These will enable investigators to intercept, capture, and analyze mobile traffic closer to the source in real time. Hooks will provide timestamped data, session metadata, and encrypted channel activity logs.

- **Contextual metadata enrichment**: Through integration with 5G base stations, investigators can extract device behavior profiles, location shifts, and app usage patterns while maintaining low-latency access to data streams.

- **Event-triggered logging**: Systems will be configured to generate forensic logs upon predefined triggers—such as security alerts, geo-fence violations, or protocol anomalies—without waiting for external intervention.

2027–2029: Standardized Immersive Data Capture and Reconstruction

With AR/VR platforms becoming commonplace in social, educational, and commercial domains, the focus will shift toward developing standards for immersive evidence capture and interaction reconstruction.

Key developments will include

- **Standard schemas for spatial and biometric data logging**: Capturing data such as eye movement, spatial orientation, gesture trails, and avatar interactions in forensically sound formats.

- **Time-synchronized event reconstruction engines**: These will allow analysts to simulate the entire immersive experience, providing 3D replays with metadata overlays. It will enable courts and stakeholders to "witness" events in virtual timelines.

- **Cross-platform avatar tracking systems**: Identity persistence will be enhanced by linking user behavior across multiple immersive platforms, aiding in suspect identification and pattern analysis.

2030 and Beyond: Global Federated Forensic Networks and AI-Driven Detection

The final phase of the roadmap envisions a globally federated forensic intelligence infrastructure underpinned by AI, blockchain, and privacy-preserving machine learning. By 2030, forensic capabilities will extend to

- **Federated forensic networks**: Cross-border, interoperable platforms that allow secure sharing of evidence artifacts, AI models, and incident insights without violating data sovereignty. These networks will be regulated by international treaties and supported by distributed ledger technologies.

- **AI-driven behavioral detection**: Real-time models capable of detecting fraud, impersonation, or coordinated criminal behavior across mobile ecosystems. These systems will continuously learn from anonymized metadata, device logs, and contextual indicators.

- **Augmented forensic analysis**: Analysts will use AI-powered tools for automatic triaging of evidence, anomaly highlighting, and recommendation of investigative directions. Mixed-reality tools will enable immersive evidence walkthroughs for decision-makers.

CHAPTER 14 FUTURE TRENDS AND EMERGING TECHNOLOGIES IN MOBILE FORENSICS

This phase represents a maturity point in mobile forensics, where technological, legal, and ethical frameworks converge to form a cohesive global security fabric.

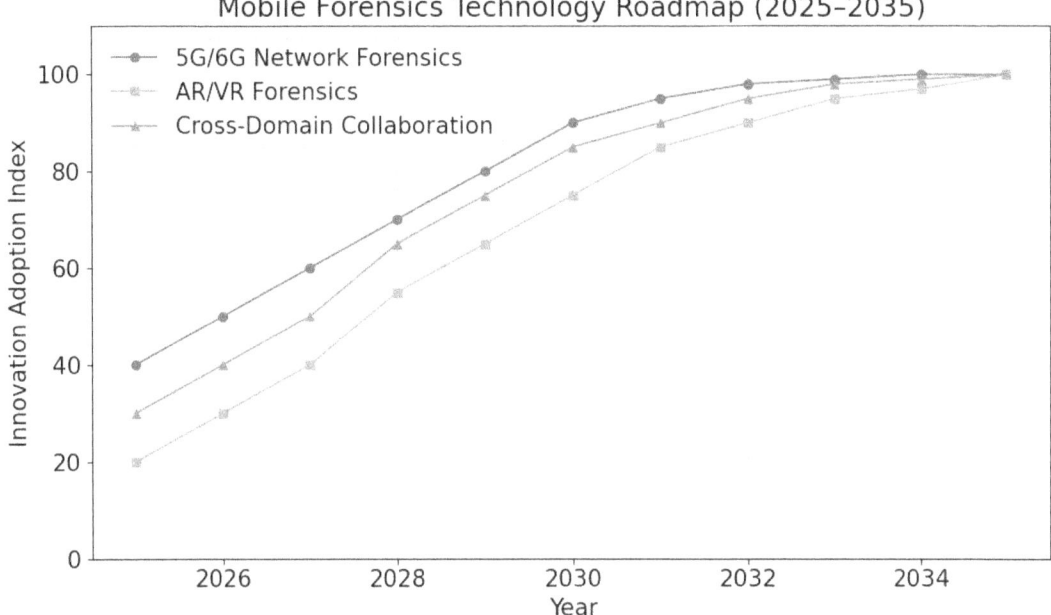

Figure 14-4. Mobile forensics technology roadmap (2025–2035)

Figure 14-4 projects the innovation adoption index for key mobile forensics technologies over a decade. It shows steady growth in 5G/6G network forensics, accelerated adoption of AR/VR forensic methods, and gradual integration of cross-domain collaboration workflows.

Parameters: Innovation Adoption Index scaled 0-100; timeline spans 2025 to 2035; layers correspond to network forensics, AR/VR forensics, and cross-domain collaboration.

Research and Policy Recommendations

In order to transition from vision to implementation, focused research and progressive policymaking are essential. The following recommendations are categorized by their relevance to standardization, technological advancement, and regulatory evolution.

Emphasis on Standardization and Cross-Border Legal Harmonization

As mobile forensics becomes a globally distributed discipline, lack of **standardized practices** and **jurisdictional incompatibilities** will hinder collaborative investigations. Key steps include

- **Establishing universal evidence formats** for mobile logs, AR/VR traces, and edge-captured metadata to ease interoperability

- **Aligning digital evidence admissibility standards** across international courts through treaties or model legislation

- **Creating trusted timestamping and validation protocols**, ensuring that forensic data from one jurisdiction is verifiable in another

Standardization must also address procedural elements like chain of custody, hash verifications, digital signatures, and logging of investigator actions. Organizations like EUROPOL, ITU, and ISO must drive these initiatives in collaboration with regional agencies.

Need for Datasets Representing Immersive and 6G Environments

Research in forensic AI is currently limited by **inadequate datasets**—especially for immersive, mobile edge, or 6G scenarios. Investment is needed in

- **Synthetic data generation frameworks** for simulating AR/VR crimes, multi-device attacks, or coordinated fraud events.

- **Public-private sandbox environments** where researchers can test forensic techniques on anonymized but realistic data.

- **Data repositories for network behavior patterns** in 6G simulations (e.g., device-to-device interference, network slicing anomalies).

These datasets will not only accelerate algorithmic development but also serve as benchmarks for certification of forensic tools.

Development of Visual Analytic Tools for Rapid Evidence Interpretation

Given the volume and complexity of future forensic data, visual analytics will become central to **evidence exploration and interpretation**. Priority should be given to

- **Interactive dashboards** that display timeline views, spatial data overlays, behavioral clustering, and anomaly maps.

- **Immersive forensics interfaces** where analysts can interact with evidence in 3D space, navigating through user sessions or network paths.

- **Visualization-driven triaging systems**, enabling investigators to rapidly isolate high-risk evidence segments without wading through all logs.

Conclusion

Mobile forensics is standing at an important crossroads. Recent developments in mobile communications, immersive technologies, and distributed computing have changed the way in which end users experience digital environments, but also forensic evidence is created, acquired, and processed. This chapter investigated some of the most prevalent developing trends in mobile forensics—5G/6G network forensics, augmented and virtual reality forensics, the importance of cross-domain collaboration and information sharing. Thanks to the ever-increasing complexity of mobile ecosystems and the increasing variety of forms of distributed data sources, a classic forensic approach observed around isolated devices, with a static data acquisition, is currently not enough.

Summary of Explored Trends and Their Implications

The move to 5G (and one day we will be saying 6G) presents a double-edged sword for forensic examiners. On the other hand, those networks also pose new challenges, such as the above-defined timing requirement for supporting edge IoT apps and crimes (e.g., drug cartel's) that can be forensically analyzed with enriched metadata available in 5G networks in real time and with context-awareness. Federation hooks deployed at base stations, mobile edge computing (MEC) nodes, and network slices can realize evidence acquisition near sources and expedient interception of communications. This could greatly expedite the reliable establishment of forensic evidence.

However, network slicing virtualization, end-to-end security, and device mobility also bring issues of complexity. Live-managed IP addresses, temporary network sessions, and jurisdictional scattering of data trails necessitate investigation techniques that are both flexible and able to join together diverging digital tracks across diversified infrastructures. The advent of 6G is predicted to further compound these complexities with the inclusion of AI-based network orchestration and a very high degree of data decentralization.

Second, the application context has been deeply modified by the appearance of Augmented Reality (AR) and Virtual Reality (AR), which add a new layer of digital interaction overlayed on the real world. Solving crime forensically in such immersive worlds demands new methods to evoke and deduce evidence. Conventional usage logs and metadata must be enriched with gesture traces, gaze paths, spatial posture, avatar identities, and haptic event traces. Analytical models need to include time-based interaction mapping and biometric association between avatars and users to replicate immersive events appropriately.

Such forensic reconstructions of immersive environments will allow analysts to re-visit virtual interactions with the detailed scrutiny of a physical crime scene. But that also adds serious concerns around privacy, consent, and how you normalize evidence formats for new types of data. The forensic community should work with industry players to create vendor-neutral standards and secure ways to handle evidence in immersive space.

Third, the chapter emphasizes the crucial importance of cross-domain cooperation and information exchange. Mobile digital evidence is not under the control of one agency or entity, but rather falls under many forms and is owned by many people.

Forensics workflows: The successful forensic workflow needs to interconnect with law enforcement, telecom operators, application vendors (developers), cloud solutions, and regulators. Orchestrated forensic request templates, secured API interfaces, and cloud forensic automation processes are needed to expedite evidence collection and preserve chain-of-custody.

Also, privacy considerations and regulation have also promoted the emergence of novel privacy-preserving approaches, including blockchain-based immutable logging and federated learning in order to facilitate collaborative model development without data leaks. All these technologies provide the transparency and auditability for forensic investigations, while respecting private data of users. Cross-domain forensics ecosystems need to be built with privacy as the underlying principle, where the investigative necessity can be met comfortably with the rights of the people.

Emphasis on Shifting from Device-Centric to Ecosystem-Aware Forensics

One theme bringing together these future directions is the transition from device-centric to ecosystem-aware forensics. In the past, mobile forensics was largely concentrated in the ability to extract and analyze the data saved on or in transit by a single physical device. The investigator's purview was largely limited to the contents of the confiscated phone, tablet, or wearable. It worked back in the early days of mobile, but in an age dominated by ubiquitous cloud sync, edge-compute-fueled services, virtualized networks, and immersive digital identities, it's no longer enough.

The modern forensic examiner must function in a holistic ecosystem; one that accepts the implicit interconnectedness of user assets, network infrastructure, cloud services, and virtual environments. Evidence may be spread over several entities, each having its own policies and technical architectures. Devices are only small parts of user activities; pertinent context may be present in network telemetry, app servers, or immersive platform logs.

This ecosystem-aware mindset implies that forensic tools and procedures need to be distributed, to interoperate, and to be dynamic. Investigation holders must gain expertise in network protocols, cloud APIs, AR/VR data formats, and privacy-preserving analytics. Nutritional forensics is the scientific study of food, supplements, and dietary goods to make sure they are safe, real, and that the labels are correct. This is done by looking for contaminants, adulterants, toxins, or substances that aren't listed on the label. It uses forensics, toxicology, and food science together to find fraud, track down the sources of foodborne sicknesses, and make sure that rules are followed. This protects public health and helps people who have been misled or hurt in court. Nutritional Forensics also needs to ensure that appropriate frameworks are in place to allow different partners to work side by side, with well-defined, secure, and auditable workflows and legally sound data sharing agreements.

Call for Adaptable, Privacy-Conscious, Real-Time Forensic Frameworks

In order to address these challenges and to take advantage of potential opportunities offered by new mobile paradigms, the forensics field should transition toward mobile adaptive, privacy-respectful, and real-time forensics frameworks.

Flexibility is a solid plus, as the world of mobile technology is ever-changing. The forensic tools and techniques shall be modular and extendable to be able to include new data sources, communication standards, and network architectures. Flexible frameworks will allow researchers to adapt quickly to the challenges of new forensic problems, such as those associated with new 6G capabilities or also next-generation immersive platforms.

This notion of Privacy by Design must be hardwired. The data needs to be investigated in a way that respects all of the various privacy laws around the world, such as GDPR, HIPAA, and various country- or region-based data protection laws. Simply put, cryptographic logs, blockchain audits, and federated learning goes a long way to ensuring that when we have to go to forensics, we minimize unnecessary exposure of personal data, maintain a clear sense of accountability. Privacy-sensitive protocols help to maintain public faith in forensic procedure and protect civil freedoms at a time when electronic monitoring is becoming more intense.

CHAPTER 14   FUTURE TRENDS AND EMERGING TECHNOLOGIES IN MOBILE FORENSICS

The real-time forensics is the cutting edge of the mobile forensics space. This near real-time analysis of potential evidence is helpful in decreasing the time gap between incident time and investigative response. For example, this is the case for cyberterrorism, child exploitation, and organized crime. Real-time platforms will take advantage of edge computing, AI-driven anomaly detection, and automated evidence triaging to speed up investigations without sacrificing forensic integrity.

Besides the above fundamental features, emerging forensic frameworks also should incorporate visual analytics, evidence reconstruction in an immersive environment, and collaboration interface to better support investigators. They should also promote multistakeholder collaboration with evident protocols and the ability to interoperate across borders and organizations.

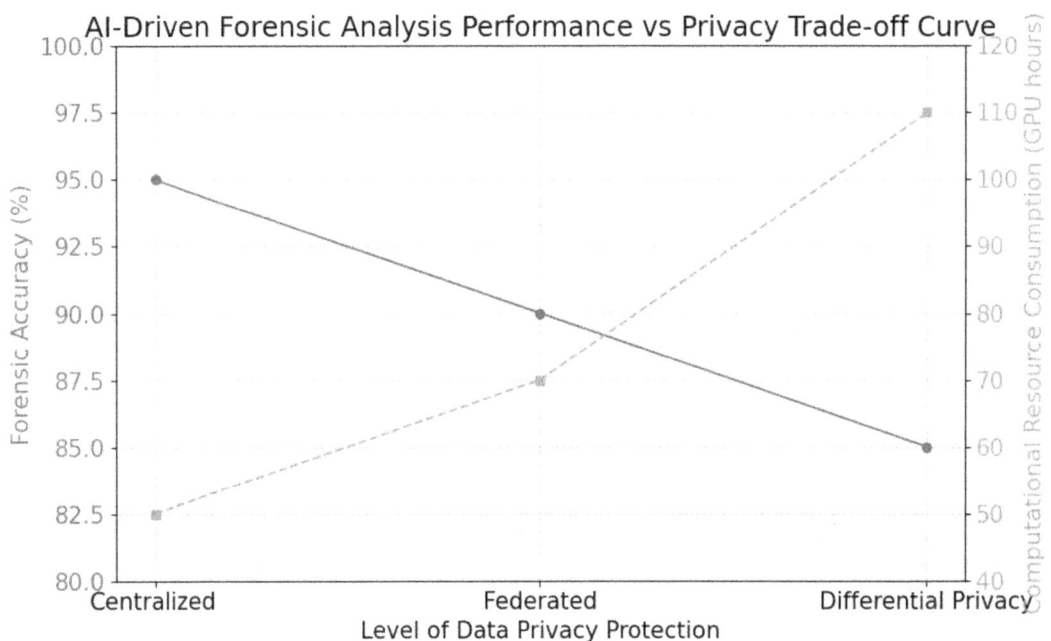

Figure 14-5. *AI-driven forensic analysis performance vs. privacy trade-off curve*

Figure 14-5 illustrates the trade-offs between data privacy preservation levels and forensic analysis costs. On one side, when privacy protection becomes stronger from centralized to differential, performance of the forensics slightly decreases and the computing resources requirements increase, which just represent a reasonable trade-off that should be considered in a reasonable future for forensic techniques. Parameters: Privacy protection levels include Centralized, Federated, and Differential Privacy; forensic accuracy measured in percentage detection rate; resource consumption shown in GPU hours.

References

[1] Papadoudis, N., Vasilaras, A., Panagiotopoulos, I., & Rizomiliotis, P. (2022). Mobile Forensics and Digital Solutions: Current Status, Challenges and Future Directions. *Special Issue 6 Eur. L. Enf't Rsch. Bull.*, 211.

[2] Khan, A. A., Shaikh, A. A., Laghari, A. A., Dootio, M. A., Rind, M. M., & Awan, S. A. (2022). Digital forensics and cyber forensics investigation: security challenges, limitations, open issues, and future direction. *International Journal of Electronic Security and Digital Forensics*, 14(2), 124–150.

[3] Alam, M. N., & Kabir, M. S. (2023, May). Forensics in the Internet of Things: application specific investigation model, challenges and future directions. In *2023 4th International Conference for Emerging Technology (INCET)* (pp. 1–6). IEEE.

[4] Malik, A. W., Bhatti, D. S., Park, T. J., Ishtiaq, H. U., Ryou, J. C., & Kim, K. I. (2024). Cloud digital forensics: Beyond tools, techniques, and challenges. *Sensors*, 24(2), 433.

[5] Kebande, V. R., & Awad, A. I. (2024). Industrial internet of things ecosystems security and digital forensics: achievements, open challenges, and future directions. *ACM Computing Surveys*, 56(5), 1–37.

[6] Waseem, Q., Alshamrani, S. S., Nisar, K., Wan Din, W. I. S., & Alghamdi, A. S. (2021). Future technology: Software-defined network (SDN) forensic. *Symmetry*, 13(5), 767.

[7] Olaniyi, O. O., Omogoroye, O. O., Olaniyi, F. G., Alao, A. I., & Oladoyinbo, T. O. (2024). CyberFusion protocols: Strategic integration of enterprise risk management, ISO 27001, and mobile forensics for advanced digital security in the modern business ecosystem. *Journal of Engineering Research and Reports*, 26(6), 31–49.

[8] Pallangyo, H. J. (2022). Cyber security challenges, its emerging trends on latest information and communication technology and cyber crime in mobile money transaction services. *Tanzania Journal of Engineering and Technology*, 41(2), 189–204.

[9] Liao, Z., Pang, X., Zhang, J., Xiong, B., & Wang, J. (2021). Blockchain on security and forensics management in edge computing for IoT: A comprehensive survey. *IEEE Transactions on Network and Service Management, 19*(2), 1159–1175.

[10] Shahzad, F., Javed, A. R., Zikria, Y. B., Rehman, S., & Jalil, Z. (2021). Future smart cities: requirements, emerging technologies, applications, challenges, and future aspects. *TechRxiv, 1*, 14.

[11] Ali, M. I., & Kaur, S. (2021). Next-Generation Digital Forensic Readiness BYOD Framework. *Security and Communication Networks, 2021*(1), 6664426.

[12] Javed, A. R., Shahzad, F., ur Rehman, S., Zikria, Y. B., Razzak, I., Jalil, Z., & Xu, G. (2022). Future smart cities: Requirements, emerging technologies, applications, challenges, and future aspects. *Cities, 129*, 103794.

[13] A. Bhardwaj and K. Kaushik, "Metaverse or Metaworst with Cybersecurity Attacks," IT Prof., vol. 25, no. 3, pp. 54–60, May 2023, doi: 10.1109/MITP.2023.3241445.

[14] K. Kaushik and A. Bhardwaj, "Zero-width text steganography in cybercrime attacks," Comput. Fraud Secur., vol. 2021, no. 12, 2021, doi: 10.1016/S1361-3723(21)00130-5.

[15] Upadhyay, D., Sharma, K. B., Gupta, M., Upadhyay, A., & Venu, N. (2024, December). Deep Learning for Channel Prediction in Non-Stationary Wireless Fading Environments. In *2024 IEEE International Conference on Intelligent Signal Processing and Effective Communication Technologies (INSPECT)* (pp. 1–6). IEEE.

[16] K. Kaushik, "Investigation on Mobile Forensics Tools to Decode Cyber Crime," Secur. Anal., pp. 45–56, May 2022, doi: 10.1201/9781003206088-4.

[17] A. Bhardwaj and K. Kaushik, "Investigate Financial Crime Patterns Using Graph Databases," IT Prof., vol. 24, no. 4, pp. 27–36, Jul. 2022, doi: 10.1109/MITP.2022.3157029.

[18] Upadhyay, D., Sharma, K. B., Gupta, M., Upadhyay, A., & Venu, N. (2024, December). Sparse Channel Estimation in Massive MIMO Systems Using Compressed Sensing and Neural Networks. In *2024 IEEE International Conference on Intelligent Signal Processing and Effective Communication Technologies (INSPECT)* (pp. 1–6). IEEE.

CHAPTER 15

Ethical and Legal Frameworks in Mobile Forensics

Introduction

Mobile forensics has evolved to become a key player in contemporary criminal and civil proceedings, ESP, and incident response. Mobile devices are digital lifelines—they store communication logs (calls, messages, etc.), personal media (photos, videos), app data, geolocation records, payment data [1], and frequently represent the most important sources of digital evidence available. But pulling, parsing, and presenting that data raises complicated ethical and legal issues. The goal of this chapter is to give a basic introduction of the ethical obligations and legal considerations that regulate the field of mobile forensics. It demonstrates how certain domains of sensitive information must be navigated by practitioners without loss of individual rights, procedural fairness, or admissibility [2].

The most up-to-date forensic tools are becoming much more effective as mobile technologies mature. Today's forensic suites can recover deleted messages, passwords, pull application metadata, and decrypt some content in certain scenarios. Although this forensic opportunity is essential for crime investigation or national security, it raises serious concerns about consent of users, violation of privacy [3], excessive collection of irrelevant people's information, and its possible misuse. Ethical responsibility thus becomes as broad as technical ability, guaranteeing the investigative state apparatus uses its force in the right way as far as possible, according to the right promises, and with the corollary of accountability [4].

The essence of mobile forensic ethics is not only to ensure investigation results, but also to ensure justice, human dignity, and public confidence in the justice system. In the same way, laws try to make these boundaries explicit—what's allowed, what needs to be in place in order to protect, what levels of proof are required that must be met for any digital object to be admissible as evidence in a court of law. The goal of this chapter is to frame these two pillars—ethics and the law, and how they are intertwined in forming solid and fair mobile forensic practice [5].

Ethical Foundations in Mobile Forensics

Ethics in mobile forensics: Ethical considerations in the field of mobile forensics revolve around the concepts of honesty, impartiality, respect, and integrity, of course, with the added requirement to consider the highest standards of diligence [6], proportionality, and confidentiality. Unlike other digital investigations, mobile forensics are often of the most personal nature, including photos, voice notes, health records, and intimate chats. This requires that forensic examiners treat that information with the appropriate level of sensitivity, that only case-specific content is examined and that all irrelevant or unspecific content is somehow protected from examination or ignored [7].

Professional ethical codes, such as from the International Society of Forensic Computer Examiners (ISFCE), ACM, and IEEE hold that examiners should not have a conflict of interest, should report findings objectively, and should not claim certainty beyond what is warranted by the evidence. Additionally, mobile forensics practitioners must demonstrate ethics by not succumbing to outside pressures, whether by law enforcement, lawyers, or political actors, that may seek to interpret or demand access that is not warranted [8].

The sanctity of privacy is the cornerstone of ethical practice. Fortunately, forensic investigators can use data minimization principles to extract and examine only what they must in order to investigate the alleged offense and do not rely on bulk collection programs that are likely to sweep in unwarranted personal information. In cases that concern vulnerable or protected populations—for example, minors, journalists, or patients—we need to be more careful from an ethical point of view, often needing to consult with a lawyer before analyzing the content [9].

Crucially, ethical considerations are not just academic; they are decisions in how we act and relate to evidence. Unjust behaviors can compromise not only the image of a professional or institution but also the admissibility of relevant evidence and even the investigation as a whole. Thus, ethics must be viewed as an essential standard of operation—not just as an additional code of morality [10].

Legal Mandates and Evidentiary Integrity

The judicial framework in mobile forensics shapes the procedural limits around the investigative work that can be carried out. These consist of procedures for accessing data, rules for seizures, guidelines on handling evidence, and standards for admissibility [11]. There is a legal requirement in many jurisdictions for authorized access to, or search of, mobile devices (e.g., search warrants, user consent). The slightest variation from this procedure will cause the testimony to be inadmissible and the accused to have been deprived of his constitutional right not to be the subject of an unreasonable search and seizure [12].

The legal custody chain is a significant safeguard. This includes the documentation of the collection, storage, transfer, and examination of digital evidence. This also means that the evidence cannot have been tampered with or modified and can be safely presented in court [13]. Any weak link in that chain can also open the door for a legal challenge, so it is important for forensic teams to keep meticulous records from the first step to the last.

The state rules of evidence for digital evidence can be different depending on location, but generally will follow evidentiary models such as the Daubert or Frye tests. These criteria consider as to whether the applied strategies are scientifically valid, reproducible, and supported by the professional community involved. For mobile forensics, it implies the use of proven tools, well-documented techniques, and skilled handlers. Failure to take such steps may result not only in exclusion of the evidence but also could result in legal liability of the examiner [14].

Jurisdictional issues are further addressed by legal prescriptions, particularly when investigating cross-border. When it comes to cloud storage and worldwide service providers, the data involved in a case could be located in another country. In such circumstances, Mutual Legal Assistance Treaties (MLATs) or international cooperation guidelines should be utilized. Forensic scientists should know these legal geographies to avoid international juridical infractions or diplomatic implications.

In regulated markets like healthcare, banking, or critical infrastructure, industry-specific legislation such as HIPAA, PCI-DSS, or GDPR also requires to be addressed. These laws can layer on further restrictions for accessing, processing, and even retaining data—particularly in forensic contexts. Legal literacy involving other noncriminal laws (civil, regulatory, international, etc.) is therefore necessary for forensic experts as well.

Human Rights and Public Trust

A frequently undiscussed element of mobile forensics is the social effect, particularly with regard to human rights and public trust. Unchecked forensic tooling can almost instantly become abuse for surveillance, or profiling. Thus, tolerance of human rights—in terms of privacy, free speech, fair trial and more—is not a barter in either ethic or law.

International human rights treaties, such as the Universal Declaration of Human Rights (UDHR) and the International Covenant on Civil and Political Rights (ICCPR), emphasize that any interference with personal privacy must be "lawful, necessary and proportionate. Those experts working in democracies will need to think hard about the principles they embed into their work: namely, that technological authority should not be bought at the cost of basic freedoms [15].

Media coverage and highly publicized court cases have influenced the public's perception of digital forensics, and that of mobile forensics in particular, as have revelations regarding surveillance by governments. Ethical transgressions, such as taking a phone without permission or leaking sensitive material, can also have reputational implications for institutions and governments that linger for many years. Keeping ethics and law protections in place therefore has a strategic goal as well: maintaining public trust in digital justice technologies.

The Need for an Integrated Framework

Indeed the ethical and legal aspects are not distinct, rather two aspects of a single forensic perspective. Moral conduct frequently antedates legality, and legal norms often change as a result of questions posed by budding ethical debates. A good mobile forensic strategy brings them together, to the point where professionals are not only taught how to use these tools, but are also trained in rights-honoring investigative practices.

All personnel associated with mobile forensics need to be trained and certified and continue to receive education in the legal–ethical aspects. Standard operating procedures should consist not only of technical guides but also of ethical decision trees, legal checklists. In addition, oversight tools–like audits, peer reviews, and legal advisory panels–should be institutionalized to catch potential missteps and keep them from becoming liabilities.

CHAPTER 15   ETHICAL AND LEGAL FRAMEWORKS IN MOBILE FORENSICS

Thus, the introduction sets the premise that a comprehensive umbrella underpinning both ethical obligation and legislative demand must guide mobile forensics. In the rest of this chapter, we will explore these dimensions in detail.

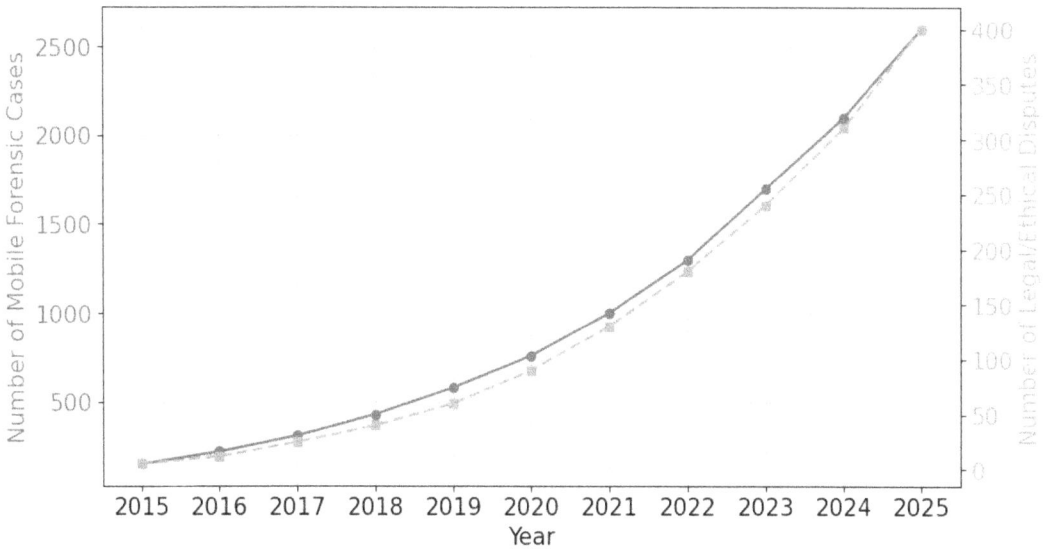

Figure 15-1. *Rise in mobile forensic cases*

This two-axis plot shows the rise of both mobile forensic cases and the legal or moral disputes that go along with them from 2015 to 2025. It shows how hard it is to balance technological research with legal and moral concerns. According to the data, the number of mobile forensic cases will rise from 150 in 2015 to over 2,600 by 2025. At the same time, the number of legal disputes will rise from 5 to 400. This shows that there is a clear link between the rise in forensic cases and the increase in legal disputes they cause. To improve interpretation, it is important to make it clear if the dataset includes global statistics, regional trends, or case studies from a single country. This is because the type of data greatly affects the results' applicability and scope.

Ethical Guidelines for Forensic Practitioners

Ethical considerations of practitioners have garnered much more attention as mobile forensics has been increasingly integrated into law enforcement, civil investigations, and intelligence practice. The ability to take and then examine and interpret such personal and often-sensitive data from mobile devices carries with it great moral responsibilities. This chapter also discusses key ethical tenets to which mobile forensics practitioners should adhere with particular focus on ethical behavior, data trustworthiness, privacy protection, and data minimization [16]. These principles impact not only the admissibility of evidence at trial, they inform public confidence and protect individual freedoms.

Professional Conduct and Data Integrity

At the heart of the paradigm of ethical practice in mobile forensics is the practitioners' pledge to professionalism, honesty, and integrity. Forensic scientists are held to an extremely high standard of behavior since the results of their work frequently shape judicial ruling and administrative outcomes. Several international and professional organizations, including the International Society of Forensic Computer Examiners (ISFCE), Scientific Working Group on Digital Evidence (SWGDE), the IEEE, the ACM, and the Digital Forensics Certification Board (DFCB), have developed their own codes of ethics and best practices to guide responsible conduct in the discipline.

At the heart of these principles are a few fundamental tenets: honesty, objectivity, transparency and accountability. Honesty is the process of giving true, unprejudiced, fact-based opinions without adulterations. In practical terms, forensic reports would therefore need to present the facts behind the results obtained when examining the mobile devices in question, avoiding speculation or hyperbole. Investigators have a natural tendency to give in to the urge—almost inevitably created by investigative strains—to reach conclusions that exceed that which the evidence supports.

A forensic interpretation is neutral for objectivity. Mobile forensic practitioners regularly collaborate with the cops, but they don't owe their ethical allegiance to the prosecution or the defense. They are supposed to be independent fact-finders. Conscious or unconscious bias could undermine the reliability of the conclusions/findings and erode the trustworthiness of the forensic science. To practice ethically, it would also mean not having personal stakes in cases to recuse oneself when impartiality is an issue.

Transparency: The clear evidence of the use of tools, methods adopted, decisions made, and problems faced during the forensic analysis. An ethical practitioner should be able to explain the basis for their case in a framework that can be readily examined by colleagues. This is especially critical when results are subject to legal cross-examination. Clarity in process and the ability to replicate results inspire confidence while reducing ambiguity.

Accountability requires that practitioners should follow standard protocols, validate their tools, and process the evidence correctly. Mistakes made in digital forensics can result in innocent people going to jail, or the key piece of evidence in a case being thrown out. Consequently, mobile forensic examiners must regularly check and validate their processes to ensure stability and fidelity [7].

Reporting restrictions: One of the "real world" aspects of professional practice. Ethical therapists respect the limitations of what the evidence allows. For example, metadata from a phone may indicate the exchange of messages between two parties but cannot definitively indicate the message content or meaning without accompanying context. Additionally, timestamps can be manipulated by system glitches and tools can generate false positives. Ethical researchers always come with a disclaimer and make obvious distinctions between established fact and reasonable inference.

Apart from the conduct of particular cases, professional ethics relates to training, certification, and ongoing professional development. Satisfied professionals do understand that mobile forensics and mobile technology are moving targets. They continue to learn, stay certified, and avoid using a method that is outdated or obsolete. Certifying authorities such as ISFCE and DFCB also have compulsory recertification process in place, making them refresh ethical and technical skill sets constantly.

Privacy Respect and Minimization Practices

The security of personal privacy is one of the most important ethical issues in mobile forensics. Smartphones are particularly intimate items, storing our most private calls and messages, health records, and photographs, as well as the content of your travels (like where you've been) and, if you use a mobile wallet, even payment details and biometric data. Accordingly, forensic professionals must strike a balance between the requirement for data acquisition and the responsibility not to overstep the mark.

Primarily, it is the principle of data minimization that is at play here. It implies that no data should be collected and analyzed than is required for purposes of the investigation. The combination of over-collecting data makes it risk-prone–not just for privacy violations but for legal risk, particularly in the age of privacy-facing regulations like the GDPR (General Data and Privacy Regulation) and California Consumer Privacy Act (CCPA), or in industries like healthcare dealing heavily with HIPAA. These regulations require institutions and professionals to data with the principle of purpose limitation and to safeguard user privacy unless otherwise ordered by the court workaround.

As a practical matter, privacy-protective forensics starts with sensible scoping. Investigators and analysts need to specify what are the relevant data types for an investigation and confine their tools accordingly. For instance, if it's a message app harassment case, access must be confined to that app and associated logs, and not all logs of the device storage. Legitimate use of forensic tools (e.g., Cellebrite, XRY or Oxygen Forensic Suite) includes setting up the tools to not capture irrelevant data where possible.

The second central concern is consent. In civil matters or internal investigations, you may need the owner's consent to forensically image or extract data from the device. Good faith providers have the duty to describe the level of analysis, possible perils, and retention rules. In law enforcement, there can be a replacement, a court-issued warrant, but the accordance of the warrant additionally restricts what can be accessed. Violations of medical ethics take place when they operate beyond this area of practice, and when they presume, without challenge, vague or overly general mandates.

There's a surveillance risk as well. With the advancements on forensics, deleted data can be recovered, the device log can be passively monitored, and cloud backups can be obtained. Unchecked, this power could be abused in the direction of surveillance, especially when wielded by authoritarian governments or without judicial oversight. Ethical standards need to have built-in audit and checks and balances, similar to a requirement for dual control on carrying out of invasive extraction methods or maintaining audit trails of what has been done.

Third parties must also be considered in this privacy consideration. Frequently the data extracted from a mobile device includes text or email messages or pictures of other persons who are not the subjects of the investigation. And there is some fact analysis—the ethical practitioner needs to decide if any such information is material to the case. If it's not, then it needs to be redacted or anonymized or not included in the report. This is especially important when dealing with vulnerable populations, for example, children, victims of violence, informants.

There is a more ethical component—the issue of data storage and preservation. Forensic data should be saved safely, possibly in an enciphered and access-restricted form, after it has been collected and examined. Unauthorized access, data leaks, or breaches in storage could cause unacceptable destruction and break a number of privacy legislations. And clinicians are also supposed to follow policies about how long to keep data, deleting it when an investigation or trial ends, unless they are legally required to keep it.

In conclusion, ethical standards for forensic professionals are not just about attitudes, but they need to be reflected in daily professional conduct. Ethical decision-making, ranging from professional behavior, respect for an individual's privacy, and data minimization, is at the core of the legitimacy and societal acceptance of mobile forensics. These norms of behavior need to keep pace with technological advancement and they should be supported through training, oversight, and regulation. Without them the formidable tools of mobile forensics potentially become tools not of truth but of injustice.

Legal and Regulatory Frameworks While Handling Evidence

Mobile forensics investigation has to be within legal jurisdiction to make sure the evidence is acceptable to court and following the regional and international laws. This includes accounting for the complexities of jurisdiction, procedural due process, and maintaining the authenticity of digital evidence form acquisition through storage. Legal regulations are of the utmost importance in determining the scope, depth, and procedure or practice of mobile forensic analysis, as the data storage, personal device usage, and remote communications become more and more globalized.

There are, of course, legal mistakes in the development of evidence that can render evidence inadmissible in court or breach of a fundamental right, and perhaps, worse for the investigator, personal legal liability. Thus, forensic experts must be technically proficient and legally knowledgeable. This part examines two issues of great importance; namely, jurisdictional limits in mobile forensic investigations and the chain of custody relating to evidence preservation.

CHAPTER 15   ETHICAL AND LEGAL FRAMEWORKS IN MOBILE FORENSICS

Jurisdictional Boundaries and Legal Access

Mobile data is a transnational entity, and as such is one of the most significant legal oddities and paradoxes in the digital forensic landscape today. It's not that these records are really stored on a device, which is the concern of law enforcement on-device evidence is not cloud evidence, but the account the device has to connect to can open the floodgates to documents on more than the device used at the time the warrant application is made. Which is to say that although the piece may be in one country, the part that it retrieves from a cloud datacenter could be in a data center halfway around the world.

Questions in these instances can pertain to what legal standards are relevant. Should Country A's law enforcement be entitled to access data held in Country B? Can international laws about data sovereignty trump a country's right to investigate? These questions have become even more important as services like iCloud, Google Drive, and WhatsApp cloud backups have become more popular. Obtaining information such as this without proper legal authority could not only violate international conventions, it could also make the evidence inadmissible in court.

This is a problem, and therefore forensic investigators are frequently forced to use Mutual Legal Assistance Treaties (MLATs), or similar arrangements such as the Budapest Convention on Cybercrime, for the ability to cooperate across international borders under certain legal conditions. But such treatments are typically slow, bureaucratic, constrained by political factors. Meanwhile, in pressing circumstances, investigators may be tempted to bypass useful cloud data stored using methods of a more technical nature, potentially deviating from principles of data protection law and evidence integrity.

For user's devices that travel across countries, even for them, owner's rights vs. territorial law issues may arise. For example, if a target is a foreign national, and their phone is encrypted or associated with a foreign number, lawful access may need to be subject to multiple levels of judicial authorization. In terrorism or organized crime cases, where multination teams are involved in the investigation, mobile forensics teams may work with Interpol, Europol, or their counterparts in the regions for cybercrime units.

A necessary requirement for obtaining mobile devices is a judicial approval, generally in the form of a search warrant. These warrants issuing authorities should define the extent to which collection is to be made, in which target application and at which time frame, to prevent vast or intrusive searches. Overbroad or loosely specified warrants may be challenged with motions to suppress in court on the basis that the warrant does not serve its intended purpose and that it infringes on the defendant's constitutional rights (in jurisdictions using a charter or a bill of rights), requiring that the evidence obtained be excluded.

Chain of Custody and Evidence Integrity

The chain of custody is the formalized procedure used to document the seizure, transfer, access, and storage of digital evidence in a way that ensures the integrity of the evidence. Any mobile forensic work is just as strong as its record trail. Without proven and continuous evidence authentications, even the most damning of digital evidence could potentially be disputed or left out by the court.

Chain of custody: An unbroken record of evidence is important to show that evidence has not been arbitrarily changed or tampered with or contaminated in any way during the forensic examination. When it comes to mobile devices, it means documenting the following at every step along the way: who acquired it, how it was transported, what tools were used forensically, what they found and where that was kept, who had access to it at the time of analysis, etc.

The digital evidence is liable to the invisible sabotage or the unintentional change. Merely powering on a cell phone can modify metadata or timestamp metadata. For this reason, the primary seizure should be performed as carefully as possible, such as in Faraday bags or in airplane mode in attempts to keep the device away from any network. In making the seizure, they should photograph the equipment, record serial numbers where possible, and describe the condition of the equipment.

Forensic tools log audit during analysis. Those logs need to capture everything from bypassing the lock screen to parsing SMS records, each with a timestamp, the examiner's ID, and the tool version used. Most of the forensic tools like Cellebrite UFED and Magnet AXIOM have internal logging mechanism to support this accountability.

Storage-related procedures also have to adhere to legal and forensic rules. Evidence needs to be securely maintained, often encrypted and access-restricted. The physical devices could be enclosed and labeled, while their digital representations are hashed to produce the verification values for example, using the SHA-256 algorithm. These hashes can then be presented in court as evidence that the evidence has not been altered since it was obtained.

CHAPTER 15   ETHICAL AND LEGAL FRAMEWORKS IN MOBILE FORENSICS

In order for digital evidence to be admitted in court, it must comply with the local rules of evidence that a court is following. Many of those local standards were drawn from Indian federal rules already in place in 2008, but courts have decided the rules apply to digital data. This standard gauges the acceptability of the forensic techniques being used, as well as their peer review. Courts may also review whether the analyst is qualified, and the results reproducible.

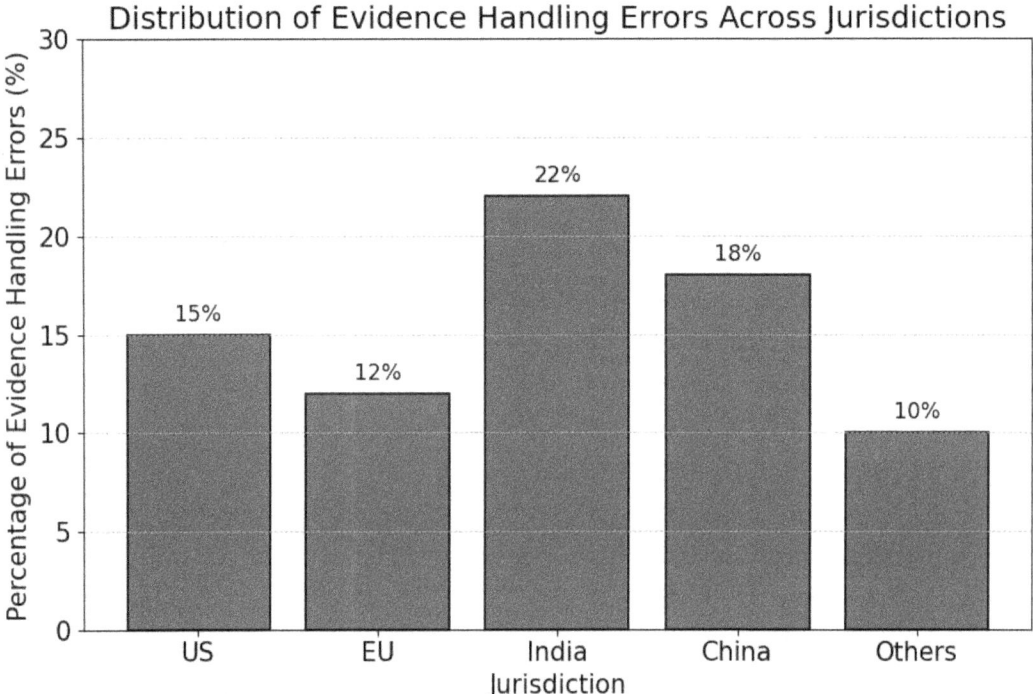

Figure 15-2. *Distribution of evidence handling errors across jurisdictions*

This bar chart shows how evidence handling mistakes are spread across different jurisdictions. India (22%) and China (18%) have the highest rates of mistakes like broken chains of custody, incomplete paperwork, and bad management of digital evidence, compared to the United States (15%), the EU (12%), and other regions (10%). These differences show how unevenly standardized forensic procedures are being used and how hard it is to make sure that evidence is always correct during investigations around the world. To put these results in a better context, it's important to say when the data were gathered, since mistake rates can change over time depending on the laws and technologies in place. Adding links to studies or reports that look into how forensic evidence is handled would also make the results more reliable and give readers ideas for more reading on forensic issues that are unique to their area.

Legal Standards, Continuous Monitoring, and Regulatory Compliance

As practices of mobile forensics are firmly cemented into disciplines such as law enforcement, corporate governance, healthcare, and finance, it is imperative to conform to legal standards and ensure continued compliance with regulatory obligations. The legal landscape pertaining to electronically stored information has developed in an attempt to keep pace with the pervasiveness of mobile technology and the ever-increasing landscape of data collection. Legal compliance is not just bureaucratic requirements, but a fundamental necessity that defines how forensic practices are performed in mobile, documented, and defended in litigation (civil and criminal) or corporate litigation situation.

This chapter elaborates on two key requirements—the requirement to be in compliance with geo-political, sectoral regulations like (GDPR, HIPAA, IT Act, etc.) and the importance of setting up monitoring systems and audit mechanisms to institutionalize the legal and procedural accountability.

Compliance with Legal Frameworks and Industry Regulations

Legal compliance guarantees forensic investigation respect for individual's rights, organizational obligations, and admissibility of digital evidence. Field mobile forensic practitioners are working in an environment that is determined by region, sector, and application. Different legal mechanisms exist in different jurisdictions for regulating the access to, privacy and security of, digital data. Similarly, various industries have specific requirements for managing confidential or classified information.

General Data Protection Regulation (GDPR)

GDPR is the most far-reaching data privacy law in the world, as it applies to all companies processing the personal data of EU citizens. In mobile forensics, being GDPR-compliant represents the lawful, fair, and transparent processing of the data and the collection being necessary and proportionate. Principles like the limitations of purpose, data minimization, accuracy, storage limitation, accountability are all in GDPR's Article 5.

In practice, this means that as a mobile forensics examiner they should: limit data acquisition to those which are pertinent to the case, not to load unnecessary personal information and to provide means to ensure data security in treatment and post-treatment process. Forensic tools should be constructed to carry out selective extraction while, if such anonymization is technically feasible, reporting should anonymize non-essential user data.

If you are involved in corporate or civil EU investigations, you may need to obtain either consent or establish a legitimate interest before extracting evidence from employee- or client-owned devices. GDPR breaches can lead to severe penalties (including up to €20 million or 4% of global turnover), thus organizations using the services need to have procedures of its own staff that are driven by GDPR.

Health Insurance Portability and Accountability Act (HIPAA)

HIPAA regulations in the United States must be followed on digital forensic investigation of a mobile device in medical institutions. HIPAA Requires: Protecting of PHI (Protected Health Information)—This means any health data that is contained, accessed, or transmitted in any format on mobile devices.

For forensic investigations of mobile devices that belong to healthcare providers and for devices that contain health-related information, the practice of digital forensics should involve secure handling, data-encryption, and data-minimization with disclosure limitation. Charges for mishandling a case are no joke and can be applied irrespective of intent, even if a set is exposed while being handled forensically. This is especially important in situations of incident response to ransomware attacks, or insider threats at medical facilities.

Information Technology Act (India) and Regional Laws

The Information Technology Act, 2000 (as revised in 2008) is the principal statute in India that regulates mobile forensics. This act deals with issues including cybercrime investigation, admissibility of digital evidence, and unlawful access of data. Investigators must adhere to due process when handling digital evidence, and the Act makes it a crime to gain illegal access to or compromise data (Sections 43 and 66). An increasingly explicit chain of custody demonstrating the integrity of confiscated digital evidence is being required by Indian courts as proof of investigator compliance with the Indian Evidence Act. In addition, new provisions pertaining to digital evidence have been included in the Bharatiya Nyaya Sanhita (BNS) and Bharatiya Nagarik Suraksha Sanhita (BNSS), which have established standards for its acceptance in criminal cases and strengthened its legal recognition. Digital records are emphasized as fundamental evidence in these frameworks, which also strengthen responsibilities regarding forensic validation and

CHAPTER 15 ETHICAL AND LEGAL FRAMEWORKS IN MOBILE FORENSICS

chain of custody. Similar to how the Privacy Act and the Surveillance Devices Act govern the institutional and legal bounds for digital forensic efforts in the United States, the United Kingdom, Australia, and Canada, sector-specific laws like FERPA and SOX do the same in other jurisdictions.

Sector-Specific Customization

For banking, financial forensics of mobile banking applications must adhere to anti-fraud laws enforced under PCI-DSS and AML directives, as well as to the national financial data laws. Forensics access to client banking data need to incorporate layers of authority and encrypted informing.

In government and national security, mobile forensics of classified devices must comply with internal compliance procedures that are often regulated by national intelligence or defense cyber units. Data breach or improper logging can be considered a national security violation. For instance, when conducting forensic investigation of unauthorized disclosures (whistleblowing or insider threat, for example), you need to get the secrecy law in order and also comply with national data policies.

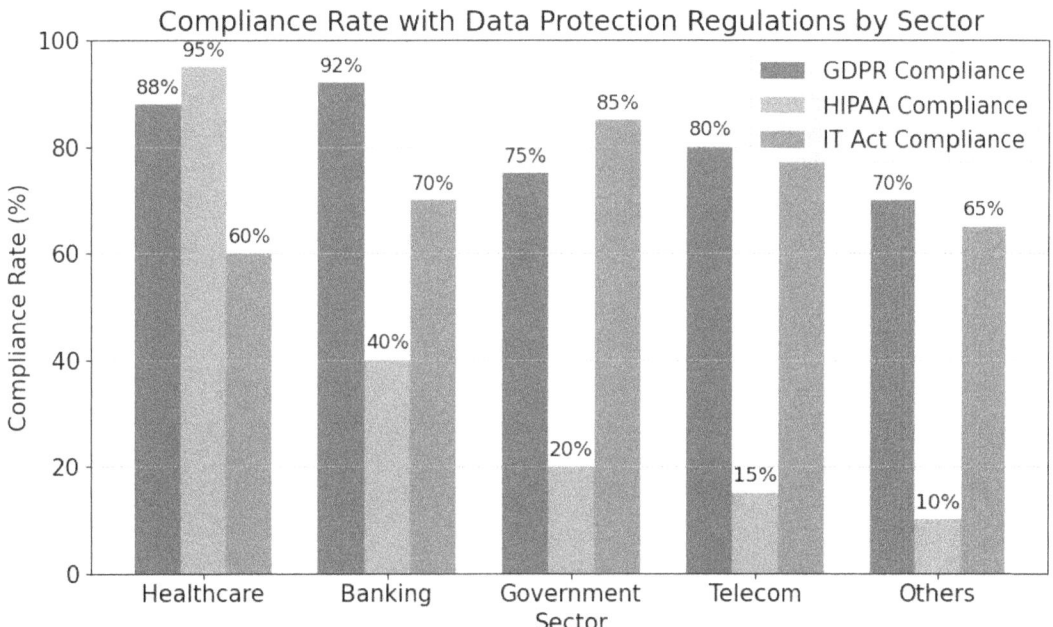

Figure 15-3. *Compliance rate with data protection regulations by sector*

This grouped bar chart shows how well five important industries—healthcare, banks, government, telecom, and others—agree with GDPR, HIPAA, and IT Act rules. Healthcare is the most compliant with HIPAA (95%), while banks and telecom are also very compliant with GDPR (92% and 80%, respectively). Government and "Others" have more even compliance rates, but they are slower to accept HIPAA. This shows that sector-specific enforcement needs to be stronger. Compliance rates range from as low as 10% to as high as 95%, which shows how different businesses are when it comes to how regulations are aligned. To improve interpretation, it is important to be clear about the time frame of the data shown, such as whether it comes from compliance polls that were done in a single year or over a number of years. This is because the time frame has a big impact on how readers understand how regulations are changing. Adding links to the dataset source would make it even more reliable and let readers look into sectoral compliance trends in more depth.

Monitoring and Process Audits

While establishing a compliant forensic workflow is important, maintaining it over time is equally critical. This is where continuous monitoring, internal quality assurance, and legal accountability mechanisms come into play.

Internal Monitoring Systems

To adhere to both external laws and internal policies, mobile forensics labs should have standard operating procedures (SOPs) and compliance test procedures in place and, carry out regular compliance checks. Active monitoring involves the use of audit logs, activity logs and version controlled tools. This helps to guarantee that everything you do, such as unlocking a device, exporting logs, or viewing of media files, is recorded and can be traced.

Legal Audits and Compliance Reviews

Periodic legal audits help forensic departments assess whether their tools, methods, and training remain legally defensible. These audits should examine the following:

- Tool validation and versioning (ensuring tools meet legal standards)
- Chain of custody documentation
- Data minimization adherence
- Consistency of reporting and metadata logging
- Staff certifications and ethical training completion

CHAPTER 15   ETHICAL AND LEGAL FRAMEWORKS IN MOBILE FORENSICS

In sensitive cases, external legal experts or compliance officers may be brought in to review processes before evidence is submitted to court. In corporate settings, compliance audits are often initiated by the internal risk or legal department to ensure that digital evidence, especially from employee-owned devices, has been collected in accordance with HR policies, labor laws, and privacy regulations.

Accountability Mechanisms

Strong accountability frameworks should include

- **Access control logs**: Showing who accessed what data and when

- **Forensic reporting trails**: Versioned, time-stamped reports with signatures

- **Escalation protocols**: For reporting misuse or procedural violations

- **Retention and disposal logs**: Ensuring data isn't stored longer than legally permissible

Establishing **legal liaisons** between forensic units and organizational legal departments is also recommended. This ensures prompt legal guidance during unclear cases and facilitates collaboration during compliance assessments or court trials.

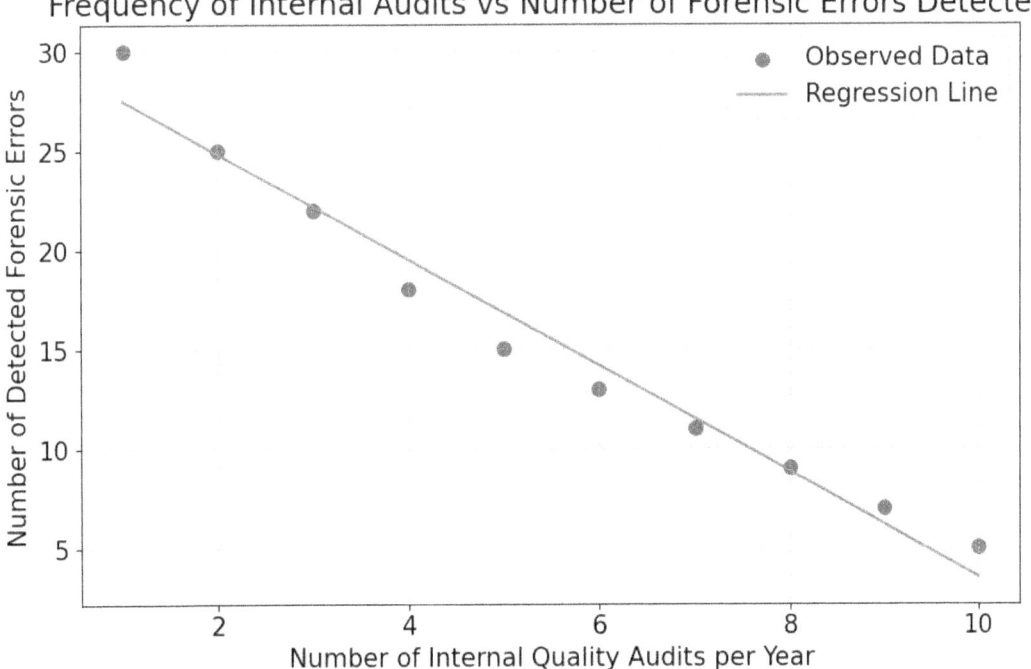

Figure 15-4. *Frequency of internal audits vs. number of forensic errors detected*

This scatter plot with regression line shows a negative correlation between the number of internal quality audits conducted annually and the number of forensic errors detected. It highlights that increasing audit frequency significantly reduces procedural, documentation, and data breach errors in mobile forensics. Parameters include audit frequency from 1 to 10 per year and forensic errors ranging from 30 to 5, with a fitted linear regression to emphasize the inverse relationship.

Conclusion

Mobile device forensics is coalescing as a fundamental part of present-day digital investigation, accompanied by the huge responsibility to serve justice through authentic and objective investigation, while at the same time protecting the rights, honor, and privacy of individuals. This chapter has shown how forensic professionals are faced with the increasing need to weigh these responsibilities under an intricately woven network of ethical considerations, legislative requirements, and regulatory standards.

The search for truth is the essence of mobile forensic practice—based on science, transparent methodologies, and respect for due process. Whether they are probing criminal conduct, regulatory violations, or threats from the inside, the results of inquiries conducted by practitioners generally become the foundation of litigation or policy. In view of this, forensic investigations should be above reproach. This continues with the first link in the chain of custody and ends with running down an investigation even under resistance to maintain the integrity of the process.

But the ability to review and exploit mobile data, the deeply personal archive of modern life, must be used with great care. As illustrated throughout the chapter, the ethical concerns of data minimization, privacy, informed consent, and third-party data use are core, not peripheral, practice imperatives. Forensic experts must not just "do it right" but "do the right thing," especially when technological capacities exceed parliamentary precision.

At the same time, the legal side of mobile forensics becomes more and more challenging, distributed, and multi-jurisdictional. In the era of cloud computing, communications encryptions, and privacy-first regulations such as GDPR, HIPAA, and the data protection acts of various countries that were recently introduced, we find ourselves in a legal minefield. They will have to be aware of and sensitive to diverse regulatory realms, acquire lawful authority for every step of the process and keep up with the progress of admissibility and accountability. It requires an incessant legal enlightenment—not only in your own country's legal frame, and in the international treaties, but also digital sovereignty concerns and data localization.

The chapter also highlighted the significance of regulation compliance and in-house control. Highly regulated industries like healthcare, finance, and government, which need to work in line with certain data handling procedures, are no exception. Failing to do so can compromise the legitimacy of an investigation as well as expose institutions to legal liability and public rebuke. Ongoing process audits, quality reviews, and audit trails are necessary to ensure that forensic activity within the organization remains in line, transparent, and ethical.

Just as with mobile devices and forensic tools, the guidance for their examination should advance. For mobile forensics to be effective in its practice, it needs to be fair as well. Global standardization of forensic processes, responsive legislation and a coherent ethical framework is mandatory in order to guarantee the fairness and justice in the process of mobile forensics. This is especially important as the international cooperation in cybercrime investigations becomes more and more significant (L. Flannery and N. Edelhoff (2007) work the limestone by ISO/IEC 27037/ 27042).

Finally, and contributing to the prior point, there is the notion that our mobile forensics professionals hold great power and carry great responsibility. Not only are they technical, they are profoundly ethical, and at a foundational level, legal. As protectors of justice and the rights of the individual, they must adopt the most fitting levels of behavior, stay abreast of the law, and constantly adapt to new technologies and public norms. It is only through this united effort that mobile forensics can maintain its power and authority in the digital era.

References

[1] Vasilaras, A., Papadoudis, N., & Rizomiliotis, P. (2024). Artificial intelligence in mobile forensics: A survey of current status, a use case analysis and AI alignment objectives. *Forensic Science International: Digital Investigation, 49*, 301737.

[2] Bharati, R. "Legal and Ethical Considerations in the Use of Digital Forensics by Law Enforcement: A Multi-jurisdictional Study.

[3] Khubrani, M. M. (2023, August). Mobile device forensics, challenges and blockchain-based solution. In *2023 Second International Conference On Smart Technologies For Smart Nation (SmartTechCon)* (pp. 1504–1509). IEEE.

[4] Mykhaylova, O., Fedynyshyn, T., Sokolov, V., & Kyrychok, R. (2024). Person-of-Interest Detection on Mobile Forensics Data—AI-Driven Roadmap. *Cybersecurity Providing in Information and Telecommunication Systems 2024, 3654,* 239–251.

[5] Asasfeh, A., Al-Dmour, N. A., Al Hamadi, H., Mansoor, W., & Ghazal, T. M. (2023, November). Exploring cyber investigators: an in-depth examination of the field of digital forensics. In *2023 IEEE Intl Conf on Dependable, Autonomic and Secure Computing, Intl Conf on Pervasive Intelligence and Computing, Intl Conf on Cloud and Big Data Computing, Intl Conf on Cyber Science and Technology Congress (DASC/PiCom/CBDCom/CyberSciTech)* (pp. 0084–0088). IEEE.

[6] Upadhyay, D., Sharma, K. B., Upadhyay, A., Gupta, M., Dhondiyal, S. A., & Asudani, D. S. (2024, November). Multi-Layered Security Techniques for Protecting Cooperative Relay Networks from Eavesdropping. In *2024 International Conference on Cybernation and Computation (CYBERCOM)* (pp. 14–19). IEEE.

[7] Casino, F., Dasaklis, T. K., Spathoulas, G. P., Anagnostopoulos, M., Ghosal, A., Borocz, I., ... & Patsakis, C. (2022). Research trends, challenges, and emerging topics in digital forensics: A review of reviews. *Ieee Access, 10,* 25464–25493.

[8] Yin, Z., Wang, Z., Xu, W., Zhuang, J., Mozumder, P., Smith, A., & Zhang, W. (2025). Digital Forensics in the Age of Large Language Models. *arXiv preprint arXiv:2504.02963.*

[9] Upadhyay, D., Kumar, A., Rani, S., Asudani, D. S., Ranswal, A. S., & Venu, N. (2025, January). Eavesdropper-Aware Optimization of Cooperative Relaying Networks in the Presence of Heterogeneous Interference. In *2025 International Conference on Cognitive Computing in Engineering, Communications, Sciences and Biomedical Health Informatics (IC3ECSBHI)* (pp. 120–125). IEEE.

[10] Kidenda, S., Muchai, R., Green, L., McHale, T., Mishori, R., & Nelson, B. D. (2022). Evaluating the effectiveness of a mobile application to improve the quality, collection, and usability of forensic documentation of sexual violence. *PLoS one*, *17*(12), e0278312.

[11] Kováč, P., Jackuliak, P., Bražinová, A., Varga, I., Aláč, M., Smatana, M., ... & Thurzo, A. (2024). Artificial Intelligence-Driven Facial Image Analysis for the Early Detection of Rare Diseases: Legal, Ethical, Forensic, and Cybersecurity Considerations. *AI*, *5*(3), 990-1010.

[12] Alotaibi, F., Al-Dhaqm, A., & Al-Otaibi, Y. D. (2023). A conceptual digital forensic investigation model applicable to the drone forensics field. *Engineering, Technology & Applied Science Research*, *13*(5), 11608-11615.

[13] Upadhyay, D., Gupta, M., Sharma, K. B., & Upadhyay, A. (2024, August). Enhancing Network Function Virtualization and Service Function Chaining: Innovative Optimization Strategies and Their Impact. In *2024 First International Conference on Pioneering Developments in Computer Science & Digital Technologies (IC2SDT)* (pp. 153-157). IEEE.

[14] Upadhyay, D., Upadhyay, A., Gupta, M., Sharma, K. B., & Yadav, D. (2024, August). An Approach of Fog Computing and Edge Computing for Computing Resources Optimization Strategies. In *2024 First International Conference on Pioneering Developments in Computer Science & Digital Technologies (IC2SDT)* (pp. 142-146). IEEE.

[15] Elhaik, E., Ahsanuddin, S., Robinson, J. M., Foster, E. M., & Mason, C. E. (2021). The impact of cross-kingdom molecular forensics on genetic privacy. *Microbiome*, *9*(1), 114.

[16] Nteziryayo, D., Wang, J., Qian, H., Liang, M., Liu, H., Liu, X., ... & Joseph, P. (2024). Advancement and the existing landscape of forensic medicine in Africa: A comparison with developed countries. *Forensic Science, Medicine and Pathology*, 1-14.

Index

A

Abstract visualizations, 248
Access iCloud account
    forensic tools
        application data, 84
        calendars, reminders, and notes, 84
        contacts, 83
        data download, 84
        data extraction, 82
        data types, 82
        data type selection, 83
        iCloud backup, 83
        install and set up, 82
        login credentials, 82
        messages and SMS, 83
        photos and videos, 83
        Safari browser history and bookmarks, 83
        token-based access, 83
    manual access via browser, 81
Access Point (AP), 207
Access Points (WAPs) logs, 219
Accountability, 433, 444
Accountability frameworks, 443
Account activity logs, 152
Account takeover (ATO), 128
Acquisition of iOS devices
    challenges
        anti-forensic measures, 73, 74
        data loss, 73
        encryption, 72, 73
    cloud-based acquisition, 71, 72
    data acquisition, 75–84
    file system acquisition, 70, 71
    forensic acquisition, 68
    logical acquisition, 69, 70
    physical acquisition, 68, 69
Adam optimizer, 353, 357
ADB, *see* Android Debug Bridge (ADB) technique
ADB and rooted access (Android), 242
Address matching, 191
Advanced Encryption Standard (AES), 68, 205
Advanced persistent threats (APTs)
    definition, 225
    known and unknown vulnerabilities exploitation, 225, 226
    network intrusion, 226
    social engineering, 225
AES, *see* Advanced Encryption Standard (AES)
Ahead-of-Time (AOT) compilation, 34
AIDL, *see* Android Interface Definition Language (AIDL)
AI ecosystems, 3
AI-integrated operating systems, 3
AI-powered apps, 369
AI-powered tools, 418
Airplane mode (flight mode), 14
Algorithms flag anomalies, 325
ALM, *see* Application lifecycle management (ALM)

INDEX

AMM, *see* Automated Market Maker (AMM)
Android, 4, 5
Android 16, 50
Android-based applications, 179
Android browsers
    Firefox for Android, 234
    Google Chrome, 234
    Opera Mini and UC Browser, 234
    Samsung Internet Browser, 234
Android Debug Bridge (ADB) technique, 51, 136
Android devices
    app data
        authentication credentials, 46
        protocol buffers (protobuf) and SQLite, 45
        user preferences, 46
    file system, 55–59
    forensic acquisition
        anti-forensic techniques, 50
        device variability, 50
        encryption, 49
        file system acquisition, 48, 49
        logical acquisition, 48
        physical acquisition, 47, 48
        rooting, 49, 50
    forensic tools and methods, 51–55
    locations
        /cache, 46
        /data, 46
        /sdcard, 47
        /system, 46
    system data
        cached data, 44, 45
        geolocation data, 45
        log files, 44
        system settings, 44
    user data
        browser history, 44
        call logs, 43
        contacts, 42
        messages, 43
        pictures and videos, 43
        profiles, 42
        tablet computers, 42
Android file structure, 137, 138
Android file system
    APKs, 56, 57
    APKTool, 57
    app-specific data, 57
    EXT4, 56
    F2FS, 56
    Frida, 58
    JD-GUI, 58
    YAFFS, 56
Android forensics, architecture of Android, 28–42
Android Interface Definition Language (AIDL), 39
Android malware datasets, 353
Android operating system, 27
Android package files (APKs), 56, 57, 136
Android Runtime (ART), 28
    ALM, 35
    AOT compilation, 34
    code verification and security, 36
    vs. Dalvik, 36, 37
    description, 34
    environment, 34
    GC, 35
    JIT, 34
    memory management, 35
    memory profiling and optimization, 35, 36
Android storage paths

INDEX

artifact location comparison, 244
Chrome, 239
Firefox, 240
Opera/UC Browser/Samsung Internet, 240
Android system components
device control, 29
Linux kernel, 30–42
memory management, 29
multitasking, 29
overview, 29
Android Virtual Device (AVD), 141
Anomalies, 313
Anomaly detection, 292, 310, 325, 329
mobile phones
common mobile data anomalies, 311, 312
cybersecurity, 310
definition, 310
mobile forensics, 310
role, 310
semi-supervised learning, 314
supervised learning, 312
types, 310
unsupervised learning, 313
Anti-forensic techniques, 24, 50
iOS devices, 73, 74
AOT, *see* Ahead-of-Time (AOT) compilation
APFS, *see* Apple File System (APFS)
API-based integration, 414
APKTool, 57
Apple File Conduit framework, 69
Apple File System (APFS), 70, 88, 89, 247
Apple iCloud Backup, 183
Apple Pay, 337
Apple's iPadOS, 5
Application framework, Android

activities, 37
APIs, 37
app's life cycle and interactions, 37
broadcast receivers, 38
connectivity services, 38
content providers, 38
core libraries, 37
data storage, 39, 40
flexibility and scalability, 41
getSystemService () method, 39
IPC, 39
life cycle management, 40
location services, 38
notification services, 38
power management, 38
security and permissions, 41
service locator pattern, 39
services, 38
system services, 38
telephony services, 38
testing and debugging, 41, 42
Application lifecycle management (ALM), 35
Application security, 117
App sandboxing, 63, 64
App Usage Data, 346
AR and VR forensics, 411
ArcGIS, 190
Architecture of Android
ART, 28
component, 28
components, 29–42
data types, 28
Linux kernel, 28
organizational structure, 28
overview, 28, 29
Artifact-based threat detection, 251, 252
AP, *see* Access Point (AP)

451

INDEX

APKs, *see* Android package files (APKs)
ART, *see* Android Runtime (ART)
AR/VR environments, 410, 411, 417
AR/VR forensics, 411
    data sources, 410
AR/VR platforms, 409, 418
Arweave, 103
ATO, *see* Account takeover (ATO)
Augmented Reality (AR), 408, 421
Australia's Privacy Act, 390
Authorization code flow, 120
Autoencoders, 313, 322, 351, 355, 365
Automated Market Maker (AMM), 93
Autopsy, 51, 69, 245
Avatar identity, 409
AVD, *see* Android Virtual Device (AVD)
Axie Infinity, 92
Axiom, 230

B

Banking application behavior identification
    application permissions, 141, 142
    misusing application permission, 143
Banking applications, 117
Base Station Controller (BSC), 202
Base stations, 202, 203
Base Transceiver Station (BTS), 202
Behavioral analytics, 119
Behavioral fingerprinting, 411
Belkasoft, 51, 74, 75
Belkasoft Evidence Center, 76–78
Big data, 198
Biometric, 411
Biometric authentication, 72
Bitcoin, 91, 95, 107
Blackberry, 4
BLE, *see* Bluetooth Low Energy (BLE)

Blockchain
    applications, 92
    app logs, 100
    authentication and privacy, 91
    backup files, 100
    and Bitcoin, 91
    blocks of data, 91
    browser data, 100
    challenges
        anonymity, 110
        data encryption, 110
        decentralized, 110
        privacy features, 110
        pseudonymity, 110
        volume and complexity of data, 111
    cryptoanalysis, 95
    cryptocurrency investigations, 107
    cryptocurrency transactions, 108
    cryptography, 91
    crypto wallets, 92
    cybercrimes, 94
    dApp investigation, 112
    dApps, 92, 93 (*see also* Decentralized applications (dApps))
    digital ledger, 91
    integration, 112
    real-time monitoring, 112, 113
    smart contract, 112
    transparent and immutable data, 111
    wallet analysis, 109
    wallet database files, 100
    wallet investigations, 107
    wallets, 107–109
Blockchain-based audit trails, 416
Blockchain data, 95
Blockchain forensic analysis
    blockchain data, 95
    blockchain explorers, 99

Cellebrite UFED, 99
challenges
    anonymity, 101
    distributed nature, 101
    encrypted data, 101
    lack of standardized, 101
    private keys, 101
    pseudonymity, 101
cryptobased application analysis, 95, 96
data acquisition
    bit-by-bit copy, 96
    crypto-related applications, 96
    data sources, 97
    extracting data, 96
    logical acquisition, 97
    physical acquisitions, 97
data correlation and reporting, 99
data parsing, 98
DeFi and NFT Apps, 98
ElcomSoft iOS Forensic Toolkit, 100
Magnet AXIOM, 99
mobile devices, 95
transaction analysis, 98, 99
Wallet Apps, 97
XRY by MSAB, 100
Blockchain technology, 415
Bluetooth, 14
Bluetooth (BR/EDR), 211
Bluetooth Low Energy (BLE), 211, 212
Bring Your Own Device (BYOD), 235
Broadcast receivers, 38
Browser artifacts analysis
    artifact location comparison
        Android *vs.* iOS, 244
    artifact storage paths
        Android storage paths, 239, 240
        iOS Storage Paths, 240
    artifact types

bookmarks, 239
browsing history, 238
cache files, 238
cookies, 238
downloads, 239
indexed databases, 239
vs. investigative relevance, 243
saved credentials and autofill data, 238
session and tab data, 239
web storage, 239
digital artifacts, 237
forensic tools
    ADB and Rooted Access (Android), 242
    Cellebrite UFED, 242
    iTunes Backup, 242
    Jailbreaking, 242
    Magnet AXIOM, 242
    oxygen forensic detective, 242
    physical analyzer, 242
    SQLite and plist Viewers, 243
mobile web browsers, 237
volatile *vs.* persistent artifacts, 241
Browser threat detection, 358
Browsing history, 238
BSC, *see* Base Station Controller (BSC)
BscScan, 103
BTS, *see* Base Transceiver Station (BTS)
Bulk Extractor tool, 160
Burp Suite and HTTP Tool kit, 140
BYOD, *see* Bring Your Own Device (BYOD)

C

Cache files, 157, 238
California Consumer Privacy Act (CCPA), 200, 390

INDEX

Call detail records (CDRs), 199
Call Logs, 346
Captioned video forensic timeline, 365
C-based API, 33
CCPA, *see* California Consumer Privacy Act (CCPA)
CDRs, *see* Call detail records (CDRs)
Cellebrite, 112, 258
Cellebrite UFED, 51, 69, 71, 74–78, 81, 99
Cellebrite UFED and physical analyzer, 242
Cellebrite UFED Cloud, 72
Cell networks, 199
Cellular devices, 384
Cellular network architecture
  base stations, 202, 203
  HLRs, 203
  mobile switching center, 203
Centralized exchanges (CEXs), 104, 109
Certification, 396, 397
CEXs, *see* Centralized exchanges (CEXs)
Chainalysis, 104, 106
Chain of custody, 376, 380, 381, 386, 437
  auditing, 382
  in court, 378
  digital data, 376
  digital evidence, 376, 377
  document, 382
  evidence, 378
  evidence-handling processes, 381
  evidence label and tags, 381
  forensic investigations, 376
  labeling, 377
  physical device, 377
  recording and maintenance, 381
  stored in environments, 382
Chain-of-custody documentation, 414

Challenges in forensic acquisition of location-based data
  cloud synchronization issues, 182
  data volatility, 182
  device fragmentation, 182
  encryption, 182
  location spoofing, 182
Charles Proxy, 140
Chip-off and JTAG methods, 2
Chip-off method, 11, 47
Circuit-Switched (CS), 204
Civil litigation, 235
Classification algorithms, 312
Classification tasks, 292
Cleared history, 246
Cloud-based acquisition, 27
  iOS devices, 71, 72
Cloud data analysis, 172
Cloud forensics, 414
Cloud forensics characteristics
  automatic backup, 168
  data accessibility, 168
  real-time modifications, 168
  versioning, 168
Cloud mining contracts, 94
Cloud security, 171
Cloud services, 28
Cloud storage, 167
Cloud storage investigation challenges
  access regulation and confidentiality, 170
  data modifications and deletion, 170
  security and encryption, 171
  storage complications, 170
Cloud storage services, 182
Cloud-synced interactions, 3
Cloud synchronization, 168
Clustering, 333

Clustering algorithms, 313, 319, 321
Clustering tasks, 292
CNNs, *see* Convolutional Neural Networks (CNNs)
Code obfuscation, 120
Collective anomalies, 311
ColorOS, 5
Common mobile data anomalies, 311, 312
Common mobile network threats
    IMSI catchers, 224
    MITM attacks, 223
    SIM jacking, 224
    Wi-Fi-linked threats
        evil twin attacks, 225
        rogue access point, 224
Communication type, 350
Community detection, 325
Completely Fair Scheduler (CFS), 30
Compound, 93
Confusion matrix, 357
Connectivity services, 38
Consent
    GDPR and CCPA, 392
    national security, 392
Containment, 228
Content classification, 325
Content providers, 38, 39
Contextual anomalies, 311
Convolutional Neural Networks (CNNs), 286, 347, 351, 365
    gallery images, 347
    object-level evidence, 347
    textual evidence embedded in images, 347
Cookies, 157, 238
Cookies falsification, 252
Core libraries, 37
Core Location API, 180
Core Network (CN), 198, 204
Correlation-based Feature Selection (CFS), 283
Corrupted cookies, 252
Credential stuffing, 128, 320
Credit card fraud detection, 337
Crime scene investigators, 7
Criminal investigations, 235
Cross-browser behavioral profiling, 255
Cross-browser forensic analysis
    artifact volume comparison across browsers, 256
    challenges, dataset merging, 256, 257
    cross-browser behavioral profiling, 255
    forensic investigation, mobile browsers, 254
    multi-browser correlation, 254
    structural and encryption differences
        Firefox, 255
        Google Chrome, 255
        Safari, 255
    user session overlap across browsers, 257
Cross-domain cooperation, 422
Cross-domain knowledge, 369
Cross-validation, 294
CRUD operations, 38
Cryptoanalysis, 95
Cryptocurrency, 95, 107, 108, 111
    wallet analysis, 109
Cryptocurrency investigations, 107
Cryptocurrency transactions, 108
Cryptocurrency wallets, 95, 107
Cryptographic hashing, 414
Cryptographic signatures, 36
Cryptography, 91, 117
Cuckoo and MOBSF Sandbox, 141

INDEX

Cybercrimes, 234
   DPRK's DMM Bitcoin Exploit, 94
   GainBitcoin, 94
Cybercrime units, 436
Cybercriminals, 344
   malware/spyware, 127
   mobile banking crimes and frauds, 126, 127
   novel attack methods, 126
   social media applications, 149
   UPI medium, 115

D

Dalvik Executable (DEX) format, 138
Dangerous permissions, 142
dApps, see Decentralized applications (dApps)
Data, 6, 8
Data acquisition, 348
   data analysis, 80
   defined, 75
   file system acquisition
      begin extraction, 79, 80
      review and save data, 80
      specify directories, 79
      start, 79
   iCloud, 81–84
   logical acquisition
      authenticate device, 76, 77
      connect device, 76
      review and save data, 77, 78
      selection, 76
   physical acquisition
      address encryption challenges, 78, 79
      authenticate device, 78
      bit-for-bit image, 78
      connect device, 78
      review and save data, 79
      selection, 78
   preparation, 75
   reporting, 80
Data analysis, 80
Data breach, 441
Data corruption/tampering, 193
Data exfiltration, 316
Data extraction/acquisition methods
   advantages and disadvantages, 8
   and analysis of mobile phones, 8
   chip-off, 11
   classification system, 8
   Hex Dump, 10, 11
   logical extraction, 10
   micro read, 12
   risk analysis in mobile forensic process, 9
   selective extraction, 9, 10
Data gaps, 184
Data generated by social media and messaging applications
   access and authentication data, 153
   additional data acquired, 154
   content interaction history, 152, 153
   geolocation data, 154
   user-generated content
      comments and responses, 150
      followers/following, 152
      interaction and engagement data— user activity details, 152
      media content, 150
      memes and gifs, 151
      messages and call, 150
      metadata, 151
      photographs, 150
      tagging and mentions, 152

text posts, 150
    user profile data—personal
        information, 151, 152
    videos, 151
  user's network and relationship
    details, 153
Data hiding analysis, 19
Data integrity, 195
  algorithms, 383
  backups and redundancy, 384
  chain of custody, 379
  contamination, 379
  digital evidence, 379, 383
  ethical and legal responsibilities, 380
  evidence, 379
  forensic investigations, 381
  forensic tools, 383
  Hash values, 383
  practice, 382
  protocols, 380
  tools and software, 383
  write-blocker, 382
Data in transit encryption, 120
Data minimization, 389
Data modification, 24
Data parsing, 98
Data recovery, 54
Data remanence, 246–248
Data security risks
  data corruption/tampering, 193
  unauthorized access, 193
Data smoothing, 274
Data splitting, 293
Data subject, 389
Data volatility, 182, 237
De-authentication attacks, 210
Decentralized applications (dApps),
    108, 111

attribution
  forensic tracing, 105
  link analysis, 105
  wallet address mapping, 105
Axie Infinity, 92
challenges and opportunities, 102
compound, 93
data acquisition
  blockchain data, 103
  blockchain explorers, 104
  blockchain transaction data, 103
  IPFS clients, 104
  Off-Chain Data, 103
  smart contract analysis tools, 104
  smart contract interactions, 103
  wallet data, 103
data analysis
  chainalysis, 104
  CipherTrace, 105
  elliptic, 104
  Maltego, 105
  privacy, 106
  smart contract interaction, 104
  Token Terminal/DeFiLlama, 105
  transaction flow analysis, 104
  transactions, 104
  wallet addresses, 104
  wallet address linking, 104
DeFi platforms, 102
distributed ledger technology, 102
features, 102
Gaming platforms, 102
investigations, 102, 103, 112
legal and jurisdictional issues
  cooperation with centralized
    exchanges, 107
  cross-border issues, 106
  lack of centralized authority, 106

457

INDEX

Decentralized applications (dApps) (*cont.*)
    requesting data from decentralized protocols, 107
    working with international partners, 106
  NFT marketplaces, 102
  nodes, 92
  OpenSea, 93
  PancakeSwap, 93
  privacy
    advanced forensic techniques, 106
    cooperation with exchanges, 106
    mixing services, 106
  and privacy-focused cryptocurrencies, 105
  social media, 102
  Uniswap, 93
Decentralized exchanges (DEXs), 93, 100
Decentralized finance (DeFi), 95, 110
Decision trees, 287, 288, 318
Deep learning, 366
Deep learning applications, 368
Deep learning models, 330, 364
Deep learning techniques
  background information, 345
  characteristic features, 342
  cloud-based analytics services, 359
  cloud logs, 344
  CNNs, 347
  conventional mobile forensic tools, 342
  digital evidence, 346
  factors, 341
  forensic investigations, 344
  forensic investigators, 343
  forensic techniques, 345
  intelligent forensic tools, 342
  law enforcement authorities, 343
  manual article screening, 343
  mobile apps, 341
  operating systems, 341
  pipelines, 343
  potency, 350
  preprocessing methods, 349
  smartphones and tablets, 341
  traditional, 345
  visualization, 345
Deep Q-Networks (DQN), 270
Defense lawyers, 392
DeFi, *see* Decentralized finance (DeFi)
Deleted data recovery
  challenges
    cleared history, 246
    incognito mode, 245, 246
  data remanence, 246–248
  digital forensics, 244
  file carving technique, 245
  file system behavior, 246–248
  journal analysis, 245
  method *vs*. effectiveness, 250
  parameter settings, 249, 250
  recovery methods comparison, 249
  recovery success rate *vs*. time since deletion, 247, 248
  SQLite recovery, 244
  WAL, 245
Denial-of-service (DoS) attacks, 271
Detection, 228
Device level security
  developer mode, 122
  FBE, 121
  FDE, 121
  identify jailbroken/rooted devices, 121
  in-app security features, 121
  minimize permissions, 121
  restrict app functionality on compromised devices, 122

DEXs, *see* Decentralized exchanges (DEXs)
Digital artifacts, 237, 309
Digital data, 438
Digital evidence
 investigations and quality, 7
 law enforcement, 18–22
Digital evidence acquiring methods, cloud storage forensics, 171
Digital forensic environments, 384
Digital forensic investigators, 195
Digital forensics, 3, 193, 233, 257
 ML (*see* Machine learning (ML))
 mobile forensics, 309
Digital investigation relevance, 233, 234
Digital ledger, 91
Digital sovereignty, 445
Dimensionality reduction techniques
 LDA, 280
 PCA, 280
 sample python code
  LDA, 281
  PCA, 281
  t-SNE, 281
 t-SNE, 281, 282
Direct memory dump, 129
Distributed ledger technologies (DLTs), 102, 110
DLTs, *see* Distributed ledger technologies (DLTs)
DPRK's DMM Bitcoin Exploit, 94
DoS attacks, *see* Denial-of-service (DoS) attacks
DQN, *see* Deep Q-Networks (DQN)
Drive-by attacks, 251
Dynamic analysis
 android emulator, 140
 Frida, 140
 mobile application security evaluations, 139
 network proxy tools, 140
 network traffic analysis, 141
 purpose, 139, 140
 real-time monitoring and examination, 139
 Xcode simulator, 140
Dynatrace, 140

E

Early stopping
 patience parameter, 298
 stopping criterion, 298
 validation monitoring, 298
 weights, 298
Ecosystem-aware forensics, 422
Elbow Method, 334
ElcomSoft Cloud Explorer, 72
Elcomsoft Phone Breaker, 81, 84
Elliptic, 106
Email forensics, mobile phones
 attachments, 331
 clustering algorithms
  hierarchical clustering, 334
  K-means clustering, 333
 communication records, 331
 cross-platform integration, 332
 digital investigation, 331
 email classification, forensic investigations
  detecting AI-generated text, 335
  email data, 336
  fraud transaction alerts, 336
  phishing emails detection, 335
  reconstruction, email threads, 335

INDEX

Email forensics, mobile phones (*cont.*)
   metadata, 331
   ML approaches
      logistic regression, 333
      Naive Bayes, 332
      NLP, 333
      SVMs, 332
   NLP techniques
      NER, 333
      sentiment analysis, 333
      tokenization, 333
   social engineering, 331
   supervised learning algorithms, 332
   unsupervised learning, techniques, 332
   usage, 331
Email Phishing, 127
Encrypted communication, 205
Encrypted data, 55
Encryption, 4, 236
Endpoint Detection and Response (EDR), 327
Endpoint vulnerabilities, 328
EPC, *see* Evolved Packet Core (EPC)
EPC elements, 204
Ethereum, 95, 107
Etherscan, 103, 108
Ethical forensic frameworks, 367
Ethical issues in geolocation data analysis
   bias and discrimination, 194
   intrusiveness of location tracking, 194
Evidence extraction and digital investigation
   archives, 18
   documentation and report, 17
   evidence intake, 12, 13
   identification
      brand, model, version, 13
      goals of the examination, 13
      legal authority/limitations, 13
      potential evidence, 14
      removable and external data storage, 13, 14
   isolation, 14, 15
   logical acquisition, 15, 16
   physical acquisition, 15, 16
   preparation, 14
   presentation, 17
   processing, 15, 16
   type of mobile phone, 12
   verification
      vs. available data in handset, 16
      different tools, 16
      hash values, 16, 17
Evolved Packet Core (EPC), 204, 216
EXT4 file system, 56, 246
External application data, 154

F

F2FS, *see* Flash-Friendly File System (F2FS)
Fabricated media/deepfakes, 174
Facial recognition, 325
Faraday bags/antistatic bags, 15, 155
Feature extraction
   dimensionality reduction techniques, 280–282
   multimedia features, 279, 280
   network traffic features, 278, 279
   text-based features, 277, 278
   timestamp extraction, 276, 277
Feature phones, 2, 6
Feature-rich devices, 2
Feature selection
   CFS, 283
   Lasso regression, 283

machine learning models, 282
mutual information, 283
output, sample code, 284
RFE, 282, 283
sample python code, 284
Federated forensic networks, 418
Federated learning (FL), 368, 415
FHSS, see Frequency-hopping spread spectrum (FHSS)
File-Based Encryption (FBE), 121, 236
File carving, 258
File carving technique, 245
File-level and full-disk encryption systems, 23
File synchronization, 167
File system acquisition, 48, 49, 70, 71
File system behavior, 246–248
File system evaluation, 172
File system performance, 247
File version verification, 172
Financial fraud, 234
Financial fraud detection
  machine learning
    autoencoders, 322
    clustering algorithms, 321
    isolation forest, 322
    one-class SVMs, 322
    self-training models, 322
    techniques, 321
  mobile contexts, 319, 322
  mobile domain
    credential stuffing, 320
    mobile payment fraud, 320
    SIM Swap Fraud, 320
    unauthorized transactions, 320
Financial technology applications, 116, 117
FIR, see First information report (FIR)
Firefox
  Android, 234
  Android storage paths, 240
  iOSBrowsers, 235
First information report (FIR), 171
Flash-Friendly File System (F2FS), 56
Flexibility, 423
"FORCEDENTRY" exploit, 225
Forensic acquisition
  iOS devices, 68
Forensic acquisition and analysis, 154
Forensic acquisition procedure, see Forensic imaging
Forensic analysis
  data, 157
  open-source and commercial software, 157
  open-source software, 159
Forensic artifacts/evidence
  internet browsing history
    cache files, 157
    cookies, 157
    deleted files, 158
    Facebook Messenger data, 158
    file metadata, 158
    social media applications, 158
    User Profile Information, 158
    web browser and application data, 158
    Windows Registry, 158
  prefetch documents, 159
Forensic challenges in analyzing geolocation data
  accuracy and precision, 188
  data volatility and loss, 188
  geolocation spoofing, 189
  multi-device synchronization, 188
  privacy concerns, 187
  from third-party services, 189

INDEX

Forensic cloning, 155
Forensic copies, 382
Forensic data layers, 410
Forensic errors, 443
Forensic examiners, 4, 8, 13, 18, 20, 25
Forensic imaging, 155
    definition, 155
    file formats, 155
    tools
        Ditto Dx, 156
        Logicube Forensics Falcon, 156
        Super Imager, 157
        Tableau TX1, 156
Forensic interpretation, 432
Forensic investigations, 390, 396, 413
Forensic investigators, 27, 202
Forensic link, 407
Forensic methods for collecting and analyzing geolocation data
    challenges, 184
    from cloud services, 182, 183
    cross-referencing, 183
    data integrity
        documentation, 182
        importance, 181
    data preservation
        cloud data extraction, 181
        cloud synchronization, 181
        data copying, 181
        file system extraction, 181
        live acquisition, 181
        logical extraction, 181
        physical extraction, 181
    from devices, 180
    digital cameras and social media, 180
    fitness trackers and smartwatches, 180
    geospatial mapping, 183
    timeline reconstruction, 183
Forensic request protocols, 413
Forensics analyst, 180
Forensics investigation challenges
    APIs and data access, 174
    data authenticity and integrity, 173
    data modification and deletion, 173
    disinformation and extensive quantity, 174
    distributed data, 172
    diverse platforms, 173
    encryption and privacy challenges, 173
    encryption is platform-specific, 174
Forensic social media analysis, 323
Forensic software, 395
Forensic strategy, 55
Forensics workflows, 422
Forensic tools, 171, 304, 427, 437
    accuracy and compliance, 397
    accuracy and reliability, 395
    compatibility, 393
    compatibility testing, 397
    credibility, 398
    data integrity, 394
    ISO/IEC 17025 standard, 398
    ISO/IEC 27001, 399
    laws, regulations, or guidelines, 395
    reliability, accuracy, and reputation, 394
    reproducibility, 396
    security features, 394
    selection, 393
    technology, 396
    use, 393
    validation, 397
Forensic tools, Android
    ADB, 51
    autopsy, 51
    Belkasoft, 51

cellebrite UFED, 51
commercial tool, 52
data extraction and analysis, 51
data recovery, 54
device preparation, 52
encrypted data, 55
hash values, 54
logical, physical, and file system acquisition, 52–54
Magnet AXIOM, 51
monitor process, 54
Oxygen Forensic Detective, 51
report generation, 54
review extracted data, 54
save and secure data, 54
Forensic workflow, 442
Fourth-generation (4G) mobile networks, 203
Fraud detection, 316
Fraud transaction alerts, 336
Frequency-hopping spread spectrum (FHSS), 211
Frida, 58
Frida toolkit, 140
FTK Imager, 230
Full-disk encryption (FDE), 55, 68, 72, 78, 121
Full file system mobile extraction, 131
Funtouch OS, 5
Fused Location Provider API, 179
Future trends and emerging technologies
  adaptive management strategies, 402
  AR and VR platforms, 404
  AR/VR forensics, 408
  conventional forensic tools, 402
  ecosystems, 402
  environments, 409
  financial transactions, 401

forensic experts, 406
forensic visibility, 407
5G/6G networks, 404, 406–408
6G scenarios, 406
5G to 6G, 405
legacy 4G networks, 406
MEC, 405
mobile applications, 403
mobile environments, 404
network slice, 406
privacy-focused architectures, 403
privacy-preserving collaboration, 404
speed, 401
stakeholders, 412
tenets, 417
tools, 403
user's operations, 403
virtual networks, 407
virtual object, 409
witnesses, 402

G

GainBitcoin, 94
3G and 4G Networks
  CN, 204
  components, 204
  forensic challenges
    data retention law, 205
    encrypted communication, 205
    handovers and mobility management, 205
  RAN, 204
Garbage collector (GC), 35
GATT protocol, 212
GC, *see* Garbage collector (GC)
GDPR, *see* General Data Protection Regulation (GDPR)

INDEX

General Data Protection Regulation (GDPR), 200, 202, 388, 439, 440
Geofencing, 122
Geographic Information Systems (GIS), 183
Geolocation artifacts, 177
    cell tower data, 179
    geotagged media, 179
    GPS logs, 178
    location history, 179
    Wi-Fi connection logs, 179
Geolocation detection methods
    anomalies detection, locational patterns, 123
    GPS-based geolocation monitoring, 122
    immediate location verification, 123
    VPN/Proxy/TOR identification, 123
Geolocation spoofing, 189
Geospatial analysis, 190
Geotagging, 178
getSystemService () method, 39
5G infrastructures, 417
GIS, *see* Geographic Information Systems (GIS)
GIS tools, 183
Global positioning system (GPS), 177
GNNs, *see* Graph Neural Networks (GNNs)
Google Chrome
    Android Browsers, 234
    Android storage paths, 239
    iOSBrowsers, 235
    iOS storage paths, 240
Google Drive, 2
Google Earth, 190
Google Location History, 182
Google Maps, 182
Google Pay, 337
Google's Play Protect, 338
GPS, *see* Global positioning system (GPS)
GPS-based geolocation monitoring, 122
GPS enabled devices, 177
GPS logs, 178
Graph-based learning techniques, 325
Graph Neural Networks (GNNs), 363
Grid search, 296

H

HAL, *see* Hardware Abstraction Layer (HAL)
Hardware Abstraction Layer (HAL)
    abstraction layer, 31
    abstracts device, 32
    basic understanding, 32, 33
    device portability and flexibility, 34
    interfacing with device drivers, 33
    modular design, 32
    standardized APIs, 33
Hash values, 16, 17, 54
Health Insurance Portability and Accountability Act (HIPAA), 389, 390, 440
    and CCPA, 390
Hex Dump, 10, 11
HFS+, *see* Hierarchical File System Plus (HFS+)
Hierarchical File System Plus (HFS+), 70, 88
Hierarchical clustering, 334
High-profile attacks
    Pegasus Spyware (2020), 227
    Simjacker Attack (2019), 227
    T-Mobile Data Breach (2021), 226
    WhatsApp Spyware Incident (2020), 227

HIPAA, *see* Health Insurance Portability and Accountability Act (HIPAA)
HLRs, *see* Home Location Registers (HLRs)
Home Location Registers (HLRs), 203
Hooks, 417
Hybrid models, 314
Hyperparameter tuning techniques
    Bayesian optimization, 297
    grid search, 296, 297
    random search, 297

I

iCloud, 2, 71, 72, 79, 82–84, 90
    drive files, 82
iCloud data acquisition
    access iCloud account, 81–84
    Apple ID, 81
    authentication tokens, 81
    cloud-based service, 81
    specific credentials/particular access, 81
    syncing device data, 81
    2FA, 71
iCloud/iOS backup analysis, 84, 85
Illicit downloads, 252
IM, *see* Instant Messaging (IM)
IMSI catchers and stingrays, 214, 224
IMSI tracing, 221
In-app security features, 121
Incident point triangulation, 411
Incognito and private modes, 237
Incognito mode, 245
Indian Evidence Act, 440
Indicators of compromise (IoCs), 252
Infra-Red, 14
Infrastructure-level traffic capture

access points and routers, 218
cellular infrastructure monitoring, 218
monitoring mobile infrastructure's ability, 218
network forensic appliances, 218
Infrastructure mode, 207
Insider threats, 317
Instant Messaging (IM), 346
Instant notifications, 119
Intents, 39
Inter-app communication, 63
Internal Monitoring Systems, 442
Internet connectivity, 116
Internet of Things (IoT), 211
Internet Protocol Security (IPsec), 205
Internet service provider (ISP), 187
InterPlanetary File System (IPFS), 103
Inter-process communication (IPC), 39
Interquartile Range (IQR), 273
IoCs, *see* Indicators of compromise (IoCs)
iOS, 5, 6
iOS 8, 72
iOSBrowsers
    Chrome, 235
    Firefox, 235
    Safari, 234
iOS devices
    acquisition (*see* Acquisition of iOS devices)
    architecture
        app sandboxing, 63, 64
        closed-source system, 62
        cocoa touch, 62
        core OS, 62
        core services, 62
        encryption, 63
        media layer, 62
        runtime, 63, 64

INDEX

iOS devices (*cont.*)
    Secure Enclave, 63
        system security and permissions, 64
    Cellebrite UFED, 74
    data recovery and decryption
        jailbreaking, 85–88
        passcode bypass, 85–87
    filesystem and app analysis, 88–90
    and iCloud backup analysis, 84, 85
    locations
        /private/var/log, 67
        /private/var/mobile/Containers/
          Data/Application, 67
        /var/mobile/Library, 67
        /var/mobile/Media, 67
    Magnet AXIOM, 74, 75
    Oxygen Forensic Detective, 75
    system data
        cache files, 66
        log files, 66
        settings and configurations, 66
    user data
        forensic analysis, 65
        location data, 66
        messages, 65
        notes and voice memos, 65
iOS file systems
    and analyzing app data, 89
    APFS, 88, 89
    DB Browser for SQLite, 90
    HFS+, 88, 89
    iExplorer, 89
    iMazing, 90
iOS forensics
    architecture, 62–64
    definition, 61
iOS storage paths
    Android storage paths, 244

    Chrome, 240
    Safari, 240
IoT, *see* Internet of Things (IoT)
IP address
    assignment logs, 220
    definition, 220
    forensic use, 220
    NAT and CGNAT correlation, 220
IP addresses and devices
        identification, 119
IP address geolocation, 187
iPadOS, 61
iPads, 61
IPA files, 89
IPFS, *see* InterPlanetary File System (IPFS)
iPhone 15, 75
IPhone malware, 145–147
iPhone OS, 5
iPhones, 2, 61
IPsec, *see* Internet Protocol Security (IPsec)
ISP, *see* Internet service provider (ISP)
ISO/IEC 17025 standard, 398
Isolation forest, 313, 322
iTunes, 84
iTunes backup, 242

J

Jailbreaking, 121, 242
    assess iOS Version and device
        compatibility, 87
    backup of original device, 86
    data extraction, 88
    device preparation, 86
    monitor progress, 87
    retrieving data, 85
    root access to iOS device, 85
    selected methodology, 87

tool selection, 86
validate access, 87
Jammers, 15
Java-based platforms, 4
Java OS, 6
JD-GUI, 58
JioOS, 5
JIT, *see* Just-In-Time (JIT)
Joint Test Action Group (JTAG), 11, 47
Journal analysis, 245
Jurisdictional incompatibilities, 420
Just-In-Time (JIT), 34
JWT-JSON web tokens, 118

K

Key analysis techniques for geolocation artifacts
    correlation with external data sources, 190
    geospatial analysis, 190
    location validation, 191
    pattern and behavior analysis, 190
    reverse geocoding, 191
    timeline analysis, 189
Keyloggers, 127
Keyword and topic detection, 324
K-means clustering, 290, 333
k-Nearest Neighbors (k-NN), 290
k-NN, *see* k-Nearest Neighbors (k-NN)
Know Your Customer (KYC), 109
KYC, *see* Know Your Customer (KYC)
KYC/AML data, 104

L

Lack of resources, 24
Lasso regression, 283

Late fusion techniques, 364
Law enforcement
    DOJ manual, 18
    evidence guidelines
        admissible, 19, 20
        authentic, 20
        believable, 20
        complete, 20
        in digital forensics, 19
        reliable, 20
    general guidelines
        application and file analysis, 19
        data hiding analysis, 19
        investigators, 18
        ownership and possession, 19
        time frame analysis, 19
    good forensic practices
        documenting the evidence and changes, 21
        preserving the evidence, 21
        reporting, 22
        secure the evidence, 20, 21
    policies and procedure standards, 18
    SWGDE, 18
Law enforcement agencies, 413
Law enforcement authorities, 415
LDA, *see* Linear discriminant analysis (LDA)
Linear discriminant analysis (LDA), 280
Linear regression, 318
Linux-based operating system, 4
Linux kernel, 28
    application framework, 37–42
    ART (*see* Android Runtime (ART))
    HAL (*see* Hardware Abstraction Layer (HAL))
    hardware drivers, 31
    low-level system operations, 30

INDEX

Linux kernel (*cont.*)
   memory management, 30
   process scheduling, 30, 31
   resource control and stability, 31
   security, 30
Live data, 161
Location-based data, 177
   cell tower triangulation, 178
   definition, 178
   geotagging, 178
   GPS data, 178
   mobile device applications, 179
   social media platforms, 187
   Wi-Fi and Bluetooth, 178
Location-based detection, 122
Location-based information, 154
Location data storing path, 184
Location events, 133
Location services, 38
Log files, 219
Logical acquisition, 15, 16, 27, 48, 51, 64, 131
   iOS devices, 69, 70
Logical extraction, 10
Logistic regression, 318, 333
Long Short-Term Memory (LSTM), 287, 314
LSTM, *see* Long Short-Term Memory (LSTM)
Lookout Mobile Security, 337

M

Machine learning (ML)
   AI, 264
   anomaly detection (*see* Anomaly detection, mobile phones)
   applications, mobile digital investigations, 302
   challenges, 304–306
   credit card fraud detection, 337
   digital artifacts types, 309
   digital forensics, 264
   evaluation, 302
   financial fraud detection (*see* Financial fraud detection)
   forensic tools and processes, 303
   Google's Play Protect, 338
   identify anomalies/malicious behaviors, mobile network, 264
   limitations, 304–306
   Lookout Mobile Security, 337
   methods, 264
   mobile banking fraud detection, 337
   mobile digital investigations, 265, 285 (*see also* Mobile digital investigations)
   mobile forensics, 309
   mobile payment fraud, 337
   predictive analysis (*see* Predictive analysis)
   reinforcement learning, 264
   Samsung's Knox, 338
   semi-supervised learning, 264
   SIM swap fraud detection, 337
   social media-based data analysis, 327
   supervised learning, 264
   techniques, 327
   unsupervised learning, 264
   validation, 302
   workflow integration, 303, 304
   Zimperium mobile threat defense, 338
Machine learning-driven threat intelligence, 331
Magnet AXIOM, 51, 74–78, 99, 112, 242, 245, 258
Magnet cyber tools, 230, 231

Malicious Apps, 226
Malicious programs, 25
Malicious redirects, 251
Maltego, 105
Malware categories
    account hijacking, 128
    counterfeit mobile banking applications, 128
    counterfeit payment links, 128
    credential stuffing, 128
    Keyloggers, 127
    loan fraud and identity theft, 128
    repercussions, 127
    social engineering, 128
    social media and messaging applications, 128
    spyware, 127
    Trojan Horse applications, 127
Malware detection, 316
Man-in-the-middle (MITM) attacks, 198, 218, 223, 224
Markov Decision Process (MDP), 270
Mass-scale disinformation campaigns, 174
MDP, *see* Markov Decision Process (MDP)
MDM, *see* Mobile Device Management (MDM)
MEC, *see* Multi-access Edge Computing (MEC)
Media type, 350
Memory carving tools, 163
Memory management, 30
    ART, 35
Metadata analysis, 172
MFA, *see* Multi-Factor Authentication (MFA)
Micro read, 12
Milestone roadmap projects, 419
MITM attacks, *see* Man-in-the-middle (MITM) attacks

ML, *see* Machine learning (ML)
MLM structure, *see* Multilevel marketing (MLM) structure
Mobile, 251
Mobile Application security Guidelines by Cert-In
    ensure secure application deployment and operations, 126
    guidelines for audit of applications, 125, 126
    secure development practices, 124, 125
    security in designing of application engagement of security, 124
        secure SDLC adoption, 124
        security by design approach, 124
Mobile applications storing GPS data
    database, 179
    file system, 179
    internal storage, 179
Mobile apps data, 272
Mobile banking
    application-based technology, 115
Mobile banking applications, 441
Mobile banking crimes and frauds
    email phishing, 127
    phishing, 126
    SMS phishing (Smishing), 127
    voice phishing, 127
Mobile banking data/malicious applications
    harmful files detection, 135, 136
    infection vector/pathway, 135
    malicious files analysis
        static analysis, 137, 139, 140
    questionaries and evidence extraction strategy, 134
    safeguarding harmful files, 136

INDEX

Mobile banking fraud detection, 337
Mobile-based applications, 116
Mobile-based banking applications, 147
Mobile-based threats, 320
Mobile Bluetooth forensic artifacts identification
   Bluetooth MAC addresses, 213
   connection history, 213
   GATT Data and BLE characteristics, 213
   MAC address spoofing, 213
   paired device logs, 212
   proximity-based evidence, 213
   short-range wireless architecture, 213
Mobile Bluetooth networks
   architecture, 212
   BLE, 211, 212
   Bluetooth versions and range, 211
   GATT protocol, 212
   low power consumption, 212
   pairing and bonding, 212
   short range wireless architecture, 211
Mobile browser forensics, 258
   acquisition strategies, 237
   Android Browsers
      Firefox for Android, 234
      Google Chrome, 234
      Opera Mini and UC Browser, 234
      Samsung Internet Browser, 234
   artifacts analysis (*see* Browser artifacts analysis)
   civil litigation, 235
   corporate investigations, 235
   criminal investigations, 235
   cross-browser forensic analysis
      artifact volume comparison across browsers, 256
      challenges, dataset merging, 256, 257
      cross-browser behavioral profiling, 255
      forensic investigation, mobile browsers, 254
      multi-browser correlation, 254
      structural and encryption differences, 254, 255
      user session overlap across browsers, 257
   deleted data recovery (*see* Deleted data recovery)
   forensic acquisition challenges
      DataVolatility, 237
      encryption, 236
      incognito and private modes, 237
      permissions and app restrictions, 236
      sandboxing, 236
   iOSBrowsers
      Chrome, 235
      Firefox, 235
      Safari, 234
   overview, 233, 234
   vital field, study under digital forensics, 233
   web-based threats
      artifact-based threat detection, 251, 252
      IoCs, 252
      mobile browser threat vectors, 251
      timeline correlation, malicious behavior, 252–254
Mobile data, 436
Mobile data acquisition, 96, 97
Mobile data anomaly detection, 329
Mobile data-collecting procedure, 129

INDEX

Mobile device, 314, 444
Mobile device data, 378
Mobile device logs, 220
Mobile Device Management
    (MDM), 69, 327
Mobile device network traffic
    infrastructure-level traffic, 218
    packet sniffing (*see* Packet sniffing)
    technical methods and tools, 213
    Test Access Point, 216, 217
    tools and software
        MITM Proxy/Burp suite, 218
        Tcpdump, 217
        tPacketCapture, 217
        Wireshark, 217
Mobile devices, 323, 331, 344, 346, 403,
    407, 432, 436, 445
    complexity, 3
    encryption and security, 386
    heterogeneity, 385
    human error, 385
    jurisdictions, 386
    layout and dynamic nature, 384
    tools and techniques, 385
    volatile data, 384, 385
    volatility, 386
Mobile device's call log, 272
Mobile digital investigations, 263
    CNNs, 286
    data cleaning
        handling missing values, 275, 276
        noise handling techniques, 273
        preparing raw data, 273
    data collection
        app usage data, 273
        browser history and cookies, 272
        call log, 272
        location data, 272

        mobile apps data, 272
        multimedia files, 272
        network traffic, 273
        SMS/MMS messages, 272
    data preparation, 285
    decision tree, 287, 288
    feature engineering, 285
    feature extraction
        dimensionality reduction
            techniques, 280–282
        multimedia features, 279, 280
        network traffic features, 278, 279
        text-based features, 277, 278
        timestamp extraction, 276, 277
    feature selection
        CFS, 283
        Lasso regression, 283
        machine learning models, 282, 285
        mutual information, 283
        output, sample code, 284
        RFE, 282, 283
        sample python code, 284
    fine-tuning and optimization
        early stopping, 298, 299
        hyperparameter tuning techniques,
            296, 297
        regularization techniques, 299–301
    k-NN, 290
    machine learning, 265, 285
    ML (*see* Machine learning (ML))
    ML applications, 302
    model selection criteria
        labeled data size, 293
        model complexity *vs.*
            interpretability, 292
        multimedia data, 291
        scalability and speed, 292
        task type, 292

INDEX

Mobile digital investigations (*cont.*)
    text data, 291
    time-series data, 291
  Naive Bayes, 290
  NNs, 286
  overfitting, 295
  Python-based code, RNN-LSTM, 287
  Random forest, 288, 289
  reinforcement learning (RL), 270, 271
  RNNs, 287
  semi-supervised learning, 268–270
  supervised learning, 265, 266
  SVM, 289
  text classification, Naive Bayes, 290
  training machine learning models
    data splitting, 293
    Stratified K-fold validation technique, 294
    training methodologies, 294, 295
  underfitting, 295, 296
  unsupervised learning, 267
  usage, CNN model, 286
Mobile ecosystems, 368, 417
Mobile email forensic research, 331
Mobile forensic
  accountability, 439
  clarity, 433
  codes, 428
  data minimization, 433
  digital evidence, 437, 438
  element, 430
  ethical and legal aspects, 430
  ethical component, 435
  ethical considerations, 428, 432
  ethics, 428
  evidence handling, 438
  healthcare, 434
  human rights treaties, 430
  investigation, 435
  investigators, 432, 436
  judicial framework, 429
  jurisdictional issues, 429
  legal compliance, 439
  legal custody chain, 429
  legal mistakes, 435
  markets, 429
  mobile, 431
  and mobile technology, 433
  obligation and legislative, 431
  operating procedures, 430
  personal privacy, 433
  practices, 439
  principles, 432
  privacy-protective, 434
  scientists, 432
  storage-related procedures, 437
  surveillance risk, 434
  therapists, 433
Mobile forensic analysis, 367
Mobile forensic investigations
  data access and processing standards, 393
  legal principles, 392
  procedural errors, 392
Mobile forensics, 129, 427
  blockchain (*see* Blockchain)
  challenges
    anti-forensic techniques, 24
    communication shielding, 25
    data modification, 24
    device alteration, 24
    dynamic nature of evidence, 24
    hardware differences, 22
    in-built security features, 23, 24
    lack of necessary tools, 25
    lack of resources, 24

legal issues, 25
malicious programs, 25
operating systems, 22, 23
password recovery, 24
crime scene investigators, 7
data, 6, 8
data extraction/acquisition methods, 8–12
desktop computers, 7
digital evidence, 18–22
digital forensics, 309
encryption, 4
evidence extraction and digital investigation, 12–18
evolution, 1–3
extracting information, 3
feature phones, 6
lack of necessary tools, 25
lack of resources, 24
ML (*see* Machine learning (ML))
operating systems, 4–6
recovering data process, mobile devices, 309
risk analysis, 9
technologies, 8
time, 7
tools, 7
traditionally investigators, 7
Mobile forensics acquisition methods, 147
examining collected data, 130
full file system mobile extraction, 131
logical acquisition, 131
physical acquisition, 129, 130
Mobile forensics domain, 359
Mobile forensics timeline analysis
application permission, 132
application scenarios, 132
chronological sequence of events, 131
data sources, 132
location events, 133
system events
chronological arrangement, 133
data aggregation, 133
timestamp organization, 133
trends and insights
connecting events, 134
cross-referencing data, 133
finding anomaly, 133
identifying deletion, 134
visual/chronological depiction of events, 133
Mobile hardware, 369
Mobile malware, 328
Mobile malware detection
accuracy *vs.* number of training epochs, 352
behavioral analysis, 352
concerns, 352
learning curve, 352
NLP models, 353
static analysis, 352
URL classifier, 354
Mobile network forensic investigation
phase 1—detection, 228
phase 2—containment, 228
phase 3—eradication and mitigation, 228
phase 3—recovery, 229
tools
axiom, 230
cellebrite UFED, 229
FTK Imager, 230
magnet cyber tools, 230, 231
MSAB XRY Cloud, 230
NetWitness, 230
oxygen forensics suite, 229

Mobile network forensic
investigation (*cont.*)
　Paraben's Device Seizure, 229
　Sleuth Kit and Autopsy, 230
　Wireshark, 230
　XRY, 229
Mobile network forensics, 201
Mobile network forensics artifacts
　evidence preparation, 222
　IP address, 220
　log files
　　mobile device logs, 220
　　router and network logs, 219
　　WAPs, 219
　SIM and IMSI tracing, 221, 222
Mobile network forensics challenges
　data volume and complexity, 201
　encryption and data
　　protection, 201
　legal and ethical limitations, 201
　mobility and dynamic network
　　environment, 201
Mobile networks, 197
　attacks and threats (*see* Common
　　mobile network threats)
　cellular networks (*see* Cellular
　　networks architecture)
　complexity and security, 198
　devices and services, 198
　3G and 4G, 203–206
　host of services, 197
　parts, 198
　significance, 199
　　discovering cybercrime, 200
　　incident response and
　　　investigation, 200
　　legal and regulatory
　　　compliance, 200
　　location-based forensic
　　　evidence, 199
　　mobile devices as evidence
　　　hubs, 199
　　ubiquity of mobile devices, 199
Mobile network traffic classification
　CNNs and RNNs, 356
Mobile operating systems, 2
　Android, 4, 5
　in data collection, examination, and
　　analysis, 4
　forensic examiners, 4
　iOS, 5, 6
　Java-based platforms, 4
　Nokia's Series 30+ (S30+), 4
　Symbian, 4
Mobile payment fraud, 320, 337
Mobile payment services, 320
Mobile phone-based social media
　analytics
　criminal intent detection, 326
　deception detection patterns, 327
　psychological and behavioral
　　insights, 327
　relationships and networks, 326
　tracking movement and locations, 326
Mobile phones, 233
　anomaly detection (*see* Anomaly
　　detection, mobile phones)
　email forensics (*see* Email forensics,
　　mobile phones)
　social media forensics
　　communication evidence, 323
　　content analysis, 323
　　geolocation information, 323
　　machine learning
　　　methods, 324–326
　　timeline reconstruction, 323

threat intelligence (*see* Threat
intelligence, mobile phones)
Mobile platforms, 263
Mobile switching center (MSC), 203
Mobile technologies, 427
Mobile web browsers, 237
Mobile Wi-Fi networks
AP
infrastructure mode *vs.* Ad hoc
mode, 207
roaming and handoffs, 207
channels, 208
channel hopping and forensic
implications, 209
selection and interference, 208
coverage, 206
data collection, 206
forensic investigations challenges, 211
key components, 207
modern wireless mobile
communications, 206
SSID
broadcast, 208
conflicts and rogue APs, 208
definition, 207
Mobile Wi-Fi traffic and forensic
collection
de-authentication attacks, 210
passive sniffing, 210
PCAP
captured data, 210
monitor mode, 209
tools, 209
rogue AP and evil twin, 210
Model training parameters, 358
Modern mobile communication, 206
Modern mobile devices, 360
Modern mobile phones, 386

Monero, 105
MSAB XRY, 69
MSAB XRY Cloud, 230
MSC, *see* Mobile switching
center (MSC)
Multi-access Edge Computing (MEC), 405
Multi-browser correlation, 254
Multi-Device Synchronization, 188
Multi-Factor Authentication (MFA), 118
Multilevel marketing (MLM) structure, 94
Multimedia Content, 346
Multimedia data, 291
Multimedia features, 279
Multimodal analysis, 363
deep learning models, 361
early fusion, 362
late fusion processes, 362
in mobile forensics, 363
Multimodal fusion models, 361
Multimodal models, 360
Mutual information, 283
MyCrypto, 108

N

Naive Bayes, 290, 332
Named entity recognition (NER), 333
NAND Flash memory, 11
National Institute of Standards and
Technology (NIST), 398
Natural language processing (NLP), 291,
324, 333, 354
Navigation apps, 180
NER, *see* Named entity recognition (NER)
NFTs, *see* Non-fungible tokens (NFTs)
NetWitness, 230
Network connections—IP resolution, 222
Network devices, 222

INDEX

Network interface card (NIC), 214
Network intrusion, 226
Network-level threats, 329
Network tap devices
    EPC, 216
    monitoring and troubleshooting
        communications, 216
    PGW, 217
    S1 and S5 interfaces, 217
    SGW, 216
    tapping traffic, 216
Network traffic, 205
Network traffic features, 278
Neural networks (NNs), 286, 312, 319
Next-generation mobile networks, 407
Next-generation models, 369
Next generation networks, 408
NIC, *see* Network interface card (NIC)
NIST, *see* National Institute of Standards
    and Technology (NIST)
NLP, *see* Natural language
    processing (NLP)
NNs, *see* Neural Networks (NNs)
Noise handling techniques
    data smoothing, 274
    outlier detection and removal, 273
        data visualization, 275
        sample code, 274
Nokia's Series 30+ (S30+), 4, 6
Non-fungible tokens (NFTs), 95
Notification services, 38
Nutshell, 61

O

onRestoreInstanceState(), 40
onSaveInstanceState(), 40
OOM, *see* Out-Of-Memory (OOM) Killer

OpenSea, 93
Open-source intelligence (OSINT), 109
Opera Mini, 234
Operating systems, 22, 23
Optimization techniques, 334
Origin OS, 5
OSINT, *see* Open-source
    intelligence (OSINT)
Outliers, 310
Out-Of-Memory (OOM) Killer, 30
Overfitting, 295
OWASP ZAP, 140
Oxygen Forensic Detective, 51, 75, 242,
    245, 258
Oxygen Forensics, 69, 71, 78
Oxygen Forensics Cloud Extractor, 72
Oxygen forensic tool, 169
OxygenOS, 5

P

Packet capture (PCAP), 209
Packet Data Network Gateway (PGW), 217
Packet data network (PDN), 216
Packet sniffing
    cellular networks (3G/4G)
        encryption protocols, 214
        IMSI catchers and stingrays, 214
        SIM cards and tracking, 215
    definition, 214
    direct capture on mobile devices
        rooting/jailbreaking, 215
        TLS/SSL interception, 216
        VPN-based capture, 215
    on Wi-Fi networks
        monitor mode, 214
        promiscuous mode, 214
Packet-Switched (PS), 204

PancakeSwap, 93
Paraben's Device Seizure, 229
Passcode bypass
    assess iOS version and device compatibility, 87
    backup of original device, 86
    data extraction, 88
    device preparation, 86
    monitor progress, 87
    security features, 86
    selected methodology, 87
    tool selection, 86
    validate access, 87
Passive sniffing, 210
Password recovery, 24
Pattern recognition, 329
PCAP, *see* Packet capture (PCAP)
PDN, *see* Packet data network (PDN)
Peer-to-peer networks, 102
Pegasus Spyware (2020), 227
Periodic legal audits, 442
Permissions-based mechanism, 41
Persistent artifacts, 241
Personally identifiable information (PII), 128
PGW, *see* Packet Data Network Gateway (PGW)
Phishing, 126, 335
Phishing campaigns, 328
Physical acquisition, 15, 16, 27, 47, 48, 51, 64
    chip-off acquisition, 130
    data retrieving, 129
    definition, 129
    forensic tools, 130
    iOS devices, 68, 69
    principal strategies, 129
Physical extraction, 10

Physical iOS acquisition, 68
Piconet, 212
PII, *see* Personally identifiable information (PII)
Point anomalies, 310
Power management, 38
Predictive analysis, 315
    data exfiltration, 316
    insider threats, 317
    machine learning
        clustering algorithms, 319
        decision trees, 318
        linear regression, 318
        logistic regression, 318
        neural networks, 319
        Random forests, 318
        SVMs, 318
    malware detection, 316
    mobile forensic investigations, 319
    mobile forensics, 315
        fraud detection, 316
        prediction user behavior, 315
        risky activity detection, 315
    modeling futures concept, 315
    unauthorized access, 317
Predictive mobile threat intelligence, 330
Pre-trained model, 349
Principal component analysis (PCA), 280
Privacy and security
    cloud services, 375
    communication and transaction, 373
    data, 374, 375
    forensic investigators, 376
    information, 375
    integrity, 374, 376
    law enforcement, 374
    legal compliance, 374

INDEX

Privacy and security (*cont.*)
    privacy laws, 376
    social media and instant messaging services, 375
Privacy and security implications of geolocation artifacts
    privacy concerns, 192
        legal boundaries and consent, 193
        media geotagging, 192
        private location access, 192
        unwarranted surveillance, 192
Privacy laws, 388, 391
    GDPR, 388
Privacy legislation, 391
Privacy risk comparison, 416
Proportionality, 389
Proprietary Systems, 6
Protected application programming interface
    API authentication and authorization
        authorization code flow, 120
        implicit flow, 121
    HTTPS
        code obfuscation, 120
        data in transit encryption, 120
        SSL pinning, 120
Protocol buffers (protobuf), 45
PS, *see* Packet-Switched (PS)

Q

Q-learning, 270
QR Code Fraud, 128

R

Radar chart, 370
Radio Access Network (RAN), 198, 204

RAM dumps
    common artifacts
        chat records and correspondence, 162
        data from social media applications, 162
        files pertaining to social media, 163
        web browser artifacts, 162
        web browser sessions, 162
    data acquiring, 160
    data acquiring tools, 161
    forensic data, 160
    importance in social media forensics, 161
    standard analysis methods
        memory carving, 163
        network connection, 163
        process analysis, 163
        reconstructing sessions, 164
        string search, 163
    standard analysis tools
        Rekall, 164
        volatility framework, 164
RAN, *see* Radio Access Network (RAN)
Random forest, 288, 289, 318
Random search, 297
Raw data, 349
Real-time forensics, 424
Real-time location analysis, 122
Real-time models, 418
Real-time monitoring, 112, 113
Recovery, 229
Recurrent Neural Networks (RNNs), 287, 347, 365
    audio transcripts, 347
    and LSTMs, 351

patterns in app usage logs, 347
sequential data, 347
Recursive Feature Elimination (RFE), 282, 283
Regularization techniques, 295
    augmentation, 301
    batch normalization, 301
    data augmentation (in deep learning), 300
    dropout, 300, 301
    Elastic Net, 299, 300
    L1 regularization (Lasso), 299
    L2 regularization (Ridge), 299
    machine learning domain, 299
Reinforcement learning (RL)
    basic concept, 270
    mobile digital investigations, 270, 271
Rekall, 164
REQUEST_INSTALL_PACKAGES android permission, 143
REQUEST_INSTALL_PACKAGES permission, 143
Reverse geocoding, 191
RFE, *see* Recursive Feature Elimination (RFE)
Risky activity detection, 315
RL, *see* Reinforcement learning (RL)
RNNs, *see* Recurrent Neural Networks (RNNs)
ROC curve, 354
Role of consent
    criminal investigation, 388
    ethical concerns, 387
    informed consent, 387
    investigation, 387
    investigator, 387

S

Safari
    iOSBrowsers, 234
    iOS storage paths, 240
Samsung Internet Browser, 234
Samsung's Knox, 338
Sandboxing, 236
Scalpel, 245
The Scientific Working Group on Digital Evidence (SWGDE), 18
SD cards, 47
Seasonal Hybrid Extreme Studentized Deviate (S-H-ESD), 314
Secure Enclave, 63, 68
Secure Software Development Life Cycle (SDLC), 124
Security features, 3
Security features by banking applications
    cryptography, 117
    device level security, 121, 122
    geolocation and IP-based identification, 122, 123
    protected application programming interface, 120, 121
    session management and timeout, 117–119
Selective extraction, 9, 10
Self-training models, 322
SELinux, 30
Semi-supervised learning, 264, 293
    anomaly detection
        hybrid models, 314
        LSTMs, 314
        mobile forensic investigations, 314
        S-H-ESD, 314
    basic concept, 269
    mobile digital investigations, 268–270

INDEX

Sentiment analysis, 324, 333
Service locator pattern, 39
Serving Gateway (SGW), 216
Session management
    mobile applications, 118
    optimal strategies
        brief session duration, 118
        MFA, 118
        session expiration, 119
        session revocation, 118
        session storage, 118
        session tokens, 118
Session timeouts, 119
S-H-ESD, *see* Seasonal Hybrid Extreme Studentized Deviate (S-H-ESD)
Sideloading, 63
Silhouette Analysis, 334
SIM and IMSI fundamentals, 221
SIM card, *see* Subscriber Identity Module (SIM) card
SIM card analysis through forensic device exams, 221
SIM cloning and fraud, 221
Simjacker Attack (2019), 227
SIM jacking, 224
SIM Swap Fraud, 320
SIM swap fraud detection, 337
Smart contract, 112
Smart devices, 177
Smartphone revolution, 115
Smartphones, 2, 433
SMS Phishing (Smishing), 127
Social engineering, 128, 225, 331
Social media, 327
Social media analysis, 149
Social media and messaging applications, 150
Social media applications, 154, 323
Social media artifacts extraction from social media
    application utilization data, 165
    browsing history/cookies, 165
    connections and social networks, 165
    credentials and account details, 165
    deleted content, 165
    file system acquisition method, 166
    forensics tool, 166
    messages and posts, 164
    metadata, 164
    multimedia files, 165
    notifications/alerts, 165
    web browser information, 165
Social media artifacts from cloud, 167
Social media data, 327
Social media digital artifacts, 158
Social media forensics, mobile phones
    communication evidence, 323
    content analysis, 323
    geolocation information, 323
    machine learning methods
        anomaly detection, 325
        community detection, 325
        content classification, 325
        emotion and behavioral profiling, 326
        facial recognition, 325
        graph-based learning techniques, 325
        keyword and topic detection, 324
        multimedia sharing/posting, 324
        network analysis, 325
        NLP, text analysis, 324
        sentiment analysis, 324
    social media analytics
        insights, 326, 327

value, 326, 327
   timeline reconstruction, 323
Social media platforms, 95
Social networking platforms, 157
Social network scraping, 154
Source code review, 125
Spyware, 127
SQLite, 45, 244
SQLite Database, 40
SQLite recovery, 244, 258
SSL pinning, 117
Standardization, 420
Standardized APIs, 33, 34, 42
Standardized practices, 420
Stratified k-fold cross-validation, 294, 295
Stratified sampling, 294
String search, 163
Subscriber Identity Module (SIM) card, 215
Supervised learning, 264
   advantage, 266
   algorithms, 329, 332
   anomaly detection
      classification algorithms, 312
      neural networks, 312
   basic concept, 266
   mobile digital investigations, 265, 266
Supply Chain Attacks, 225
Support vector machine (SVM), 289, 318, 332, 336
SVM, see Support vector machine (SVM)
SWGDE, see The Scientific Working Group on Digital Evidence (SWGDE)
Symbian, 4, 6
Syscalls/driver interfaces, 33

T

TCC framework, see Transparency, Consent, and Control (TCC) framework
tcpdump, 141, 217
t-Distributed Stochastic Neighbor Embedding (t-SNE), 281
Telecom operators, 413
Telecoms, 337
Telephony services, 38
Text-based features, 277, 278
Text data, 291
Text Messages, 346
Text posts, 150
TF-IDF, 334
The Information Technology Act, 440
Third-generation (3G) mobile network, 203
Threat detection, 329
Threat intelligence, mobile phones
   endpoint vulnerabilities, 328
   forensic acquisition, mobile device images, 327
   machine learning
      mobile data anomaly detection, 329
      NLP-driven phishing attack detection, mobile applications, 330
      pattern recognition, 329
      predictive mobile threat intelligence, 330
      threat detection, 329
      typology, mobile malware, 330
   mobile malware, 328
   network-level threats, 329
   phishing campaigns, 328
   role, 327
   telemetry sources, 327

INDEX

Time frame analysis, 19
Time-honored forensic tools, 359
Timeline correlation, malicious behavior, 252–254
Timeline graph maps, 412
Timeline reconstruction, 172
Time-series data, 291
Timestamp data, 159
Timestamp extraction, 276, 277
Timestamp normalization, 414
T-Mobile Data Breach (2021), 226
Tokenization, 333
Token Terminal/DeFiLlama, 105
tPacketCapture, 217
Traditional methodology, 402
Traffic analysis, 141
Training parameters, 352
Transaction authentication, 119
Transaction authorization, 119
Transformer-based models, 358
Transformer models, 347, 348
Transparency, Consent, and Control (TCC) framework, 64
Trojan Horse Applications, 127
T2 Security Chip, 68
tvOS, 5
2FA, *see* Two-factor authentication (2FA)
Two-factor authentication (2FA), 81

U

UC Browser, 234
UI framework, *see* User Interface (UI) framework
Unauthorized access, 317
Unauthorized transactions, 320
Underfitting, 295, 296
Unified Payment Interface (UPI), 116
Uniswap, 93
Unix-like operating system, 5
Unsupervised learning, 264
    anomaly detection
        autoencoders, 313
        clustering algorithms, 313
        Isolation Forests, 313
    basic concept, 268
    mobile digital investigations, 267
    techniques, 332
UPI, *see* Unified Payment Interface (UPI)
User behavior analysis
    2D t-SNE plot, 355
    LSTMs and Autoancoders, 355
    t-SNE facilitates visual identification, 356
User-generated vontent, 150
User Interface (UI) framework, 37

V

Virtual environments, 414
Virtual Reality (VR), 408, 421
Visitor Location Register (VLR), 203
Visual analytic tools
    forensics interfaces, 420
    timeline view, 420
    visualization-driven triaging systems, 421
VLR, *see* Visitor Location Register (VLR)
Voice interactions, 346
Voice Phishing (Vishing), 127
Volatile artifacts, 241
Volatile data, 161
Volatility framework, 164
VPN-based capture, 215
VPNs proxy/TOR identification, 123

VR, *see* Virtual Reality (VR)
Vulnerability assessment, 125

W

WAL, *see* Write-ahead log (WAL)
Wallet analysis, 109
Wallet investigations, 107
Wallets
    cold wallets, 107
    cryptocurrency, 107
    hot wallets, 107
    identified, 108
    linking, 109
watchOS, 5
Web-based cloud applications, 233
Web-based threats
    artifact-based threat detection, 251, 252
    IoCs, 252
    mobile browser threat vectors, 251
    timeline correlation, malicious behavior, 252–254
WhatsApp Desktop, 158
WhatsApp Spyware Incident (2020), 227
Wi-Fi, 14
Wi-Fi connection logs, 179
Windows Registry, 158
Wireshark, 141, 217, 230
Write-ahead log (WAL), 245, 258

X

Xcode Simulator, 140
Xposed Framework, 140
XRY, 229

Y

YAFFS, *see* Yet Another Flash File System (YAFFS)
Yet Another Flash File System (YAFFS), 56

Z

Zcash, 105
Zero-day vulnerabilities, 225
Zero-knowledge proofs (ZKPs), 106
Zimperium mobile threat defense, 338
ZKPs, *see* Zero-knowledge proofs (ZKPs)

GPSR Compliance

The European Union's (EU) General Product Safety Regulation (GPSR) is a set of rules that requires consumer products to be safe and our obligations to ensure this.

If you have any concerns about our products, you can contact us on

ProductSafety@springernature.com

In case Publisher is established outside the EU, the EU authorized representative is:

Springer Nature Customer Service Center GmbH
Europaplatz 3
69115 Heidelberg, Germany